COSMOLOGY AND ELEMENTARY PARTICLE PHYSICS

Other Related Titles from AIP Conference Proceedings

616 Experimental Cosmology at Millimetre Wavelengths: 2K1BC Workshop
Edited by Marco DePetris and Massimo Gervasi, May 2002, 0-7354-0062-8

586 Relativistic Astrophysics: 20th Texas Symposium
Edited by J. Craig Wheeler and Hugo Martel, October 2001, 0-7354-0026-1

575 Astrophysical Sources for Ground-Based Gravitational Wave Detectors
Edited by Joan M. Centrella, July 2001, 0-7354-0014-8

564 Quantum Electrodynamics and Physics of the Vacuum: QED 2000, Second Workshop
Edited by Giovanni Cantatore, May 2001, 0-7354-0000-8

555 Cosmology and Particle Physics: CAPP 2000
Edited by Ruth Durrer, Juan Garcia-Bellido, and Mikhail Shaposhnikov, March 2001, 1-56396-986-6

541 Theoretical High Energy Physics: MRST 2000
Edited by C. R. Hagen, November 2000, 1-56396-966-1

540 Particle Physics and Cosmology: Second Tropical Workshop
Edited by José F. Nieves, October 2000, 1-56396-965-3

523 Gravitational Waves: Third Edoardo Amaldi Conference
Edited by Sydney Meshkov, June 2000, 1-56396-944-0

493 General Relativity and Relativistic Astrophysics: Eighth Canadian Conference
Edited by C. P. Burgess and R. C. Myers, November 1999, 1-56396-905-X

456 Laser Interferometer Space Antenna: Second International LISA Symposium on the Detection and Observation of Gravitational Waves in Space
Edited by William M. Folkner, December 1998, 1-56396-848-7

453 Particles, Fields, and Gravitation
Edited by Jakub Rembielinski, December 1998, 1-56396-837-1

452 Toward the Theory of Everything: MRST'98
Edited by James M. Cline, Marcia E. Knutt, Gregory D. Mahlon, and Guy D. Moore, November 1998, 1-56396-845-2

444 Particle Physics and Cosmology: First Tropical Workshop/High Energy Physics: Second Latin American Symposium
Edited by José F. Nieves, September 1998, 1-56396-775-8

COSMOLOGY AND ELEMENTARY PARTICLE PHYSICS

Coral Gables Conference on Cosmology and Elementary Particle Physics

Fort Lauderdale, Florida 12–16 December 2001

EDITORS

Behram N. Kursunoglu
Global Foundation, Inc., Coral Gables, FL

Stephan L. Mintz
Florida International University, Miami, FL

Arnold Perlmutter
University of Miami, Coral Gables, FL

SPONSORING ORGANIZATIONS
Global Foundation, Inc.
Alpha Omega Research Foundation

AMERICAN INSTITUTE OF PHYSICS

Melville, New York, 2002

AIP CONFERENCE PROCEEDINGS ■ VOLUME 624

Editors:

Behram N. Kursunoglu
Global Foundation, Inc.
6200 Leonardo Street
Coral Gables, FL 33146
USA

E-mail: kursungf@attbi.com

Stephan L. Mintz
Department of Physics
Florida International University
Miami, FL 33199
USA

E-mail: mintz@helios.fiu.edu

Arnold Perlmutter
Department of Physics
University of Miami
Coral Gables, FL 33124
USA

E-mail: perlmutter@physics.miami.edu

L.C. Catalog Card No. 2002107359
ISBN 0-7354-0073-3
ISSN 0094-243X
Printed in the United States of America

Contents

III. ELEMENTARY PARTICLE PROCESSES

IV. NEUTRINOS AND CP VIOLATION

V. STUDENT PAPERS

PREFACE

The Thirtieth Coral Gables Conference on Cosmology and Elementary Particle Physics can be recorded as the conference which was attended by the largest number of young physicists ever and emphasized those subjects of high energy physics dominated by gravitational processes. These include the expansion of the universe with increasing acceleration, a new understanding of the gravitational force, not only attraction but repulsion too; dark energy, improved understanding of the nature of multiplicity of gravitational force, the concept of quintessence, and the origin of the dark matter content of the universe. These still remain as a challenge to further descriptions of this aspect of the universe. Perhaps, we also need to understand the neutrino distributions in the universe with improved new theories.

We expect the interest of the youngest physicists in the Coral Gables Conferences will increase further and we shall even incorporate graduate student participation in this series of Higher Energy Physics and Cosmology Conferences. We do not, of course, intend to over do this student participation but we would like to balance it out by bringing about direct interaction between graduate students and experts in the field. We thank Edward and Mrs. Maria Bacinich for their generous contributions. Edward has chosen to pursue with love and dedication the subject of high energy physics and cosmology for the past thirty years in the Alpha Omega Research Foundation in Palm Beach, Florida. We thank Lady Blanka Rosenstiel who is Honorary Consul of the Republic of Poland in Florida, Founder and President of The Chopin Foundation of the U.S., and The American Institute of Polish Culture for her continuing support which has allowed the Coral Gables Conferences to take place on an annual basis. The Trustees, the Chairman of the Global Foundation, and the Editors wish to extend special thanks to these enlightened citizens for contributing to the progress of science on the frontiers.

Behram N. Kursunoglu
Stephan L. Mintz
Arnold Perlmutter

INTRODUCTION

One of the most important developments of the past couple of years is the fact that gravitational force is both attractive and repulsive. Matter, as we have known it, acts as a source of attractive gravity but there is another kind of matter which is referred to in the current literature as quintessence whose gravitational effect consists of repulsive force. The competition between these two fundamental forces is the cause for the expansion of the Universe where the repulsive force generated by quintessence exceeds the attractive force generated by ordinary matter.

The important question at this point is to ask what is quintessence? We cannot leave it in the air subject to misunderstanding and we must find out the nature of this substance. An important question to ask is, "What is it that we call quintessence?" The pursuit of the answer is a keystone of the agenda for the 30th meeting of the Coral Gables Conference.

In view of the presence of international terrorism threatening our civilization it has been decided that in every conference of the Global Foundation the agenda will include a session on international terrorism as it may be affecting our established modus vivandi. A simple example, not yet practiced by the terrorist, is the safe operation of an electricity generating nuclear reactor. It is known that there are ways of increasing fuel density and hence the rate of fission leading to chain reaction, thereby, igniting a colossal nuclear explosion. The conference program in this case could contain some scenarios of this variety.

ABOUT THE GLOBAL FOUNDATION, INC.

The Global Foundation, Inc. was established in 1977 and utilizes the world's most important resource... people. The Foundation consists of senior men and women in science and learning, outstanding achievers and entrepreneurs from industry, governments, and international organizations, and promising, enthusiastic young people. These people form a unique and distinguished interdisciplinary entity, and the Foundation is dedicated to assembling all the resources necessary for them to work together. The distinguished senior members of the Foundation convey their expertise and accumulated experience, knowledge, and wisdom to the younger membership on important global issues and frontier problems in science.

Our work is a common effort, employing the ideas of creative thinkers with a wide range of experiences and viewpoints.

GLOBAL FOUNDATION BOARD OF TRUSTEES

Behram N. Kursunoglu, *Global Foundation, Inc., Chairman of the Board, Coral Gables*

M. Jean Couture, *Former Secretary of Energy of France, Paris*

Dr. Manfred Eigen*, *Max-Planck-Institut, Göttingen*

Willis E. Lamb, Jr.*, *University of Arizona*

Louis Néel*, *Université de Grenoble, France*

Henry King Stanford, *President Emeritus Universities of Miami & Georgia*

Richard Wilson, *Harvard University*

Arnold Perlmutter, *Secretary of Global Foundation, University of Miami*

Sevda A. Kursunoglu, *Vice President, Global Foundation*

Isabel Y. Tamindzija, *Secretary to the Chairman, Global Foundation, Inc.*

FORMER TRUSTEES

Robert Herman, University of Texas

Robert Hofstadter,* Stanford University

Walter C. Marshall, Lord Marshall of Goring

Frederick Reines,* Irvine, California

Abdus Salam,* Trieste, Italy

Glenn T. Seaborg,* Berkeley California

Eugene P. Wigner,* Princeton University

Lord Solly Zuckerman, London, UK

*Nobel Laureate

GLOBAL FOUNDATION'S RECENT CONFERENCE PROCEEDINGS

Recent Developments in High-Energy Physics
Edited by: Behram N. Kursunoglu, Arnold Perlmutter and Linda F. Scott
Plenum Press, 1980.

Making the Market Right for the Efficient Use of Energy
Edited by: Behram N. Kursunoglu
Nova Science Publishers, Inc., New York, 1992.

Unified Symmetry in the Small and in the Large
Edited by: Behram N. Kursunoglu and Arnold Perlmutter
Nova Science Publishers, Inc., New York, 1993.

Unified Symmetry in the Small and in the Large – 1
Edited by: Behram N. Kursunoglu, Stephan Mintz and Arnold Perlmutter
Plenum Press, 1994.

Unified Symmetry in the Small and in the Large – 2
Edited by: Behram N. Kursunoglu, Stephan Mintz and Arnold Perlmutter
Plenum Press, 1995.

Global Energy Demand in Transition: The New Role of Electricity
Edited by: Behram N. Kursunoglu, Stephan Mintz and Arnold Perlmutter
Plenum Press, 1996.

Economics and Politics of Energy
Edited by: Behram N. Kursunoglu, Stephan Mintz and Arnold Perlmutter
Plenum Press, 1996.

Neutrino Mass, Dark Matter, Gravitational Waves, Monopole Condensation, and Light Cone Quantization
Edited by: Behram N. Kursunoglu, Stephan Mintz and Arnold Perlmutter
Plenum Press, 1996.

Technology for Global Economic and Environmental Survival and Prosperity
Edited by: Behram N. Kursunoglu, Stephan Mintz and Arnold Perlmutter
Plenum Press, 1997

High-Energy Physics and Cosmology
Celebrating the Impact of 25 Years of Coral Gables Conferences
Edited by: Behram N. Kursunoglu, Stephan Mintz and Arnold Perlmutter
Plenum Press, 1997.

Environment and Nuclear Energy
Edited by: Behram N. Kursunoglu, Stephan Mintz, and Arnold Perlmutter
Plenum Press, 1998.

Physics of Mass
Edited by: Behram N. Kursunoglu, Stephan Mintz and Arnold Perlmutter
Plenum Press, 1999.

Preparing the Ground for Renewal of Nuclear Power
Edited by: Behram N. Kursunoglu, Stephan Mintz and Arnold Perlmutter
Plenum Press, 1999.

Confluence of Cosmology, Massive Neutrinos, Elementary Particles, and Gravitation
Edited by: Behram N. Kursunoglu, Stephan Mintz and Arnold Perlmutter
Plenum Press, 1999.

International Energy Forum 1999
Edited by: Behram N. Kursunoglu, Stephan Mintz and Arnold Perlmutter
Plenum Press, 2000.

International Conference on Orbis Scientiae 1999
Quantum Gravity, Generalized Theory of Gravitation, and Superstring Theory-Based Unification
Edited by: Behram N. Kursunoglu, Stephan Mintz and Arnold Perlmutter
Plenum Press, 2000.

The Challenges to Nuclear Power in the Twenty-First Century
Edited by: Behram N. Kursunoglu, Stephan Mintz and Arnold Perlmutter
Plenum Press, 2000.

Global Warming and Energy Policy 2000
Edited by: Behram N. Kursunoglu, Stephan Mintz and Arnold Perlmutter
Plenum Press, 2001.

International Conference on Orbis Scientiae 2000
The Role of Attractive and Repulsive Gravitational Forces in Cosmic Acceleration of Particles, The Origin of the Cosmic Gamma Rays
Edited by: Behram N. Kursunoglu, Stephan Mintz and Arnold Perlmutter
Plenum Press, 2001

The Role of Neutrinos, Strings, Gravity, and Variable Cosmological Constant in Elementary Particle Physics
Edited by: Behram N. Kursunoglu, Stephan Mintz and Arnold Perlmutter
Plenum Press, 2001

THE 30th CONFERENCE ON HIGH ENERGY PHYSICS AND COSMOLOGY
CORAL GABLES CONFERENCE ON COSMOLOGY
AND ELEMENTARY PARTICLE PHYSICS
DECEMBER 12 – 16, 2001
ORGANIZED BY GLOBAL FOUNDATION, INC.

Distinguished Guests, Ladies and Gentlemen:

Thank you Behram for your generous words. I am truly sorry I was unable to attend the entire conference, and I want to thank you for asking me to speak at this lovely dinner.

Since I know Behram and many of you, for at least 25 years, you might remember that in the past I had intention to arrange, during one of your conferences, a special presentation of Poland's scientific contributions to the world sciences. Tonight in my speech I will try to give you an overview – and as brevity is the soul of wit, I will try to be brief.

But let me say a word or two about the contribution to science, indeed to the evolvement of our civilization, of men and women of Poland, the country of my birth.

First I like to mention the names of Polish scientists whose important works have become known in our lifetime" Kazimierz Funk, called father of vitamins – who of us does not take at least one vitamin a day?

Then, anthropologist Bronislaw Malinowski, a founder of cultural anthropology. You might recall how some forty years ago, inspired by Malinowski's theories, Jacob Bronowsky produced an immemorial BBC series, which led millions around the world to become interested in science, culture and history.

And Stanislaw Ulam, whom I met thanks to Behram, a mathematician who was part of the Los Alamos team that conceived and built the Atomic Bomb. Later, as you know, Stan, using only his mathematical brain, proved that building a hydrogen bomb, with the A-bomb as a trigger, was possible and would not result in an uncontrolled chain reaction, as some scientist had feared. Stan Ulam, therefore, was the father of the H-bomb.

But, I would like to dwell a bit longer on the accomplishments of two Polish scientists whose discoveries truly changed the course of history: Mikolaj Kopernik and Maria Skodowska Curie, best know as Madame Curie.

When Kopernik was born in Torun in 1473, the entire world firmly believed in Ptolemy's geocentric conception of the universe. Almost 1700 years earlier, Ptolemy postulated that our planet was the center of the universe around which everything revolved. Questioning that pseudo scientific and religious belief was an act of courage, then even heresy.

But Kopernik had the courage of a true scientist and the conviction that truth would not be a detriment to faith. After many astronomers and physicians – in 1543, Kopernik published his fundamental work, *De revolucionibus orbium celestium.* The book enshrined forever the heliocentric concept of universe.

You may ask, why the Vatican did not castigate Kopernik when almost a century later it persecuted Galileo for expounding Kopernik's theory and making it universally accepted? First, because the Polish astronomer died shortly after the publication of his seminal work, second, because the sinister tentacles of the inquisition did not reach then, and ever, to Poland. Incidentally, the Vatican took 500 years to admit its error in excommunicating and jailing Galileo. It was done by John Paul II, the Polish pope.

The second Polish scientist I would like to mention is Marie Skodowska. I am proud to talk about her as a Pole and as a woman.

Marie was born in Warsaw in 1867. At the time when very few women even in Europe had any formal schooling, Marie received an early training in sciences from her father, a scientist of his own. But she became involved in students' revolutionary organizations and had to flee the czarist-ruled Warsaw to avoid arrest, first to Krakow and later to Paris.

In Paris, Marie took a science degree at Sorbonne where she was a student of a physics professor Pierre Curie. The two married in 1895 and became the best-known science couple in France, if not in Europe.

The Curies began research into radioactivity and discovered two new elements: polonium, named in honor of Marie's native land, and radium. In 1903, Pierre and Marie were awarded the Nobel Prize for the discovery of radioactivity. Two years later, Pierre was elected to the Academy of Sciences. Shortly thereafter he was run over by a dray in Paris and killed instantly.

For the next twenty-six years, until her death, Madame Curie carried her scientific work alone. In 1911, she was awarded the Noble Prize for the discovery of radium and its proprieties, the first person to receive the Nobel Prize twice. Her work in radium eventually led to the splitting of the atom and is a precursor of today's nuclear science.

In 1921, President Harding, on behalf of the women of the United States, presented her one-gram of uranium in recognition of her service to the world science. When she died in Savoy, France, in 1934, Marie Skodowska-Curie was one of the most honored scientists in the world, and one of the most famous women scientists ever.

One year later, the Curie's daughter, Irene Curie and her husband, Frederic Joliot, were also awarded the Nobel Prize in chemistry.

I could continue by giving you many examples of Polish scientists contribution, but let me just add, without comment, a few names you might recognize: Mieczyslaw Bekker (moon rover), Walter Golaski (vascular protesis), Hilary Kiproski (biology of cells), Albert Michelson (speed of light), Tadeusz Reichstein (hormones of the adrenal cortex) Tadeusz Sendzimir (innovations in mining and metallurgy), Joseph Tykocinski (first sound-on-film), Alexander Wolszcan (first to identify a new planetary system beyond our own), Stephen Wozniak (Apple Computer, Inc).

Thank you,

Blanka Rosenstiel
Honorary Consul of the Republic of Poland in Florida
Founder and President of:
The Chopin Foundation of the U.S. (1977)
The American Institute of Polish Culture, Inc. (1972)

I. PROGRESS IN GRAVITATIONAL THEORY

New Facts on the Nature of Gravitational Force And Nonlinear Oscillations of Space

Behram N. Kursunoglu
Global Foundation, Inc.
P.O. Box 249055
Coral Gables, Florida 33124-9055
e-mail:kursun@globalfoundationinc.org

Abstract

This paper discusses the letters received by this author from Albert Einstein, Erwin Schrödinger, and Paul Adrian Maurice Dirac, about fifty years ago which comment on my nonsymmetrical generalization of Einstein's general relativistic theory of gravitation. The writing of this paper, because of the dates of the letters, seems to have been delayed by half a century. Of the three versions of the nonsymmetrical theory (Einstein, Schrödinger and Kursunoglu Theories) my own paper contains results obtained as solutions of Generalized Theory of Gravitation field equations. In this paper it is shown that the field equations yield space nonlinear oscillations; a quartet of gravitational forces, quintessence, and replace Einstein's Cosmological Constant by two invariant parameters r_o and q related according to $r_o^2 q^2 = c^4/2G$, where r_o is a length varying between zero and infinity and where q^2 has the dimensions of energy density. These parameters govern the expansion of the universe with increasing acceleration and their existence yield four different solutions at each space-time point.

Introduction

In the 1999 Coral Gables Conference I had reported on the latest observations, with regard to the expansion of the universe with increasing acceleration which theoretically was explained in terms of the existence of repulsive gravitational forces predicted in 1991 from the unified theory[2] (see page 493 in reference 2) and further that space itself was subject to some nonlinear oscillations with variable frequencies. These predictions of the unified theory were published in the proceedings of the 1999 Coral Gables Conference[1]. Contrary to the existing paradigm amongst most cosmologists and physicists, I strongly agree with Einstein's regret (though for different reasons) for his introduction of the cosmological constant into his general relativistic theory of gravitation. In the unified theory the 16 component non-symmetric tensor $\hat{g}_{\mu\nu} = g_{\mu\nu} + \varepsilon q^{-1} \Phi_{\mu\nu}$ where $\varepsilon = \pm 1, \pm i$ contains the invariant parameter q, where q^2 is an energy density and is related to an intrinsic fundamental length parameter r_o by the relation

$$r_o^2 q^2 = c^4/2G \quad . \tag{1}$$

CP624, *Cosmology and Elementary Particle Physics*, edited by B. N. Kursunoglu et al.

DUBLIN INSTITUTE FOR ADVANCED STUDIES
SCHOOL OF THEORETICAL PHYSICS
64-65 MERRION SQUARE, DUBLIN.

15. V. 1951.

Dear Mr. Kursunoglu,

I have no time to study your paper
now. I am amazed to the extent of finding
it difficult to believe that the modification
you suggest should entail

a) the equations of motion for a charged mass-
point to be about what is to be expected from
older theories, and

b) the electromagnetic field acting as a source
of the gravitational potential in much the same
way as one would expect from Einstein's theory
+ Maxwell's theory.

I advise you to check upon this very
carefully to see whether your astonishingly brief
reasoning is really valid.

Born's value 9.18 · 10^{15} e.s.u. was marred
by a mistake that remained for a long time until
noticed. The mistake is a factor $(1.2361)^4$ de
~~too~~. which, when corrected, gives 3.957 · 10^{15} e.s.u.
(See my paper Proc. Roy. Ir. Acad. 49, 135,
1943.) – More important is that on your page
72, value of g_m, the division by c is incorrect, fortunately! Stronger fields than 200.000 Gauss

have been produced. Moreover if Born's non-linearity
would come into play at the order of
50.000 Gauss this would have been noticed
long ago.

Yours truly

E. Schrödinger

Typed "transcription" of E. Schrödinger's handwritten letter

DUBLIN INSTITUTE FOR ADVANCED STUDIES
SCHOOL OF THEORETICAL PHYSICS
64-65 MERRION SQUARE, DUBLIN

15.5.1951

Dear Mr. Kurşunoğlu,

I have no time to study your paper now. I am amazed to the extent of finding it difficult to believe that the modification you suggest should entail:

a) The equations of motion for a charged mass point to be about what is to be expected from older theories, and

b) The electromagnetic field acting as a source of the gravitational potential in much the same way as one would expect from Einstein's theory + Maxell's theory.

I advise you to check upon this very carefully to see whether your astonishingly brief reasoning is really valid.

Born's value $9.18.10^{15}$ e.s.u. was marred by a mistake that remained for a long time unnoticed. The mistake is a factor $(1.2361)^4$ detracts, which when corrected gives $3.957.10^{15}$ e.s.u. (See my paper Proc. Roy. Ir. Acad. 49.135. 1943.) More important is that on your page 72, value of q_m the division by $\underline{c}$ is incorrect, fortunately! Stronger fields than 200,000 Gauss have been produced. However, if Born's non-linearity would come into play at the order of 50,000 Gauss this would have been noticed long ago.

Yours truly,

E. Schrödinger

DUBLIN INSTITUTE FOR ADVANCED STUDIES

SCHOOL OF THEORETICAL PHYSICS

64-65 MERRION SQUARE, DUBLIN

29th June, 1951.

Mr. Behram Kurşunoğlu,
Fitzwilliam House,
<u>C A M B R I D G E</u>

Dear Mr. Kurşunoğlu,

Supposing all your computations are correct, your suggested field equations are decidedly a possibility. It would be interesting to know whether the right-hand side of equations (15) would yield in first approximation the conventional Einstein Tensor as sources of the gravitational field. I suppose you know that neither in Einstein's nor in my version of the non-symmetric theory it does.

You did not say what the "B"'s mean in equations (17) and (18).

With all that, I must confess that I have no great confidence in your version. The way in which the symmetric parts of both the covariant and contravariant g^{ik} intervene seems to me rather artificial, as can be seen particularly well from the Lagrangian $\mathcal{L}$ you mention at the end of your paper. To modify a frequently quoted remark of Pauli's: if you do decide to join together what God has separated (viz. a symmetric and an anti-symmetric tensor) you should at least not have to divorce them again in order to get the fundamental equations.

Yours truly,

E. Schrödinger

E. Schrödinger.

THE INSTITUTE FOR ADVANCED STUDY

PRINCETON, NEW JERSEY

January 29, 1952

Behram Kursunoglu
Fitzwilliam House
The University
Cambridge, England

My dear Mr. Kursunoglu:

I have to excuse myself for not answering your interesting communication of June 14.

I must confess that the argument for equating $R_{\alpha\beta}$ to $\lambda g_{\alpha\beta}$ is not convincing to me, for the simple reason that I believe that Minkowski space (flat space) should be a possible field.

But this is not the main point. The essential thing is that the existence of the Bianchi identities is not enough to make the system

$$\left.\begin{aligned} g^{\underset{+}{\alpha}\underset{-}{\beta}}{}_{;\gamma} &= 0 \\ R_{\alpha\beta} &= \lambda g_{\alpha\beta} \\ \Gamma_{\alpha} &= 0 \end{aligned}\right\} \qquad (11)$$

compatible in the usual sense. If you, for simplicity, assume that Γ^{i}_{kl} have been expressed by the $g_{\alpha\beta}$ (from the first equation), you have 16 + 4 equations (really only 16 + 3 independent equations), for the 16 $g_{\alpha\beta}$. This means you have 3 equations too many, which are not balanced by identities.

There is, therefore, the risk that the manifold of solutions may be too small. I have taken the same risk in the Appendix to my book on Relativity, but later considerations make it very improbable that such a risk is justified from a physical standpoint (too few solutions)

Sincerely yours,

A. Einstein

Albert Einstein

P. S. The variation principle $\delta\left(\int g^{ik} R_{ik}\, d\tau\right) = 0$ does not yield $g^{\underset{+}{i}\underset{-}{k}}{}_{;s} = 0$.

Department of Physics

The Florida State University
Tallahassee, Florida 32306

Oct 22 1973

Dear Behram

I have been comparing my new field theory with yours In both there are the 10 Einstein $g_{\mu\nu}$ and the four electromag. potentials My theory has one extra field function Yours has six The extra field function in my theory energy or momentum It is of the nature of a mathematical parametrization rather th. physical field. What are the corresponding results with your theory?

It would be interesting if you could find solutions of your field equa corresponding to plane waves. You could then see whether your extra field functions give waves carry energy and momentum My guess is that they do. In that case you must find physical meaning for them Presumably the waves all travel with the velocity of light So they will not be connected with short-range forces, and you ill have a problem

I could talk about this theory of mine at your January meeting, if you think it a suitable I am sorry I cannot come to the Onsager celebrations

wishes from

Paul Dirac

The relation (1) above was obtained[6] in 1950 while I was a graduate student in Cambridge University and it is the condition for the non-symmetric theory to reduce to general relativity in the limit of r_o=0, the correspondence limit. The parameter q has the dimensions of an electric field and therefore the field tensor $\hat{g}_{\mu\nu}$ is dimensionless. The relation (1) can also be regarded as an equation of state. The role of r_o, which varies between zero and infinity, in this theory is as important as the role of Planck's constant ħ in quantum theory. This is the missing link in Einstein and Schrödinger versions[5] of the non-symmetric theory, which differ from my own version of non-symmetric theory in the fact that they have no parameters in their theories. Both Einstein[4] (see the Appendix II page 146) and Schrödinger have commented on the subject of constants in a physical theory, as indicated in the following letters.

The four letters included in this paper are presented with comments:

- Schrödinger's letter dated May 15, 1951,
- Schrödinger's letter dated June 29, 1951,
- Einstein's letter dated January 29, 1952,
- Letter dated October 22, 1973 of Paul Adrian Maurice Dirac my professor in Cambridge University.

Schrödinger's letter dated May 15, 1951.

Comments:

I have received from Erwin Schrödinger about six letters from 1950 to 1952. I would like to include here two of his letters and comment on them.

Differences between my own version that of Schrödinger's are contained in the presence of two invariant parameters in my formulation of the non-symmetric theory while Schrödinger's version contained only a cosmological constant first introduced to general relativity by Einstein. On the other hand in Einstein's version of the unified theory there are no constants. Schrödinger's letter above comments on my results favorably but has not fully made up his mind about my theory. First he says that he finds it difficult to believe that the modification I suggested should entail:

a.) The equations of motion for a charged mass point to be about what is to be expected from the old theories and,

b.) The electro magnetic field acting as a source of the gravitational potential in much the same way as one would expect from Einstein theory plus Maxwell's theory, I had actually provided complete proof on the validity of my statement and yet Schrödinger chose to advise me as he says in his letter "I advise you to check upon this very carefully to see whether your astonishingly brief reasoning is really valid". However, despite Schrödinger's interjection of some doubt I knew and I have verified over many times on the complete validity of my statements.

This state of affairs to yield the laws of motion from the field equations in a unified field theory was the first time that a unified theory yields the fundamental laws of motion of a charged particle in an external electromagnetic field. The situation in Einstein's version because of the absence of any constant is worse than the case of Schrödinger's theory.

Continuing with Schrödinger's theory as compared with mine I would like at this point to comment on his June 29, 1951 letter to me.

Schrödinger's letter dated June 29, 1951

Comments:

This letter was typed and his comment on my version of the non-symmetric theory, as seen from the first paragraph. is quite encouraging "Your suggested field equations are decidedly a possibility," and then he continues "It would be interesting to know whether the right –hand side of equations (15) would yield in the first approximation the conventional Einstein Tensor as source of the gravitational field. I suppose you know that neither in Einstein's nor in my version of the non-symmetric theory it does."

His second paragraph consists of arguments contradicting those of the first paragraph. It is remarkable to see a great man like Schrödinger uses "Pauli's remark: if you do decide to join together what God has separated (viz. a symmetric and an anti-symmetric tensor) you should at least not have to divorce them again in order to get the fundamental equations". Schrödinger is totally wrong. I wish he were alive and see the consequences my version of the non-symmetric unified theory. In this case Pauli was trying to look funny by what he said and Schrödinger quoting it was committing a sin right from the beginning.

Einstein's letter dated January 29, 1952

Comments:

During my three-year stay in Cambridge I have received answers from Einstein to five letters that I wrote to him. In his present letter he is addressing to me as "My Dear Mr. Kursunoglu" which indicates that I have been communicating with him for a number of years. I never wrote to him about a formulation of my version of the non-symmetric theory but somehow I preferred to write to him about Schrödinger's formulation, and his negative comments do therefore refer to Schrödinger's version of the non-symmetric theory, but when I visited him in person while working in Cornell University as a Post Doctoral fellow from Cambridge I managed to arrange a visit to Princeton just to confer with Einstein. When I explained to him about the fundamentals of my work he was very generous in his remarks to the extent that his parting statement at the end of our four hours discussions of his and of my works was:"Your theory because of the existence of the fundamental length is more general than mine, but only time will show which one of us is right". I was very happy to hear Einstein comparing my work with that of his theory.

Letter dated October 22, 1973 of Paul Adrian Maurice Dirac my professor in Cambridge University

Comments:

I was most fortunate being a graduate student in Cambridge, where Dirac was the most distinguished professor, and interested at that time in Einstein's proposals. Very frequently I had the opportunity to see him in his office and telling him the latest in the non-symmetric field theory. He often asked me difficult questions to which I was not able always to give the right answers. From his inquiries about the theory and frequent visits to his office I had, of course, benefited greatly. At this point I would like to quote Dirac when he was visiting at the Center for Theoretical Studies at the University of Miami, which was an annual affair. We were invited by a well-to-do citizen of Miami, Mr. and Mrs. Kirk Landon to dinner in their home. One of the guests was Mrs. Lee Bloomberg who asked Dirac "What do you think of Behram's work?" Dirac immediately answered, "Well, we are rivals!" I was very pleased to hear this comment from Dirac and I thanked him for his generosity of making me his rival. At that time Dirac as a distinguished member of the Center for Theoretical Studies was also working on unconventional unified field theories, as the reader can see from the above letter Dirac continued the same work when he moved to Florida State University in Tallahassee. I was again very much flattered when he compared, as can be seen from his above letter, my theory with his own theory. I have answered his question with a paper in Physical Review, which he agreed with my answer to his question.

Quartet of Gravitational Forces

The expansion of the universe is now linked up with the predicted existence of the *quartet of gravitational forces*. This multiplicity of the gravitational forces can be seen by studying the sign of the gravitational constant G in the equation of state (1). The four possible substitutions in the relation (1) for the two invariant parameters r_o and q are

$$G_1^+ = [\pm r_o\,,\ \pm q\,],\quad G_1^- = [\pm i r_o\,,\ \pm q\,],\ \text{and}$$

$$G_2^- = [\pm r_o\,,\ \pm iq\,],\ G_2^+ = [\,\pm i r_o\,,\ \pm iq\,], \tag{2}$$

Where $G^+ = G$ (attractive gravity), and $G^- = -G$ (repulsive gravity). Thus, the quartet of gravitational forces consists of two attractive G_1^+, G_2^+ and two repulsive G_1^-, G_2^- forces.

This multiplicity of the gravitational forces is due to the existence of the relation (1) in the unified field theory, which multiplicity is not possible in the general theory of relativity.

Spherically Symmetric Field Equations

I would like, for the use of the field equations of unified theory, to introduce an invariant function f, which was already contained in the original field equations, and a new variable β by the relation

$$dr = f\, d\beta. \tag{3}$$

In terms of the new variable β, the spherically symmetric field equations can be written as

$$\frac{1}{2} r_o^2 \frac{d}{d\beta}[S \exp(\rho_{nT})\, \dot{\Phi}_{nsT}] = R^2 \cos \Phi_{nsT} + \ell_o^2 \sin \Phi_{nsT}, \tag{4}$$

$$\frac{1}{2} r_o^2 \frac{d}{d\beta}[S \exp(\rho_{nT})\, \dot{\rho}_{nsT}] = R^2 \sin \Phi_{nsT} + \ell_o^2 \cos \Phi_{nsT} + \exp(\rho_{nT}), \tag{5}$$

$$\frac{1}{2} r_o^2 \frac{d}{d\beta}[S \exp(\rho_{nT})] = \exp(\rho_{nT}) \left[\left(1 - \frac{\exp(\rho_{nT}) \sin \Phi_{nsT}}{R^2 + r_o^2} \right) \right], \tag{6}$$

$$2\, \ddot{\rho}_{nT} + \dot{\rho}^2_{nsT} + \dot{\Phi}^2_{nsT} = 0\,, \tag{7}$$

where

$$\dot{\rho} = \frac{d\rho}{d\beta} \tag{8}$$

and where now S,ρ , and Φ can be regarded as functions of the new variable β. The equation (7) is derivable from the three equations (4), (5), (6) which can be written as

$$\left(\frac{d^2}{d\beta^2}+\omega^2\right)\exp\left(\frac{1}{2}\rho\right)=0,\ \omega=\frac{1}{2}\dot{\Phi}, \tag{9}$$

implies nonlinear oscillations of the *orbiton* magnetic charge layers, which in the special case of β-independent ω (= ω_o) the oscillations are linear. Hence we see that the structure of an orbiton, in view of the variable frequency ω , oscillates like a pendulum whose length is changing. An orbiton, which confines all the monopoles, is formed as a result of a condensation[2] of infinite number of magnetic charges or monopoles into alternating and decreasing thickness of layers of positive and negative charges (with zero sum) which by itself is a patch. At distances large compared to r_o the function exp(½ ρ)→ r. The functions S and R are defined by

$$f=\upsilon\cosh\Gamma,\quad S=\frac{\exp(u)}{\cosh^2\Gamma}, \tag{10}$$

$$\cosh\Gamma=(R^2+r_0^2)\exp(-\rho),\ [(\exp(2\rho)+\lambda_0^4)]^{1/2}=R^2+r_0^2, \tag{11}$$

$$\ell_0^2=q^{-1}\mid g\mid,\lambda_0^2=q^{-1}\mid e\mid, \tag{12}$$

where e and g represent electric and magnetic charges, respectively.

If we assume Φ is a constant then there are no oscillations and we can calculate r_0 in the form

$$r_0^2=(\ell_0^4+\lambda_0^4)^{1/2},\ \text{where}\ \ell_0^2=(2G/c^4)N^2\mid g\mid(e^2+g^2)^{1/2}, \tag{13}$$

$$\lambda_0^2=(2G/c^4)N^2\mid e\mid(e^2+g^2)^{1/2}, \tag{14}$$

where, λ_o and ℓ_o are constants of integration.

The dynamics of space in the presence of matter is not simple. The existence of magnetic charge carrying quintessence with repulsive gravitational force versus ordinary matter with attractive gravitational force are in a state of competition which drives the expansion of the universe with increasing acceleration . The oscillations are not confined to a point but, as in the case of super -strings theory, they behave like a patch where the oscillating system has , as in the general relativity, a Schwarzschild radius. The size of a patch can be of the order of r_o and its mass can be defined as

$$M=(c^2/2G)\,r_o, \tag{15}$$

Where r_o is equal to the Schwarzschild radius of the mass M. If we treat a proton as a patch then its Schwarzschild radius is of the order of 10^{-54} cm. The largest oscillating patch is the universe itself with its Schwarzschild radius being equal to its size $r_o = 10^{28}$ cm. In this case i.e., $r_o \sim \infty$ and the field equations (4)-(7) would yield flat space-time solutions which would indicate the fact that universe is flat.

In this paper I have thus shown that the fundamental parameter r_o differs from Einstein's concept of cosmological constant in a fundamental way since in this case a single parameter is part of the non-symmetric structure and prevails over the entire evolution of the universe. Why did Einstein not consider the necessity and highly visible existence of r_o ? Was he mostly influenced by the disappointment with his own cosmological constant ? Actually Einstein answers this question, presumably, based on my correspondence with him during 1950-1952 on the subject matter of constants in the field equations, in his famous book[4] , page 146, "all such additional terms bring a heterogeneity into the system of equations, and can be disregarded, provided that no strong physical argument is found to support them" . I must point out that without r_o we can not even obtain the classical laws of motion of a charged particle in an electromagnetic field. I obtained many additional results to refute entirely Einstein's claim quoted above. The most striking fact can be found in the definition of mass: elementary particles, the sun, the earth, a black hole, neutron star etc. obtained as

$$M = (c^2/2G)\, r_o = (1/2)\, (r_o/r_p)\, m_p, \qquad (16)$$

Where r_p and m_p represent Planck length and Planck mass, respectively.

Finally, with regard to nonlinear oscillations, there is the possibility that space itself consists of fluctuating bubbles as often referred to as *quantum foam or* in this case we may prefer a description like *nonlinear foam.*

References:

(1) Behram N. Kursunoglu ,Coral Gables 1999 Conference Proceedings "Confluence of Cosmology, Massive Neutrinos, Elementary Particles, and Gravitation" pages 5-23, Edited by Behram N. Kursunoglu, Stephan L. Mintz, Arnold Perlmutter

(2) Behram N. Kursunoglu, *After Einstein and Schrödinger, A New Unified Field Theory* Journal of Physics Essays, Vol.4, pp439-518, (1991)

(3) Behram N. Kursunoglu , Phys.Rev. D6. Vol.13, Number 6, 1538 (1976)

(4) Albert Einstein, *The Meaning of Relativity,* Princeton University Press, (1953)

(5) Abraham Pais, *Subtle is the Lord* .....Science and the Life of Albert Einstein, Oxford University Press, (1982)

(6) Behram N. Kursunoglu, Phys.Rev.88, 1369 (1952); Phys. Rev. D9, 2723(1974)

A Conjecture on the Existence of Attractive and Repulsive gravitational Forces in the Generalized Theory of Gravitation

Arnold Perlmutter
Physics Department
University of Miami
Coral Gables, Florida 33124-8046

Abstract

Following the conjecture of Kursunoglu that the fundamental relation, $r_0^2 q^2 = \frac{c^4}{2G}$, implies the existence of four types of gravitational interactions (two attractive and two repulsive), we write down explicitly the resulting four sets of spherically symmetric field equations that hold at each point of space-time.

In the previous paper, Kursunoglu discusses the possible simultaneous existence of attractive and repulsive gravitational forces in his *Generalized Theory of Gravitation.*[1] The four spherically symmetric field equations are written in terms of a variable β, where

$$dr = f d\beta, \tag{1}$$

where

$$f = v \cosh\Gamma, \ S = \frac{\exp(u)}{\cosh^2\Gamma},$$

$$\cosh\Gamma = (R_0^2 + r_0^2)\exp(-\rho),$$

$$[\exp(2\rho) + \lambda_0^4]^{1/2} = R_0^2 + r_0^2, \ \ell_0^2 = \mid g \mid q^{-1}, \ \lambda_0^2 = \mid e \mid q^{-1}.$$

The parameters r_0^2 and q^2 are related through the fundamental relation

$$r_0^2 q^2 = \frac{c^4}{2G}. \tag{2}$$

It is easily seen that the universal gravitational constant, G, can take on a negative value through the substitutions,

$$r_0 \to i r_0 \text{ or } q \to i q, \tag{3}$$

CP624, *Cosmology and Elementary Particle Physics,* edited by B. N. Kursunoglu et al.

and hence

$$r_0^2 \to -r_0^2,\ q^2 \to -q^2. \tag{4}$$

There are then four possible different sets of field equations that can be generated, which we will characterize by

$$\begin{aligned} &G_1^+[\pm r_0, \pm q],\ \ G_1^-[\pm ir_0, \pm q], \\ &G_2^-[\pm r_0, \pm iq],\ \ G_2^+[\pm ir_0, \pm iq]. \end{aligned} \tag{5}$$

Clearly, there should be a quartet of different gravitational forces, G_1^+ and G_2^+ being attractive, and G_1^- and G_2^- being repulsive.

In order to make this conjecture explicit, I will now write each of the sets corresponding to the substitutions (5).

$\underline{G_1^+[\pm r_0,\ \pm q]\ .\ r_0^2 \to r_0^2, q^2 \to q^2, \lambda_0^2 \to \lambda_0^2, \ell_0^2 \to \ell_0^2}$

$$\frac{1}{2}r_0^2\frac{d}{d\beta}\left[S\exp(\rho_{n\tau})\frac{d\Phi_{ns\tau}}{d\beta}\right] = \left\{\left[\exp(2\rho_{n\tau})+\lambda_0^4\right]^{\frac{1}{2}} - r_0^2\right\}\cos\Phi_{ns\tau} + \ell_0^2\sin\Phi_{ns\tau}, \tag{6a}$$

$$\frac{1}{2}r_0^2\frac{d}{d\beta}\left[S\exp(\rho_{n\tau})\frac{d\rho_{n\tau}}{d\beta}\right] = \left\{\left[\exp(2\rho_{n\tau})+\lambda_0^4\right]^{\frac{1}{2}} - r_0^2\right\}\sin\Phi_{ns\tau} \quad \ell_0^2\cos\Phi_{ns\tau} + \exp(\rho_{n\tau}), \tag{6b}$$

$$\frac{1}{2}r_0^2\frac{d}{d\beta}\left[\frac{dS}{d\beta}\exp(\rho_{n\tau})\right] = \exp(\rho_{n\tau})\left\{1 - \frac{\exp(\rho_{n\tau})\sin\Phi_{ns\tau}}{\left[\exp(2\rho_{n\tau})+\lambda_0^4\right]^{\frac{1}{2}}}\right\}, \tag{6c}$$

$$\left[\frac{d^2}{d\beta^2} + \frac{1}{4}\left(\frac{d\Phi_{ns\tau}}{d\beta}\right)^2\right]\exp(\frac{1}{2}\rho_{n\tau}) = 0. \tag{6d}$$

$\underline{G_1^-[\pm ir_0, q]\ \ r_0^2 \to r_0^2, q^2 \to q^2, \lambda_0^2 \to \lambda_0^2, \ell_0^2 \to \ell_0^2}$

$$\frac{1}{2}r_0^2\frac{d}{d\beta}\left[S\exp(\rho_{n\tau})\frac{d\Phi_{ns\tau}}{d\beta}\right] = -\left\{\left[\exp(2\rho_{n\tau})+\lambda_0^4\right]^{\frac{1}{2}}+r_0^2\right\}\cos\Phi_{ns\tau} - \ell_0^2\sin\Phi_{ns\tau}, \qquad (7a)$$

$$\frac{1}{2}r_0^2\frac{d}{d\beta}\left[S\exp(\rho_{n\tau})\frac{d\rho_{n\tau}}{d\beta}\right] = -\left\{\left[\exp(2\rho_{n\tau})+\lambda_0^4\right]^{\frac{1}{2}}+r_0^2\right\}\sin\Phi_{ns\tau} - \ell_0^2\cos\Phi_{ns\tau} - \exp(\rho_{n\tau}), \qquad (7b)$$

$$\frac{1}{2}r_0^2\frac{d}{d\beta}\left[S\exp(\rho_{n\tau})\right] = -\exp(\rho_{n\tau})\left\{1-\frac{\exp(\rho_{n\tau})\sin\Phi_{ns\tau}}{[R_0^2-r_0^2]}\right\}, \qquad (7c)$$

$$\left[\frac{d^2}{d\beta^2}+\frac{1}{4}\left(\frac{d\Phi_{ns\tau}}{d\beta}\right)^2\right]\exp\left(\frac{1}{2}\rho_{n\tau}\right) = 0. \qquad (7d)$$

$G_2^-\ [\pm r_0, \pm iq]$, $r_0^2 \to r_0^2, q^2 \to -q^2, \lambda_0^2 \to \lambda_0^2, \ell_0^2 \to -\ell_0^2$

$$\frac{1}{2}r_0^2\frac{d}{d\beta}\left[S\exp(\rho_{n\tau})\frac{d\Phi_{ns\tau}}{d\beta}\right] = \left\{\left[\exp(2\rho_{n\tau})+\lambda_0^4\right]^{\frac{1}{2}}-r_0^2\right\}\cos\Phi_{ns\tau} - \ell_0^2\sin\Phi_{ns\tau}, \qquad (8a)$$

$$\frac{1}{2}r_0^2\frac{d}{d\beta}\left[S\exp(\rho_{n\tau})\frac{d\rho_{n\tau}}{d\beta}\right] = \left\{\left[\exp(2\rho_{n\tau})+\lambda_0^4\right]^{\frac{1}{2}}-r_0^2\right\}\sin\Phi_{ns\tau} - \ell_0^2\cos\Phi_{ns\tau} + \exp(\rho_{n\tau}), \qquad (8b)$$

$$\frac{1}{2}r_0^2\frac{d}{d\beta}\left[\frac{dS}{d\beta}\exp(\rho_{n\tau})\right] = \exp(\rho_{n\tau})\left\{1-\frac{\exp(\rho_{n\tau})\sin\Phi_{ns\tau}}{[\exp(2\rho_{n\tau})+\lambda_0^4]^{\frac{1}{2}}}\right\}, \qquad (8c)$$

$$\left[\frac{d^2}{d\beta^2}+\frac{1}{4}\left(\frac{d\Phi_{ns\tau})^2}{d\beta}\right]\exp\left(\frac{1}{2}\rho_{n\tau}\right) = 0. \qquad (8d)$$

$G_2^+\ [\pm ir_0, \pm iq]$ $r_0^2 \to -r_0^2, q^2 \to -q^2, \lambda_0^2 \to \lambda_0^2, \ell_0^2 \to -\ell_0^2$

$$\frac{1}{2}r_0^2\frac{d}{d\beta}\left[S\exp(\rho_{n\tau})\frac{d\Phi_{ns\tau}}{d\beta}\right] = -\left\{\left[\exp(\rho_{n\tau})+\lambda_0^4\right]^{\frac{1}{2}}+r_0^2\right\}\cos\Phi_{ns\tau} + \ell_0^2\sin\Phi_{ns\tau}, \qquad (9a)$$

$$\frac{1}{2}r_0^2\frac{d}{d\beta}\left[S\exp(\rho_{n\tau})\frac{d\rho_{n\tau}}{d\beta}\right] = -\left\{[\exp(2\rho_{n\tau}) + \lambda_0^4]^{\frac{1}{2}} + r_0^2\right\}\sin\Phi_{ns\tau} + \ell_0^2\cos\Phi_{ns\tau} - \exp(\rho_{n\tau}), \tag{9b}$$

$$\frac{1}{2}r_0^2\frac{d}{d\beta}\left[\frac{dS}{d\beta}\exp(\rho_{n\tau})\right] = -\exp\rho_{n\tau}\left\{1 - \frac{\exp(\rho_{n\tau})\sin\Phi_{ns\tau}}{[R_0^2 - r_0^2]^{\frac{1}{2}}}\right\}, \tag{9c}$$

$$\left[\frac{d^2}{d\beta^2} + \frac{1}{4}(\frac{d\Phi_{ns\tau}}{d\beta})^2\right]\exp(\frac{1}{2}\rho_{n\tau}) = 0. \tag{9d}$$

At every point of space and time, each of the above sets of equations (6), (7), (8), (9), describes the proposed quartet of forces. The fundamental rules of r_0 and q are manifestly revealed by the fact that when $r_0 \to \infty$, we obtain flat space - time. This implies that the Universe began from a very high energy density. As it expanded, this density decreased to a very small magnitude, which is, in totality, the story of the big bang creation of the Universe. The behavior of the solutions to these equations will be investigated.

References

1. B.N. Kursunoglu, previous paper (These Proceedings)

Localization Issues for Robertson-Walker Branes

Philip D. Mannheim

Department of Physics, University of Connecticut, Storrs, CT 06269
mannheim@uconnvm.uconn.edu

Abstract. We discuss some of the localization issues associated with the embedding of Robertson-Walker type Randall-Sundrum branes in a bulk AdS_5. Specifically, we show that of the branes which are embeddable in AdS_5 the geometry associated with M_4 and dS_4 branes warps away from the brane while that associated with AdS_4 and RW branes of any spatial 3-curvature antiwarps away from the brane. We discuss the gravitational fluctuations around an M_4 brane and analyze the specific role played by a delta function singularity at the brane. We show how a bulk sine-Gordon scalar field can without any fine-tuning naturally lead to localization of gravity around an M_4 brane.

THE RANDALL-SUNDRUM SET-UP

Recently Randall and Sundrum [1, 2] showed that in the presence of a 5-dimensional anti-de Sitter (AdS_5) bulk it is possible for gravity to localize to a lower dimensional brane embedded in it, and that such localization could be achieved even if the bulk extra dimension was infinite. With such an AdS_5 bulk the probability for propagation of gravitational signals can fall off exponentially away from the brane, with an observer on the brane then effectively seeing only 4-dimensional rather than 5-dimensional gravity despite the presence of the infinite extra dimension that the bulk possesses, to thus enable us to be living in a universe with a macroscopically sized fifth dimension. Given such an intriguing possibility it is thus necessary to explore just how general it might be and to ascertain in what way it might even be amenable to experimental testing. In this paper we shall therefore explore these issues.

It is useful to begin first with a discussion of AdS_5 spaces themselves and to subsequently then discuss the embedding of branes (viz. lower dimensional surfaces) in them. AdS_5 is a maximally symmetric 5-space of constant negative curvature $-b^2$. As such its Riemann tensor is given by

$$R_{ABCD} = b^2(g_{AC}g_{BD} - g_{AD}g_{BC}) \tag{1}$$

(here we use $A, B = 0, 1, 2, 3, 5$ to denote the five bulk coordinates t, x, y, z, w with $\mu, \nu = 0, 1, 2, 3$ denoting the ordinary 4-dimensional spacetime coordinates on the brane), so that the 5-dimensional Weyl tensor C_{ABCD} vanishes identically while the 5-dimensional Einstein tensor is given by $G_{AB} = -6b^2 g_{AB}$. Given Eq. (1) it is possible to construct an explicit form for the metric on the 5-space, with the most convenient one being given by a 4-dimensional Minkowski (M_4) sectioning of the 5-space, viz.

$$ds^2 = e^{-2bw}(-dt^2 + dx^2 + dy^2 + dz^2) + dw^2 \ . \tag{2}$$

CP624, *Cosmology and Elementary Particle Physics*, edited by B. N. Kursunoglu et al.

(We discuss other possible sectionings below, with each such sectioning then being associated with 4-dimensional surfaces which can in fact be embedded in AdS_5.) With the fifth coordinate w ranging from $-\infty$ to ∞, the metric of Eq. (2) has two interesting aspects. First, with the metric falling away from the brane (viz. warping) in the $w \geq 0$ region, null geodesic (viz. $dt/dw = \exp(bw)$) signals emitted at $w = \infty$ will take an infinite amount of time to reach $w = 0$, with $w = \infty$ thus being a horizon. However, with the metric rising away from the brane (viz. anti-warping) in the $w \leq 0$ region, geodesic signals emitted at $w = -\infty$ will be able to reach $w = 0$ in the finite time $t = 1/b$. Consequently new information can come in from the edge of AdS_5 in a finite time, thus making it impossible to unambiguously specify the forward propagation of Cauchy data on an initial spacelike hypersurface. AdS_5 spaces are thus globally non-hyperbolic.

As first suggested by Randall and Sundrum, if we could somehow get rid of the anti-warping region while retaining only the warping one, we would then have localization of the geometry around $w = 0$. To achieve this Randall and Sundrum therefore suggested to replace Eq. (2) by the $w \to -w$ Z_2 invariant metric

$$ds^2 = e^{-2b|w|}(-dt^2 + dx^2 + dy^2 + dz^2) + dw^2 = e^{-2b|w|}\eta_{\mu\nu}dx^\mu dx^\nu + dw^2 \ , \qquad (3)$$

a metric which thus warps for both positive and negative w. Operationally, Eq. (3) entails keeping only the $w \geq 0$ region of Eq. (2) while replacing the $w \leq 0$ region by a copy of the $w \geq 0$ region, to give a Z_2 doubling of the $w \geq 0$ region. While such a doubling then yields warping for all w, we also note that the removing of the antiwarping region from consideration thus now gives us good Cauchy propagation of initial data as well. The Randall-Sundrum proposal thus not only achieves localization of the geometry, it also nicely finesses the global non-hyperbolicity problem as well. Moreover, given the modification of Eq. (3), the Riemann tensor now no longer obeys Eq. (1). Rather, it instead evaluates to

$$R_{ABCD} = b^2(g_{AC}g_{BD} - g_{AD}g_{BC}) - 2b\delta^5_A\delta^\mu_B\delta^5_C\delta^\nu_D\eta_{\mu\nu}\delta(w) \qquad (4)$$

and is thus now only a pure AdS_5 metric in the bulk region away from $w = 0$.

In order to see what is dynamically required to yield Eq. (3) it is convenient to consider a slightly more general metric than it, viz.

$$ds^2 = e^{2f(w)}(-dt^2 + dx^2 + dy^2 + dz^2) + dw^2 \ , \qquad (5)$$

a metric whose Weyl tensor still vanishes (since the metric is conformal to flat), but whose Einstein tensor is given by

$$G_{00} = -G_{11} = -G_{22} = -G_{33} = 3e^{2f}f'' + 6e^{2f}f'^2 \ , \ G_{55} = -6f'^2 \ . \qquad (6)$$

If we consider $f(w)$ to now be a function of $|w|$, on noting that $d|w|/dw = \theta(w) - \theta(-w) = \epsilon(w)$, $d^2|w|/dw^2 = 2\delta(w)$, we see (as may be anticipated from Eq. (4)) that all four of the G_{00}, G_{11}, G_{22}, G_{33} components of the Einstein tensor must now contain a delta function term, while G_{55} must not. (With dG_{55}/dw being a second derivative function of w as is required by the Bianchi identities, G_{55} itself can only contain first derivatives of $f(w)$.) If we now impose the 5-dimensional Einstein equations, viz.

$$G_{AB} = -\kappa_5^2 T_{AB} \ , \qquad (7)$$

we thus see that all four of the T_{00}, T_{11}, T_{22}, T_{33} components of the energy-momentum tensor must contain a $\delta(w)$ term while T_{55} must not. We are thus led to introduce a source of energy-momentum at $w=0$, and it is thus at $w=0$ that we must locate a lower dimensional surface or brane (viz. one which does not contribute to T_{55}), a matter bearing membrane which is thus confined to the $w=0$ region. To this end we thus set

$$T_{AB} = T^{bulk}_{AB} + \delta^{\mu}_{A}\delta^{\nu}_{B}T^{brane}_{\mu\nu}\delta(w) \ , \tag{8}$$

and find that with the introduction of bulk and brane cosmological constants

$$T^{bulk}_{AB} = -\Lambda_5 g_{AB} \ , \ T^{brane}_{\mu\nu} = -\lambda\eta_{\mu\nu} \ , \tag{9}$$

where Λ_5 and λ are both required to be positive, the metric of Eq. (3) then emerges as the exact solution to the Einstein equations provided only that

$$6\Lambda_5 + \kappa_5^2\lambda^2 = 0 \ , \tag{10}$$

with the bulk (viz. the $w \neq 0$ region) then being found to be the desired AdS_5 with its curvature being given by

$$b^2 = -\Lambda_5\kappa_5^2/6 \ . \tag{11}$$

As we see, in order to implement the solution we thus need a relationship between Λ_5 and λ, the so-called Randall-Sundrum fine-tuning condition, a condition without which Eq. (3) could not otherwise have been obtained. Having now obtained our desired warping geometry, in order to gain further insight into it we find it very convenient to consider the embedding aspects of the problem.

EMBEDDING A BRANE IN A BULK

In order to discuss the embedding of our 4-dimensional universe into a 5-dimensional bulk space with some initially completely general metric g_{AB}, it is particularly convenient [3] to base the analysis on the purely geometric Gauss embedding formula

$$^{(4)}R^{\alpha}{}_{\beta\gamma\delta} = R^{A}{}_{BCD}q_{A}{}^{\alpha}q^{B}{}_{\beta}q^{C}{}_{\gamma}q^{D}{}_{\delta} - K^{\alpha}{}_{\gamma}K_{\beta\delta} + K^{\alpha}{}_{\delta}K_{\beta\gamma} \ , \tag{12}$$

a formula which relates the 4-dimensional Riemann tensor $^{(4)}R^{\alpha}{}_{\beta\gamma\delta}$ on a general 4-dimensional surface (one not yet Z_2 doubled) to the Riemann tensor $R^{A}{}_{BCD}$ of a 5-dimensional bulk (one not necessarily AdS_5) into which it is embedded via a term quadratic in the extrinsic curvature $K_{\mu\nu} = q^{\alpha}{}_{\mu}q^{\beta}{}_{\nu}n_{\beta;\alpha}$ of the 4-surface. Here $q_{AB} = g_{AB} - n_A n_B \equiv q_{\mu\nu}$ is the metric which is induced on the 4-surface by the embedding and is thus the one with which $^{(4)}R^{\alpha}{}_{\beta\gamma\delta}$ is calculated, while n^A is the embedding normal. Equation (12) thus shows that the 4-dimensional Riemann tensor on the surface is not simply an appropriate projection of the 5-dimensional Riemann tensor. Rather the two tensors differ by terms which explicitly depend on the extrinsic curvature of the surface. On introducing the bulk Weyl tensor

$$\begin{aligned} C_{ABCD} = R_{ABCD} &- (g_{AC}R_{BD} - g_{AD}R_{BC} - g_{BC}R_{AD} + g_{BD}R_{AC})/3 \\ &+ R^{E}{}_{E}(g_{AC}g_{BD} - g_{AD}g_{BC})/12 \ , \end{aligned} \tag{13}$$

contraction of indices in Eq. (12) immediately allows us to relate the 4- and 5-dimensional Einstein tensors according to

$$
\begin{aligned}
{}^{(4)}G_{\mu\nu} &= 2G_{AB}(q^{A}{}_{\mu}q^{B}{}_{\nu}+n^{A}n^{B}q_{\mu\nu})/3-G^{A}{}_{A}q_{\mu\nu}/6 \\
&-KK_{\mu\nu}+K^{\alpha}{}_{\mu}K_{\alpha\nu}+(K^{2}-K_{\alpha\beta}K^{\alpha\beta})q_{\mu\nu}/2-E_{\mu\nu}
\end{aligned}
\qquad (14)
$$

where

$$
E_{\mu\nu}=C^{A}{}_{BCD}n_{A}n^{C}q^{B}{}_{\mu}q^{D}{}_{\nu}\ . \qquad (15)
$$

The geometric content of Eq. (14) is, first, that of the 35 components of C_{ABCD} (viz. the 35 components of the 50 component R_{ABCD} which are independent of G_{AB}) 10 of them can be determined once the induced metric on the 4-surface is known; and second, that since the left hand side of Eq. (14) only contains derivatives with respect to the four coordinates other than the one in the direction of the embedding normal n^{A}, on the right hand side all derivative terms with respect to this fifth coordinate must mutually cancel each other identically. Thus for instance, for the metric of the form $ds^2=f(w)(-dt^2+d\bar{x}^2)+dw^2$ and for normal $n^{A}=(0,0,0,0,1)$, term by term Eq. (14) yields

$$
\begin{aligned}
{}^{(4)}G^{0}{}_{0} &= -f''/f-f'^{2}/f^{2}+f''/f+f'^{2}/4f^{2}-f'^{2}/f^{2} \\
&+f'^{2}/4f^{2}+2f'^{2}/f^{2}-f'^{2}/2f^{2}-0\ ,
\end{aligned}
\qquad (16)
$$

i.e. $0=0$ as is to be expected since ${}^{(4)}G^{0}{}_{0}$ vanishes identically in the flat 4-dimensional Minkowski space M_4. Finally, the dynamical implication of Eq. (14) is that even if G_{AB} is taken to obey the 5-dimensional Einstein equations in the 5-space, the induced 4-dimensional ${}^{(4)}G_{\mu\nu}$ would not in general be expected to obey the standard 4-dimensional ones. Consequently, the dynamical structure of embedded 4-dimensional gravity is in principle different from that of non-embedded gravity, with measurement of ${}^{(4)}G_{\mu\nu}$, viz. measurement purely within the 4-dimensional world itself, then in principle enabling us to see effects coming from higher dimensions. Equation (14) thus provides a 4-dimensional window on a higher dimensional world.

In order to extend the purely geometric Eq. (14) to the Randall-Sundrum case of interest, we note that once some energy density is placed on the $w=0$ surface, there will then be a discontinuity in the extrinsic curvature of the surface as it is crossed from one side to the other. And for the situation in which the Einstein equations hold in the bulk it can be shown very generally that this discontinuity takes the form [4]

$$
K_{\mu\nu}(w=0^{+})-K_{\mu\nu}(w=0^{-})=-\kappa_{5}^{2}[T^{brane}_{\mu\nu}-q_{\mu\nu}(T^{brane})^{\alpha}{}_{\alpha}/3]\ . \qquad (17)
$$

As such these Israel junction conditions constitute the general relativistic generalization of the discontinuity in a Newtonian gravitational field as a sheet of non-relativistic matter is crossed (viz. the direction of the field is always toward the matter distribution). While there is a discontinuity in the extrinsic curvature it is important to note that there is no such discontinuity in the induced metric itself so that $q_{\mu\nu}(w=0^{+})=q_{\mu\nu}(w=0^{-})$. To implement the Z_2 doubling we now take the 5-space metric to be a function of $|w|$, and with the extrinsic curvature being related to a first derivative of the normal, $K_{\mu\nu}(w)$

then behaves as a discontinuous $\theta(w)-\theta(-w)$ type function, so that $K_{\mu\nu}(w=0^-) = -K_{\mu\nu}(w=0^+)$. In the presence of Z_2 doubling we thus obtain

$$K_{\mu\nu}(w=0^+) = -\kappa_5^2[T^{brane}_{\mu\nu} - q_{\mu\nu}(T^{brane})^{\alpha}{}_{\alpha}/3]/2 \ . \tag{18}$$

at the brane. Now since, as we noted earlier, ${}^{(4)}G_{\mu\nu}$ involves no derivatives with respect to w (i.e. like $q_{\mu\nu}$ it is continuous at the brane), even in the event that we take g_{AB} to be a function of $|w|$, it follows that ${}^{(4)}G_{\mu\nu}$ cannot acquire any $\delta(w)$ term. Consequently, the right hand side of Eq. (14) must also contain no net $\delta(w)$ dependent term either. However, given a generic brane matter density

$$T^{brane}_{\mu\nu} = -\lambda q_{\mu\nu} + \tau_{\mu\nu} \tag{19}$$

it follows from Eqs. (7) and (8) that as far as the delta function terms are concerned, the Einstein tensor terms in Eq. (14) make a contribution

$$2G_{AB}(q^A{}_\mu q^B{}_\nu + n^A n^B q_{\mu\nu})/3 - G^A{}_A q_{\mu\nu}/6 = -\kappa_5^2[4\tau_{\alpha\beta}q^\alpha{}_\mu q^\beta{}_\nu - \tau^\alpha{}_\alpha q_{\mu\nu}]\delta(w)/6 \tag{20}$$

on the brane. Since the extrinsic curvature terms contain no $\delta(w)$ terms ($T^{brane}_{\mu\nu}$ is defined as the coefficient $\delta(w)$ in Eq. (8)), it then follows that on the brane $E_{\mu\nu}$ must contain a discontinuous delta function term of the form [5]

$$E^{disc}_{\mu\nu} = -\kappa_5^2[4\tau_{\alpha\beta}q^\alpha{}_\mu q^\beta{}_\nu - \tau^\alpha{}_\alpha q_{\mu\nu}]\delta(w)/6, \tag{21}$$

a quantity that need not vanish even if the Weyl tensor vanishes in the bulk.

With the $\delta(w)$ terms in Eq. (14) thus taking care of each other, we can now isolate the continuous non $\delta(w)$ terms in Eq. (14), and on noting that any product of any two of the components of the extrinsic curvature is itself continuous at the brane ($[\theta(w)-\theta(-w)]^2 = 1$), we find that on the brane [3]

$${}^{(4)}G_{\mu\nu} = \Lambda_4 q_{\mu\nu} - 8\pi G_N \tau_{\mu\nu} - \kappa_5^4 \pi_{\mu\nu} - \bar{E}_{\mu\nu} \tag{22}$$

where

$$\Lambda_4 = \kappa_5^2(6\Lambda_5 + \kappa_5^2\lambda^2)/12 \ , \quad 8\pi G_N = \lambda\kappa_5^4/6 \ ,$$
$$\pi_{\mu\nu} = -\tau_{\mu\alpha}\tau_\nu{}^\alpha/4 + \tau^\alpha{}_\alpha\tau_{\mu\nu}/12 + q_{\mu\nu}\tau_{\alpha\beta}\tau^{\alpha\beta}/8 - q_{\mu\nu}(\tau^\alpha{}_\alpha)^2/24 \ , \tag{23}$$

and where $\bar{E}_{\mu\nu} = [E_{\mu\nu}(w=0^+) + E_{\mu\nu}(w=0^-)]/2$ is the piece of $E_{\mu\nu}$ which is continuous at the brane. As such Eq. (22) is the equation obeyed by the Einstein tensor on the brane, and through the presence of the $\bar{E}_{\mu\nu}$ and $\pi_{\mu\nu}$ terms we thus see an explicit departure from the standard 4-dimensional Einstein equations associated with a gravitational coupling constant $8\pi G_N = \lambda\kappa_5^4/6$.[1] Now while Eq. (22) is completely general and does

[1] The presence of the quadratic $\pi_{\mu\nu}$ term was first noted in [6], while the emergence of an effective Newton constant through the cross terms in bilinear products of the $K_{\mu\nu}$ was first given in [7].

not require any a priori assumptions regarding the geometry in the bulk, in the event that the bulk is taken to be AdS_5, the continuous piece of the Weyl tensor will then vanish and we will be able to drop the $\bar{E}_{\mu\nu}$ term altogether, to then yield

$$^{(4)}G_{\mu\nu} = \Lambda_4 q_{\mu\nu} - 8\pi G_N \tau_{\mu\nu} - \kappa_5^4 \pi_{\mu\nu} \ , \tag{24}$$

with the only departure from standard gravity then being through the presence of the term quadratic in the energy density, a term which could potentially be of major concern in the early universe when the energy density is large.

Now even if we start off with an AdS_5 bulk, as soon as we put some additional matter density on the brane, that matter density will immediately set up a new gravitational field in the bulk to not only potentially modify the bulk geometry but to also possibly delocalize gravity as well. To avoid this we must thus only put matter densities on the brane for which Eq. (24) then yields a 4-metric which is embeddable in AdS_5, i.e. a metric which can be associated with a sectioning of AdS_5. As we will see below this precisely can occur for de Sitter, anti de Sitter and Robertson-Walker (collectively Robertson-Walker type) branes, viz. those highly symmetric branes of relevance to cosmology. To see how severe a constraint the very structure of the embedding actually imposes, we note that Eqs. (7) and (8) actually admit of the exact solution [8]

$$ds^2 = e^{-2b|w|}[-(1-2MG/r)dt^2 + dr^2/(1-2MG/r) + r^2 d\Omega] + dw^2 \ , \tag{25}$$

when the brane is taken to have a Ricci flat Schwarzschild geometry. Moreover, in this solution every single term in Eq. (22) vanishes identically on the brane. However, inspection of the bulk geometry in this solution shows that the bulk Weyl tensor does not vanish off the brane (cf. $C_{0101} = 2MGe^{-2b|w|}/r^3$). Thus even though the $C_{5\mu5\nu}$ components of the Weyl tensor needed for $\bar{E}_{\mu\nu}$ do vanish, its other components do not, with the bulk thus not being AdS_5 in this particular case, and with the Schwarzschild metric thus not being embeddable in AdS_5. In and of itself then requiring the bulk Einstein tensor to obey $G_{AB} = \kappa_5^2 \Lambda_5 g_{AB}$ is thus not sufficient to force the bulk Weyl tensor to vanish, and thus not sufficient to ensure that the bulk be AdS_5. Finally, we also note that since we cannot embed the Schwarzschild metric in AdS_5, if we therefore consider a general fluctuation due to the addition of a static mass source to a background brane whose geometry does embed in AdS_5, we will find that in general the fluctuation will generate a non-zero contribution to the Weyl tensor, to thus potentially not only modify the geometry in the bulk but to also induce a Weyl tensor contribution on the brane as well. As we thus see, even in the event that the background Eq. (24) is of the form of the standard 4-dimensional Einstein equations (i.e. cases in which the $\pi_{\mu\nu}$ term is negligible), nonetheless the brane fluctuations around such a background will not in fact be standard.[2] (Moreover, according to Eq. (14) fluctuations in $K_{\mu\nu}$ are also able to contribute to the fluctuations in the brane $^{(4)}G_{\mu\nu}$.) While we shall return to a discussion of the structure of the associated fluctuation equation below, we turn first to a discussion of brane backgrounds which are in fact embeddable in AdS_5.

[2] The remarks presented here were developed in collaboration with A. H. Guth, D. I. Kaiser and A. Nayeri.

EMBEDDING OF ROBERTSON-WALKER BRANES IN AdS_5

For metrics which are maximally 4-symmetric in the ordinary spacetime coordinates (viz. metrics for which ${}^{(4)}G_{\mu\nu} = \Lambda_4 q_{\mu\nu}$) the most general possible 5-dimensional metrics take the form

$$ds^2 = e^{2f(w)}[-dt^2 + e^{2Ht}(dx^2 + dy^2 + dz^2)] + dw^2 = e^{2f(w)} q_{\mu\nu} dx^\mu dx^\nu + dw^2 \quad (26)$$

and

$$ds^2 = e^{2f(w)}[e^{2Hz}(-dt^2 + dx^2 + dy^2) + dz^2] + dw^2 = e^{2f(w)} q_{\mu\nu} dx^\mu dx^\nu + dw^2 \;, \quad (27)$$

metrics which respectively correspond to dS_4 and AdS_4 sectionings of an otherwise initially general 5-space. Since both of these 5-dimensional metrics just happen to be conformal to flat for any $f(w)$, requiring their Einstein tensors to obey

$$G_{AB} = -\kappa_5^2[-\Lambda_5 g_{AB} - \delta^\mu_A \delta^\nu_B \lambda q_{\mu\nu} \delta(w)] \quad (28)$$

will then actually force the associated bulks to be AdS_5, with both the dS_4 and AdS_4 branes thus being embeddable in AdS_5. Moreover, given the explicit form of the brane energy-momentum tensor in Eq. (28) the $f(w)$ coefficients are completely determined. Thus for the dS_4 brane embedded in AdS_5 we find that [9, 10]

$$ds^2 = \sinh^2(b|w| - \sigma) \sinh^{-2}\sigma[-dt^2 + e^{2Ht}(dx^2 + dy^2 + dz^2)] + dw^2 \quad (29)$$

where $\sinh\sigma = b/H$, while for the AdS_4 brane embedded in AdS_5 we find that [9]

$$ds^2 = \cosh^2(b|w| - \sigma) \cosh^{-2}\sigma[e^{2Hz}(-dt^2 + dx^2 + dy^2) + dz^2] + dw^2 \quad (30)$$

where $\cosh\sigma = b/H$. Additionally, on the brane the residual cosmological constant is given by $\Lambda_4 = 3H^2$ in the dS_4 case and by $\Lambda_4 = -3H^2$ in the AdS_4 case. Thus we see that when the Randall-Sundrum fine tuning $\Lambda_4 = 0$ condition is not obeyed the brane becomes either de Sitter or anti de Sitter depending on the relative strengths of the input bulk and brane cosmological constants.

As brane theories both of these two metrics grow exponentially as $|w| \to \infty$ and at first sight each would appear to be of the non-localizing anti-warping type. However the dS_4 brane metric has a horizon at $b|w| = \sigma$ beyond which null geodesics can never reach the brane. Since the function $\sinh^2(b|w| - \sigma)$ falls all the way to this horizon gravity actually does localize [11] in the dS_4 brane case. For the AdS_4 brane case, while there is no such horizon ($\cosh^2(b|w| - \sigma)$ never vanishes), nonetheless the function $\cosh^2(b|w| - \sigma)$ does initially begin to fall before eventually turning round at $|w| = \sigma/b$ and then begin to rise. Consequently, for small enough H (viz. large σ) the horizon will be far away from the brane and the low energy fluctuations will be quite close to the localizing ones associated with the $H = 0$ M_4 Minkowski brane, to thus give an approximate or effective localization of low energy gravity on the brane [12]. In this sense then localization of gravity can be associated with both the dS_4 and AdS_4 brane cases, though for large H none of the above reasoning would apply in the AdS_4 case and and its localization would be lost. (For further analysis of these two cases see also [13].)

For maximally 3-symmetric RW branes [viz. branes with metrics which obey Eq. (24) with $\tau_{\mu\nu} = (\rho_m + p_m)U_\mu U_\nu + p_m q_{\mu\nu}$] their embedding in an arbitrary 5-space yields as the most general 5-space metric

$$ds^2 = -dt^2 e^2(w,t)/f(w,t) + f(w,t)[dr^2/(1-kr^2) + r^2 d\Omega] + dw^2 \ . \qquad (31)$$

However, unlike the previous dS_4 and AdS_4 brane cases, this time the 5-space metric is not automatically conformal to flat. In fact 10 of the components of the Weyl tensor do not necessarily vanish (the 6 $C_{\mu\nu\mu\nu}$ with $\mu \neq \nu$ and the 4 $C_{\mu 5 \mu 5}$), with all of them being found to be proportional to

$$C_{0505} = (4e^3 f f'' - 6e^3 f'^2 + 4e^3 f k + 6e^2 f e' f' \\ -4e^2 f^2 e'' - 2e f^2 \ddot{f} + e f \dot{f}^2 + 2 f^2 \dot{e} \dot{f})/8ef^3 \ . \qquad (32)$$

Consequently this time imposing the Einstein equations is not sufficient to make the bulk be AdS_5. Rather one must also require the Weyl tensor to vanish. Explicit calculation [14, 15] then shows that this can be done, so that maximally 3-symmetric RW metrics can indeed be embedded in AdS_5. However, while it can be done, in the static RW brane case it can only be done at a price, namely there has to be a new fine-tuning relation between the matter fields of the theory. Since the discussion is different in the static and non-static cases we shall discuss the two cases separately.

For the static case first, on solving the 5-dimensional Einstein equations and on setting the bulk Weyl tensor to zero, we find [14, 15] that the fine-tuning condition

$$\kappa_5^2(\lambda + \rho_m)(-\lambda + 2\rho_m + 3p_m) = 6\Lambda_5 \qquad (33)$$

is required of the matter fields.[3] On setting $\nu = (-2\kappa_5^2\Lambda_5/3)^{1/2}$ the most general solution is given in the $k = +1$ case by [14, 15]

$$f = (4/\nu^2)\sinh^2(\nu w_0/2 - \nu|w|/2) \ , \ e^2/f = (4/\nu^2)\cosh^2(\nu w_0/2 - \nu|w|/2) \qquad (34)$$

where $\coth(\nu\, w_0/2) = \kappa_5^2(\lambda + \rho_m)/3\nu$, and in the $k = -1$ case by [14, 15]

$$f = (4/\nu^2)\cosh^2(\nu w_0/2 - \nu|w|/2) \ , \ e^2/f = (4/\nu^2)\sinh^2(\nu w_0/2 - \nu|w|/2) \qquad (35)$$

where $\tanh(\nu w_0/2) = \kappa_5^2(\lambda + \rho_m)/3\nu$. With each of these metrics having forms which are hybrids of both of the dS_4 and AdS_4 brane case metrics which we presented above, and with both of them antiwarping far from the brane, whether or not they might lead to localization of gravity is not at all apparent. While a Karch-Randall type analysis [12]

---

[3] In passing we note that in the presence of this Eq. (33) the Einstein tensor on the brane is then given by ${}^{(4)}G_{\mu\nu} = -8\pi G_N[(\rho_m + p_m)U_\mu U_\nu + p_m q_{\mu\nu} - (\rho_m + 3p_m)q_{\mu\nu}/2] + O(\kappa_5^4\rho_m^2)$, with the leading order source acting just like a perfect fluid with energy density $\rho = 3(\rho_m + p_m)/2$ and pressure $p = -(\rho_m + p_m)/2 = -\rho/3$, i.e. acting just like negative pressure quintessence. With negative brane pressure thus potentially being able to arise due to the embedding into the bulk (the bulk stresses maintain the negative pressure on the brane), there may thus be no need to actually introduce any explicit 4-dimensional fluid with intrinsically negative pressure into cosmology at all.

has yet to be applied to either of these two metrics, we note that localization would at least appear possible in the $k=-1$ case since this metric has a horizon at $|w|=w_0$, with both the $f(w)$ and $e^2(w)/f(w)$ coefficients warping all the way to it.

In the time dependent case the AdS_5 embedded solution is found to take the form [15]

$$f(w,t)=a^2[\cosh(\nu|w|/2)-(\tau/a)\sinh(\nu|w|/2)]^2 \ ,$$
$$e(w,t)=\frac{1}{[\nu^2\tau^2-\nu^2a^2-4k]^{1/2}}\frac{df(w,t)}{dt} \ , \tag{36}$$

where the time dependent quantities $a(t)$ and $\tau(t)$ are fixed by the relevant Israel junction conditions

$$3\nu\tau=a\kappa_5^2(\lambda+\rho_m) \ ,-3\nu[\tau\dot{a}+a\dot{\tau}]=\kappa_5^2(-2\lambda+\rho_m+3p_m)a\dot{a} \ , \tag{37}$$

with Eq. (37) itself entailing the standard covariant conservation condition

$$a\dot{\rho_m}+3\dot{a}(\rho_m+p_m)=0 \ . \tag{38}$$

With a resetting of the time according to

$$dt'=\frac{2\dot{a}dt}{[\nu^2\tau^2-\nu^2a^2-4k]^{1/2}}=\frac{6da}{[12\Lambda_4a^2+\kappa_5^4(2\lambda\rho_m+\rho_m^2)a^2-36k]^{1/2}} \ , \tag{39}$$

the metric then takes the convenient form

$$ds^2=-dt'^2[\cosh(\nu|w|/2)-(d\tau/da)\sinh(\nu|w|/2)]^2+$$
$$a^2[\cosh(\nu|w|/2)-(\tau/a)\sinh(\nu|w|/2)]^2[dr^2/(1-kr^2)+r^2d\Omega]+dw^2 \ , \tag{40}$$

with the induced metric at $w=0$ now being a standard comoving RW one. For a perfect fluid source the Einstein tensor on the brane is given by

$${}^{(4)}G_{\mu\nu}=\kappa_5^2(6\Lambda_5+\kappa_5^2\lambda^2)q_{\mu\nu}/12-\lambda\kappa_5^4[(\rho_m+p_m)U_\mu U_\nu+p_mq_{\mu\nu}]/6$$
$$-\kappa_5^4[2\rho_m(\rho_m+p_m)U_\mu U_\nu+\rho_m(\rho_m+2p_m)q_{\mu\nu}]/12 \ , \tag{41}$$

with a specification of an equation of state for the fluid then enabling us to determine $a(t)$, with $\tau(t)$ then being obtainable from Eq. (37).[4] The metric of Eq. (40) thus describes the most general possible embedding of a comoving RW brane of arbitrary spatial 3-curvature k in an AdS_5 bulk,[5] and with its dependence on w being so similar to that found in the static RW case, its localization status would appear to be comparable.

---

[4] We note that the time-time component of Eq. (41) takes the form $-3(\dot{a}^2+k)/a^2=-\Lambda_4-\lambda\kappa_5^4\rho_m/6-\kappa_5^4\rho_m^2/12$, a form we immediately recognize as Eq. (39); and in passing we also note that even though Eq. (39) is not necessarily always integrable in terms of named functions, in the special quintessence case where $p_m=-\rho_m/3$, i.e. $\rho_m=B/a^2$, Eq. (39) actually admits of an exact solution, viz. $a^2(t')=A+C\cosh(\gamma t')$ where $\gamma=(4\Lambda_4/3)^{1/2}$, $A=(2k\lambda-B\nu^2)/\lambda\gamma^2-B/\lambda$, and $C=(B\nu^2-2k\lambda)^2/\lambda^2\gamma^4-B(4k\lambda-B\nu^2)/\lambda^2\gamma^2$. Since the standard purely 4-dimensional cosmology ${}^{(4)}G_{\mu\nu}=-\kappa_4^2(-\lambda q_{\mu\nu}+\tau_{\mu\nu})$ would yield $dt'=da/[\kappa_4^2(\lambda+\rho_m)a^2/3-k]^{1/2}$ for the very same 4-dimensional sources, we see that there is an intrinsic difference between standard and brane embedded cosmology.

[5] In complete analog to the situation found with regard to the M_4 and dS_4 branes, as soon as we take the brane metric to be time dependent we are immediately released from fine tuning constraints.

THE GRAVITATIONAL FLUCTUATIONS ON AN M_4 BRANE

If a small perturbative source S_{AB} is added to the background geometry associated with Eq. (7), this will induce a small change $\delta g_{AB} = h_{AB}$ in the background metric g_{AB} and lead to the fluctuation equation

$$\Delta G_{AB} = \delta G_{AB} + \kappa_5^2 \delta T_{AB} = -\kappa_5^2 S_{AB} \ , \tag{42}$$

with the associated gravitational fluctuation modes then being given as the solutions to $\Delta G_{AB} = 0$. Evaluation of Eq. (42) for fluctuations around an M_4 brane is greatly facilitated by working in the 10 condition Randall-Sundrum gauge

$$h^{5A} = 0 \ , \ h^{\mu}{}_{\mu} = 0 \ , \ h^{\mu\nu}{}_{;\nu} = 0 \ , \tag{43}$$

since ΔG_{5A} is found to vanish identically in this gauge, with the ordinary space-time components of ΔG_{AB} being found to be given by the very compact equation [1, 2]

$$\Delta G_{\mu\nu} = [\partial_w^2 - 4b^2 + e^{2b|w|}\partial_\alpha \partial^\alpha + 4b\delta(w)]h_{\mu\nu}/2 = -\kappa_5^2 S_{\mu\nu} \ , \tag{44}$$

an equation which is conveniently diagonal in the μ, ν indices. In terms of the mixed components $h^{\mu}{}_{\nu} = g^{\mu\alpha}h_{\alpha\nu} = \exp(2b|w|)h_{\mu\nu}$ Eq. (44) may be rewritten as

$$\Delta G^{\mu}{}_{\nu} = [\partial_w^2 - 4b\epsilon(w)\partial_w + e^{2b|w|}\partial_\alpha \partial^\alpha]h^{\mu}{}_{\nu}/2 = g^{-1/2}\partial_A g^{1/2}\partial^A h^{\mu}{}_{\nu}/2 = -\kappa_5^2 S^{\mu}{}_{\nu} \ . \tag{45}$$

In this gauge then each mixed fluctuation component obeys the 5-dimensional scalar Klein-Gordon equation. Moreover, for separable solutions we may simplify Eq. (45) by setting $\partial_\alpha \partial^\alpha h^{\mu}{}_{\nu} = m^2 h^{\mu}{}_{\nu}$; and thus, when we restrict $h_{\mu\nu}$ to depend on $|w|$, we find, on recalling that $d^2|w|/dw^2 = 2\delta(w)$, that Eq. (45) then yields two conditions that the allowed modes must satisfy, viz.

$$\left[\frac{d^2}{d|w|^2} - 4b\frac{d}{d|w|} + e^{2b|w|}m^2\right]\phi(|w|) = 0 \tag{46}$$

and

$$\delta(w)\frac{d\phi(|w|)}{d|w|} = 0 \ , \tag{47}$$

where we use $\phi(|w|)$ to denote each $h^{\mu}{}_{\nu}(|w|)$ component. Additionally, the allowed modes need to be properly orthonormalized. Recalling that the covariant scalar product

$$(\phi_1, \phi_2) = \int (\phi_2^* \partial_A \phi_1 - \phi_1 \partial_A \phi_2^*) n^A d\Sigma \tag{48}$$

with timelike normal n^A and spacelike hypersurface $d\Sigma$ provides a time independent norm for any modes ϕ_1 and ϕ_2 which obey the curved space Klein-Gordon equation, we see that Eq. (48) is precisely the requisite scalar product for the mixed modes $h^{\mu}{}_{\nu}$, with their finiteness thus requiring [13]

$$\int_{-\infty}^{\infty} dw e^{-2b|w|}\phi_1^*(w)\phi_2(w) < \infty \ . \tag{49}$$

Modes which obey all of Eqs. (46), (47) and (49) are readily found [1, 2, 16], with there being an isolated massless bound state graviton with wave function

$$\hat{\phi}_0(w,\bar{x},t) = Ne^{i\bar{p}\cdot\bar{x}-i|\bar{p}|t} \ , \tag{50}$$

and normalization $N = b^{1/2}$, together with a massive continuum of modes which begins at $m = 0$ with wave functions

$$\phi_m(w,\bar{x},t) = N(m)(m^2/b^2)e^{2b|w|}[Y_1(m/b)J_2(me^{b|w|}/b) - J_1(m/b)Y_2(me^{b|w|}/b)]e^{i\bar{p}\cdot\bar{x}-i(p^2+m^2)^{1/2}t} \tag{51}$$

and normalization factor [16, 13]

$$N(m) = \frac{b^2}{2m^{3/2}[J_1^2(m/b)+Y_1^2(m/b)]^{1/2}} \ . \tag{52}$$

With the fluctuation modes $h_{\mu\nu}$ being related to the mixed modes via $h_{\mu\nu} = \exp(-2b|w|)h^{\mu}{}_{\nu}$, we thus see that for all the allowed modes each associated wave function $h_{\mu\nu}$ falls off exponentially fast far way from the brane, with localization of the geometry to the brane thus entailing localization of gravity to the brane as well. Given the mode basis the retarded propagator associated with the $h_{\mu\nu}$ modes is readily calculable [16, 17], and can be written in the convenient form [13]

$$G(x,x',w,w') = e^{-2b|w|}e^{-2b|w'|}\sum_m \phi^*_m(w)\phi_m(w')\Delta(x-x',m) \tag{53}$$

where $\Delta(x-x',m)$ is the standard 4-dimensional flat Minkowski retarded propagator for a field of mass m. With a static brane source $S_{\mu\nu} = \delta^0_\mu\delta^0_\nu M\delta^3(x)\delta(w)$ at the origin of coordinates thus producing a fluctuation on the brane of the form

$$h_{00}(r,w=0) = \frac{\kappa_5^2 M|\hat{\phi}_0(w=0)|^2}{4\pi r} + \kappa_5^2 M\int_0^\infty dm\frac{|\phi_m(w=0)|^2e^{-mr}}{4\pi r} \ , \tag{54}$$

we see that the massless graviton yields the conventional $1/r$ potential on the brane with Newtonian coupling $8\pi G_N = \kappa_5^2 b$. Additionally, for large r the continuum integral gets to be dominated by the small m limit of $\phi_m(w=0)$ (viz. $\phi_m(w=0) \sim m^{1/2}$), so that the continuum integral then generates a non-leading $1/r^3$ potential [2]. Low energy brane localized gravity is thus completely standard, with the continuum of massive modes not affecting long distance low energy gravity on the brane at all.

Recalling that $b = (-\Lambda_5\kappa_5^2/6)^{1/2}$, we see that because of the Randall-Sundrum fine-tuning condition of Eq. (10) we may also set $b = \lambda\kappa_5^2/6$. We thus find that the effective 4-dimensional Newton constant defined by the propagator, viz. $8\pi G_N = \lambda\kappa_5^4/6$, is precisely that obtained in Eq. (22) via the embedding procedure. Now while this is certainly a very desirable result since it confirms the consistency of two different ways of defining G_N, the result is still somewhat puzzling since though the Eq. (22) background reduces to ${}^{(4)}G_{\mu\nu} = 0$ in the M_4 brane case, nonetheless, it is not true that fluctuations around it will obey $\Delta^{(4)}G_{\mu\nu} = -8\pi G_N \delta\tau_{\mu\nu}$ when a weak source $S_{\mu\nu} = \delta\tau_{\mu\nu}\delta(w) = \delta^0_\mu \delta^0_\nu M\delta^3(x)\delta(w)$ is introduced at $w = 0$, since, as we noted earlier, the introduction of a mass source on the brane potentially leads to changes in both C_{ABCD} and $K_{\mu\nu}$. On denoting the net effect of such potential changes by $\delta F_{\mu\nu}$, the lowest order brane fluctuations thus have to generically obey the modified

$$\Delta^{(4)}G_{\mu\nu} = -8\pi G_N \delta\tau_{\mu\nu} - \delta F_{\mu\nu} \tag{55}$$

instead. Since Eq. (55) is not a standard 4-dimensional Einstein fluctuation equation, it is not immediately clear with what strength the massless graviton then does couple, and we thus have to reconcile Eqs. (22), (54) and (55). In order to explicitly do this we have found it very instructive to monitor the $\delta(w)$ contributions to the fluctuation equation.

Since ${}^{(4)}G_{\mu\nu}$ is associated with the induced metric on the brane, and since it transforms as a rank two tensor with respect to the background geometry, we can calculate the change $\Delta^{(4)}G_{\mu\nu}$ due to the change $\delta q_{\mu\nu} = h_{\mu\nu}$ in the induced metric using standard tensor calculus techniques. In the $h^\mu{}_\mu = 0$, $h^{\mu\nu}{}_{;\nu} = 0$ gauge of interest explicit calculation then shows that $\Delta^{(4)}G_{\mu\nu} = \partial_\alpha\partial^\alpha h_{\mu\nu}/2$, so that Eq. (44) may be rewritten as

$$\Delta G_{\mu\nu} = [\partial_w^2 - 4b^2 + 4b\delta(w)]h_{\mu\nu}/2 + e^{2b|w|}\Delta^{(4)}G_{\mu\nu} = -\kappa_5^2 \delta\tau_{\mu\nu}\delta(w) \ . \tag{56}$$

On Taylor expanding $h_{\mu\nu}(|w|) = a^0_{\mu\nu} + a^1_{\mu\nu}|w| + a^2_{\mu\nu}|w|^2/2 + ...$, Eq. (56) entails that

$$a^2_{\mu\nu}/2 - 2b^2 a^0_{\mu\nu} + \Delta^{(4)}G_{\mu\nu} = 0 \ , \quad (a^1_{\mu\nu} + 2ba^0_{\mu\nu})\delta(w) = -\kappa_5^2 \delta^0_\mu \delta^0_\nu M\delta^3(x)\delta(w) \ , \tag{57}$$

so that even while $S_{\mu\nu}$ contains a $\delta(w)$ term, the equation involving $\Delta^{(4)}G_{\mu\nu}$ does not since $\Delta^{(4)}G_{\mu\nu}$ itself possesses no $\delta(w)$ term. However, on substituting for $a^0_{\mu\nu}$ in the static case of interest we obtain

$$\Delta^{(4)}G_{00} = \nabla^2 h_{00}(w=0)/2 = \nabla^2 a^0_{00}/2 = -\kappa_5^2 bM\delta^3(x) - ba^1_{00} - a^2_{00}/2 \ , \tag{58}$$

which we recognize as being of the form of Eq. (55) with $\kappa_5^2 b = 8\pi G_N$ and $\delta F_{00} = ba^1_{00} + a^2_{00}/2$. For the massless graviton exchange contribution where $h_{00}(r,w) = \exp(-2b|w|)\kappa_5^2 bM/4\pi r$, the Taylor series expansion coefficients explicitly evaluate to

$$a^0_{00} = \kappa_5^2 bM/4\pi r \ , \ a^1_{00} = -\kappa_5^2 M[b^2/2\pi r + \delta^3(x)] \ , \ a^2_{00} = \kappa_5^2 bM[b^2/\pi r + \delta^3(x)] \ , \tag{59}$$

so that δF_{00} takes the value $-\kappa_5^2 bM\delta^3(x)/2$ and is thus explicitly non-zero. Thus finally, on inserting Eq. (59) into Eq. (58) we obtain none other than

$$\nabla^2(\kappa_5^2 bM/8\pi r) = -\kappa_5^2 bM\delta^3(x)/2 \tag{60}$$

just as desired of massless graviton exchange on the brane. We thus conclude that even though the fluctuations on the brane obey the non-standard Eq. (55), nonetheless, through a delicate interplay, the resulting fluctuations turn out to still be completely canonical. Having now explored the structure of the Randall-Sundrum set-up, we now briefly discuss how such a set-up could be achieved dynamically; and shall thus explore the dynamics associated with the coupling of gravity to a bulk sine-Gordon scalar field (a model also considered in [18]), and show [19] how it naturally leads to Randall-Sundrum localization of gravity without any need for fine-tuning.

DYNAMICAL LOCALIZATION OF GRAVITY

For a scalar field with potential $V(\phi) = A^2\beta^2/8 - (A^2\beta^2/8)(1+\kappa_2^5 A^2/3)sin^2(2\phi/A)$ coupled to the metric of Eq. (5) with $T_{00} = e^{2f(w)}[\phi'^2/2 + V(\phi)]$, $T_{55} = \phi'^2/2 - V(\phi)$, there is an exact solution to the 5-dimensional Einstein equations, viz.

$$tan(\phi/A) = tanh(\beta w/2) \ , \ e^{f(w)} = [cosh(\beta w)]^{-A^2\kappa_5^2/12} \ . \tag{61}$$

Here $e^{f(w)}$ peaks at $w = 0$ while warping away from it, with the solution thus representing a thick domain wall supported by a soliton. Moreover, without assuming any input Z_2 symmetry, in the solution the output domain wall nonetheless has acquired one from the underlying symmetry structure which solitons intrinsically possess. Given the solution, if we now take the limit $A \to 0$, $\beta \to \infty$ with $A^2\beta$ held fixed, we find that [19]

$$e^{f(w)} \to e^{-(-\Lambda_5\kappa_5^2/6)^{1/2}|w|} \tag{62}$$

which is precisely of the Randall-Sundrum form. Here $\Lambda_5 = -\kappa_5^2 A^4\beta^2/24$ is the minimum value of $V(\phi)$. In this same limit we find that the scalar field energy density T_{00} develops a $\lambda\delta(w)$ component where $\lambda = A^2\beta/2$, and thus on comparing terms we naturally recover [19] the Randall-Sundrum $6\Lambda_5 + \kappa_5^2\lambda^2 = 0$ condition without fine-tuning.

ACKNOWLEDGMENTS

The author wishes to thank to Drs. A. Davidson, A. H. Guth, D. I. Kaiser and A. Nayeri for many helpful discussions. This work has been supported in part by the Department of Energy under grant No. DE-FG02-92ER40716.00.

REFERENCES

1. Randall L., and Sundrum R., Phys. Rev. Lett. **83**, 3370 (1999).
2. Randall L., and Sundrum R., Phys. Rev. Lett. **83**, 4690 (1999).
3. Shiromizu T., Maeda K., and Sasaki M., Phys. Rev. D**62**, 024012 (2000).
4. Israel W., Nuovo Cim. B**44**, 1 (1966).
5. Mannheim P. D., Phys. Rev. D**64**, 068501 (2001).
6. Binetruy P., Deffayet C., and Langlois D., Nucl. Phys. B**565**, 269 (2000).
7. Csaki C., Graesser M., Kolda C., and Terning J., Phys. Lett. B**462**, 34 (1999).
8. Brecher D., and Perry M. J., Nucl. Phys. B**566**, 151 (2000).
9. DeWolfe O., Freedman D. Z., Gubser S. S., and Karch A., Phys. Rev. D**62**, 046008 (2000).
10. Kim H. B., and Kim H. D., Phys. Rev. D**61**, 064003 (2000).
11. Garriga J., and Sasaki M., Phys. Rev. D**62**, 043523 (2000).
12. Karch A., and Randall L., J. High Energy Phys. **0105**, 008 (2001).
13. Guth A. H., Kaiser D. I., Mannheim P. D., and Nayeri A., in preparation (2002).
14. Mannheim P. D., Phys. Rev. D**63**, 024018 (2001).
15. Mannheim P. D., Phys. Rev. D**64**, 065008 (2001).
16. Garriga J., and Tanaka T., Phys. Rev. Lett. **84**, 2778 (2000).
17. Giddings S. B., Katz E., and Randall L., J. High Energy Phys. **0003**, 023 (2000).
18. Gremm M., Phys. Lett. B**478**, 434 (2000); Behrndt K., Phys. Lett. B**487**, 30 (2000).
19. Davidson A., and Mannheim P. D., Dynamical Localization of Gravity, hep-th/0009064 (2000).

Metric-Field Approach to Gravitation and the Problem of the Universe Acceleration

Leonid V. Verozub
Kharkov National University
verozub@gravit.kharkov.ua

Abstract

A metric-field approach to gravitation is presented. It is based on an idea of dependency of space-time properties on measuring instruments. Some bimetric equations that realize this idea are considered. They were tested by the binary pulsar PSR1913+16. The spherically - symmetric solution of the equations has no event horizon and no physical singularity in the center. The proper energy of a point particle is finite. There can exist supermassive compact configurations of the degenerated Fermi-gas which can be identified with observed objects in galactic centers. The problem of the Universe acceleration has a natural explanation.

1 Introduction

The geometrical properties of space-time can be described only by means of measuring instruments. At the same time, the description of the properties of measuring instruments, strictly speaking, requires the knowledge of space-time geometry. One of the implications of it is that geometrical properties of space and time have no experimentally verifiable significance by themselves but only within the aggregate "geometry + measuring instruments". We got aware of it owing to Poincaré [1]. It is a development of the Berkley - Leibnitz - Mach idea about relativity of space-time properties, which is an alternative to the well known Newtonian approach.

If we proceed from the conception of relativity of space-time, we assume that there is no way of quantitative description of physical phenomena other than attributing them to a certain frame of reference which, in itself, is a physical device for space and time measurements. But then the relativity of the geometrical properties of space and time mentioned above is nothing else but relativity of space-time geometry with respect to the frame of reference being used.[1]. Thus, it should be assumed that the concept of the reference frame as a physical object, whose properties are given and are independent of the properties of space and time, is approximate, and only the aggregate "frame of reference + space-time

[1]There is an important difference between a frame of reference (as a physical device) and a coordinate system (as the way to parameterize points of space-time) [5]

CP624, *Cosmology and Elementary Particle Physics*, edited by B. N. Kursunoglu et al.

geometry" is reasonable. The Einstein theory of gravitation demonstrates relativity of space-time with respect to distribution of matter. However, space-time relativity with respect to measurement instruments hitherto has not been realized in physical theory. An attempt to show that there is also space-time relativity to the used reference frames has been undertaken for the first time in [3], [4].

2 Fundamental Metric Form in NIFRs.

At present we do not know how the space-time geometry in inertial frames of reference (IFRs) is connected with the frames properties. Under the circumstances, we simply postulate (according to special relativity) that space - time in IFRs is pseudo-Euclidean. Next, we find a space-time metric differential form in noninertial frames of reference (NIFRs) from the viewpoint of an observer in the NIFR who proceeds from the relativity of space and time in the Berkley - Leibnitz - Mach - Poincaré (BLMP) sense.

By a noninertial frame of reference we mean the frame, whose body of reference is formed by the point masses moving in the IFR under the effect of a given force field. It would be a mistake to identify "a priori" a transition from the IFR to the NIFR with the transformation of coordinates related to the frames. If we act in such a way, we already assume that the properties of space-time in both frames are identical. However, for an observer in the NIFR, who proceeds from the relativity of space and time in the BLMP sense, space-time geometry is not given "a priori" and must be ascertained from the analysis of experimental data. We shall suppose that the reference body (RB) of the IFR or NIFR is formed by the identical point masses m_p. If the observer is at rest in one of the frames, his world line will coincide with the world line of some point of the reference body. It is obvious to the observer in the IFR that the accelerations of the point masses forming the reference body are equal to zero. Of course, this fact occurs in relativistic sense too. Let $d\eta$ and ds be denote the differential metric forms in the IFR and NIFR. Then, if $\nu^\alpha = dx^\alpha/d\eta$ is the 4- velocity vector of the point masses forming the reference body of the IFR, the absolute derivative of the vector ν^α is equal to zero,i.e.

$$D\nu^\alpha/d\eta = 0. \tag{1}$$

From the viewpoint of an observer in a NIFR who proceeds from the relativity of space-time in the BLM sense the points of his body reference are the points of his physical space, they are not exposed any forces. Consequently, their accelerations are equal to zero both in nonrelativistic and relativistic sense. This means that 4-velocity $\zeta^\alpha = dx^\alpha/ds$ of the point masses forming the reference body of the NIFR obeys equality

$$D\zeta^\alpha/ds = 0 \tag{2}$$

In other words, since for the observer in the IFR according to eq. (1) world lines of the IFR points are geodesic lines, then for the observer in the NIFR world lines of the NIFR points of the reference body are also geodesic lines in his space-time, which can be expressed by eq. (2).

This fact leads to important consequences. The differential equations of these world lines at the same time are the Lagrange equations of motion of the NIFR RB points. The Lagrange equations, describing the motion of the identical RB point masses in the IFR, can be obtained from the Lagrange action S by the principle of least action. Therefore, the equations of the geodesic lines can be obtained from the differential metric form $ds = k\ dS(x, dx)$, where k is the constant, $dSdt = L(x, \dot{x})dt$ and L is the Lagrange function. The constant $k = -(m_p c)^{-1}$, as it follows from the analysis of the case when the frame of reference is inertial. Thus, if we proceed from relativity of space and time in the BLMP sense, the differential metric form of space-time in the NIFR can be expected to have the following form [3], [4].

$$ds = -(m_p c)^{-1}\ dS(x, dx). \tag{3}$$

So, the properties of space-time in the NIFR are entirely determined by the properties of the used frame in accordance with the idea of relativity of space and time in the BLMP sense.

Consider two examples of the NIFR.

1.The reference body is formed by noninteracting electric charges moving in a constant homogeneous electric field $\mathcal{E}$. The motion of the charges is described in Cartesian coordinates by the Lagrangian

$$L = -m_p c^2\ (1 - V^2/c^2)^{1/2} + \mathcal{E}\ e\ x, \tag{4}$$

where V is the speed of the charges. According to eq. (4) the space - time metric differential form in this frame is given by

$$ds = d\eta - (wx/c^2)dx^0, \tag{5}$$

where $d\eta = (c^2dt^2 - dx^2 - dy^2 - dz^2)^{1/2}$, is the metric differential form of the pseudo - Euclidean space - time in the IFR and $w = e\mathcal{E}/m$ is the acceleration of the charges.

2. The reference body consists of noninteracting electric charges in a constant homogeneous magnetic field $\mathcal{H}$ directed along the axis z. The Lagrangian describing the motion of the particles can be written as follows

$$L = -m_p c^2(1 - V^2/c^2)^{1/2} - (m_p\Omega_0/2)(\dot{x}y - x\dot{y}), \tag{6}$$

where $\dot{x} = dx/dt$, $\dot{y} = dx/dt$ and $\Omega_0 = e\mathcal{H}/2mc$. The points of such a system rotate in the plane xy around the axis z with the angular frequency $\omega = \Omega_0[1 + (\Omega_0 r/c)^2]^{-1/2}$, where $r = (x^2 + y^2)^{1/2}$. The linear velocities of the BR points tend to c when $r \to \infty$. For the given NIFR

$$ds = d\eta + (\Omega_0/2c)\ (ydx - xdy). \tag{7}$$

In the above NIFR ds is of the form $ds = \mathcal{F}(x, dx)$ where $\mathcal{F}(x, dx) = d\eta + f_\alpha(x)dx^\alpha$, f_α is a vector field. The function $\mathcal{F}$ is a homogeneous of the first degree in dx^α. Therefore, generally speaking, the space-time in NIFR is Finslerian [2] with the sign - indefinite differential metric form.

One of the consequences of the above result is a natural explanation of the Sagnac effect and the fact of the existence of the inertial forces in NIFRs [3], [4].

3 Experimental Verification

A clock, which is in a NIFR at rest, is unaffected by acceleration in space - time of the frame. The change in rate of the ideal clock is a real consequence of the difference between the space - time metrics in the IFR and NIFR. It is given by the factor $\sigma = ds/d\eta$ from the equation $ds = \sigma d\eta$. For the rotating with the angular velocity Ω disk of the radius R the factor $\sigma = 1 - \Omega^2 R^2/2c^2$ which gives rise to the observed red shift in the well known Pound - Rebka - Snider experiments.

Another experimentally verifiable consequence of the above theory is some difference between the inertial mass m_p^{eq} of a body on the Earth's equator and the mass m_p^{pol} of the same body on the pole. It is given by

$$(m_p^{eq} - m_p^{pol})/m_p^{pol} = 1.2 \cdot 10^{-12} \tag{8}$$

The dependence of the inertial mass of particles on the Earth's longitude can be observed by the Mössbauer effect. Indeed, the change $\Delta\lambda$ in the wave length λ at the Compton scattering on particles of the masses m_p is proportional to m_p^{-1}. If this value is measured for a gamma-quantum with the help of the Mössbauer effect at a fixed scattering angle, then after transporting the measuring device from the longitude φ_1 to the longitude φ_2 we have

$$\frac{(\Delta\lambda)_{\varphi_1}^{-1} - (\Delta\lambda)_{\varphi_2}^{-1}}{(\Delta\lambda)_{\varphi_1}^{-1}} = \Theta\ [cos^2(\varphi_1) - cos^2(\varphi_2)], \tag{9}$$

where $\Theta = 1.2 \cdot 10^{-12}$.

4 Gravitation in Inertial and Proper Reference Frames

Consider a frame of reference whose reference body is formed by identical material points m_p moving under the effect of the field $\psi_{\alpha\beta}$. These frames will be called the proper frames of reference (PRFs) of the given field. Any observer, located in the PFR at rest, moves in space-time of this frame along the geodesic line of his space-time. This implies that the space-time metric differential form in the NIFR is given by eq. (3) where S is the action describing in a IFR the motion of particles forming the reference body of the NIFR. Now suppose, following Thirring [6], that in pseudo- Euclidean space-time gravitation can be described as a tensor field $\psi_{\alpha\beta}(x)$, and the Lagrangian describing motion of a test particle with the mass m_p is of the form

$$L = -m_p c[g_{\alpha\beta}(\psi)\dot{x}^{\alpha} \dot{x}^{\beta}]^{1/2}, \tag{10}$$

where $\dot{x}^{\alpha} = dx^\alpha/dt$ and $g_{\alpha\beta}$ is the symmetric tensor whose components are the function of $\psi_{\alpha\beta}$. Then, according to (3) the space-time metric differential form in the PFR is given by

$$ds^2 = g_{\alpha\beta}(\psi)\ dx^\alpha\ dx^\beta \tag{11}$$

Thus, the space-time in the PFR is a Riemannian with the curvature other than zero. Viewed by an observer in the IFR, the motion of the test particle forming the reference body of the PFR is affected by the force field $\psi_{\alpha\beta}$. But the observer located in the PRF will not observe the force properties of the field $\psi_{\alpha\beta}$ since he moves in space-time of the PRF along the geodesic line. For him the presence of the field $\psi_{\alpha\beta}$ will be displayed in another way — as space-time curvature differing from zero in these frames, e.g. as a deviation of the world lines of the neighbouring points of the reference body. For example, when studying the Earth's gravity, a frame of reference fixed to the Earth can be considered as an inertial frame if the forces of inertia are ignored. An observer located in this frame can consider motion of the particles forming the PRF reference body in flat space-time on the basis of eq. (10) without running into contradiction with experiments. However, the observer in the PFR (in a comoving frame for free falling particles) does not find the Earth's gravity as some force field. If he proceeds from the relativity of space-time, he believes that point particles, forming the reference body of his reference frame, are the point of his physical space. They are not affected by a force field and, therefore, their accelerations in his space-time are equal to zero. In spite of that, he observes a change in the relative distances of these particles. Such an experimental fact has apparently the only explanation as non-relativistic display of the deviations of the geodesic lines caused by space-time curvature. So, we observe an important fact that only in proper frames of reference we have an evidence for gravitation identification with space-time curvature.

Thus, we arrive at the following hypothesis. In inertial frames of reference, where space-time is pseudo-Euclidean, gravitation is a field $\psi_{\alpha\beta}$. In the proper frames of reference of the field $\psi_{\alpha\beta}$, where space-time is Riemannian, gravitation manifests itself as curvature of space-time and must be described completely by the geometrical properties of the latter.

Of course, eq. (3) refers to any classical field. For instance, space-time in the PRF of an electromagnetic field is Finslerian. However, since ds depends on the mass m_p and charge e of the point masses forming the reference body, this fact is not of great significance.

5 Geodesic-invariant equations of gravitation

Gravitational equations should be some kind of differential equations for the function $\psi_{\alpha\beta}$ or $g_{\alpha\beta}(\psi)$, which are invariant under a certain kind of gauge transformations of the potentials $\psi_{\alpha\beta}$. Since $g_{\alpha\beta} = g_{\alpha\beta}(\psi)$, the Einstein equations are the equations both for $g_{\alpha\beta}$ and for $\psi_{\alpha\beta}$. Under the transformation $\psi_{\alpha\beta} \to \overline{\psi}_{\alpha\beta}$ the quantities $g_{\alpha\beta}(\psi)$ undergo some transformations too and, as a consequence, the equations of the test particle motion resulting from eq. (10) and the Einstein's equations do not remain invariant. The equations of motion resulting from eq. (10) are at the same time the equations of a geodesic line of the Riemannian space-time V_n of the dimensionality n with the metric tensor $g_{\alpha\beta}(\psi)$. That is why if the given gauge transformation $\psi_{\alpha\beta} \to \overline{\psi}_{\alpha\beta}$ leaves the equations of motion invariant, then the corresponding transformation $g_{\alpha\beta} \to \overline{g}_{\alpha\beta}$ is a mapping $V \to \overline{V}$ of the Riemannian spaces leaving geodesic lines invariant, i.e. it is a geodesic, (projective) mapping. Let us assume that not only eq. (10) but also the field equations contain $\psi_{\alpha\beta}$ only in the form $g_{\alpha\beta}(\psi)$, then it becomes clear that the gauge-invariance of the equations of motion will be ensured if the field equations are invariant with respect to geodesic mappings of the Riemannian space V_n. Thus, if we start from eq. (10), then the gravitational field equations as well as the physical field characteristics must be invariant with respect to geodesic (projective) mappings of the Riemannian space-time V_n with the metric tensor $g_{\alpha\beta}\,(\psi)$.

The simplest equations of gravitation that can be considered as a realization of the above idea were proposed in paper [7]. (From another viewpoint). They are given by

$$B^{\alpha}_{\beta\gamma;\alpha} - B^{\nu}_{\beta\mu}\, B^{\mu}_{\gamma\nu} = 0\,, \tag{12}$$

In these equations

$$B^{\gamma}_{\alpha\beta} = \Pi^{\gamma}_{\alpha\beta} - \overset{\circ}{\Pi}_{\alpha\beta^{\gamma}}\,, \tag{13}$$

where $\Pi^{\gamma}_{\alpha\beta}$ and $\overset{\circ}{\Pi}{}^{\gamma}_{\alpha\beta}$ are the Thomas symbols for V_n and E_n,

$$\Pi^{\gamma}_{\alpha\beta} = \Gamma^{\gamma}_{\alpha\beta} - (n+1)^{-1}\left[\delta^{\gamma}_{\alpha}\Gamma^{\mu}_{\beta\mu} + \delta^{\gamma}_{\beta}\Gamma^{\mu}_{\alpha\mu}\right]\,, \tag{14}$$

$$\overset{\circ}{\Pi}{}^{\gamma}_{\alpha\beta} = \overset{\circ}{\Gamma}{}^{\gamma}_{\alpha\beta} - (n+1)^{-1}\left[\delta^{\gamma}_{\alpha}\overset{\circ}{\Gamma}{}^{\epsilon}_{\epsilon\beta} + \delta^{\gamma}_{\beta}\overset{\circ}{\Gamma}{}^{\epsilon}_{\epsilon\alpha}\right]\,, \tag{15}$$

$\Gamma^{\gamma}_{\alpha\beta}$ and $\overset{\circ}{\Gamma}{}^{\gamma}_{\alpha\beta}$ are the Christoffel symbols in V_n and E_n, respectively.

They are bimetric geodesic invariant equations. Each solution $g_{\alpha\beta}(x)$ of (12) refers to some coordinate system and is determined up to arbitrary geodesic mappings, which play the role gauge transformation in the theory under consideration. The physical meaning may have only geodesic invariant magnitudes, for example, the tensor $B^{\gamma}_{\alpha\beta}$. At the covariant gauge conditions $Q_{\alpha} = \Gamma^{\beta}_{\beta\alpha} - \overset{\circ}{\Gamma}{}^{\beta}_{\beta\alpha} = 0$ eqs. 12 are equivalent to the system

$$R_{\alpha\beta} = 0 \tag{16}$$

and

$$Q_{\alpha} = 0, \tag{17}$$

where $R_{\alpha\beta}$ is the Ricci tensor.

The equations do not contain the functions $\psi_{\alpha\beta}(x)$ explicitly. The simplest way of obtaining equations for such a kind of the functions $\psi_{\alpha\beta}$ is to set

$$B^{\alpha}_{\beta\gamma} = \nabla^{\alpha}\psi_{\beta\gamma} - (n+1)^{-1}\left(\delta^{\alpha}_{\beta}\nabla^{\sigma}\psi_{\sigma\gamma} + \delta^{\alpha}_{\gamma}\nabla^{\sigma}\psi_{\beta\sigma}\right), \tag{18}$$

where ∇^{α} is the covariant derivative in flat space-time. Then, at the gauge condition $\nabla^{\sigma}\psi_{\sigma\gamma} = 0$ eq. (12) are given by

$$\Box\psi_{\alpha\beta} - \nabla^{\sigma}\psi_{\alpha\gamma}\,\nabla^{\gamma}\psi_{\sigma\beta} = 0;\quad \nabla^{\sigma}\psi_{\sigma\gamma} = 0, \tag{19}$$

where $\Box$ is the covariant Dalamber operator in pseudo-Euclidean space-time. It is natural to suppose that with the presence of matter these equations are given by

$$\Box\psi_{\alpha\beta} = \varkappa(T_{\alpha\beta} + t_{\alpha\beta});\quad \nabla^{\sigma}\psi_{\sigma\gamma} = 0, \tag{20}$$

where $\varkappa = 8\pi G/c^4$, $t_{\alpha\beta} = \varkappa^{-1}\nabla^\sigma \psi_{\alpha\gamma}\nabla^\gamma \psi_{\sigma\beta}$ and $T_{\alpha\beta}$ is the matter tensor of the energy-momentum. Obviously, the equality

$$\nabla^\beta(T_{\alpha\beta} + t_{\alpha\beta}) = 0 \tag{21}$$

is valid. Therefore, the magnitude $t_{\alpha\beta}$ can be interpreted as the energy-momentum tensor of a gravitational field. [2] . At the conditions $\nabla^\sigma \psi_{\sigma\gamma} = 0$

$$t_{\alpha\beta} = \chi^{-1} B^\gamma_{\alpha\sigma} B^\sigma_{\beta\gamma} \tag{22}$$

6 Gravitational Energy of a Point Mass

If the Lagrangian of test particles is invariant under the mapping $t \to -t$, the fundamental metric form of space-time V_4 in the spherically-symmetric case can be written as

$$ds^2 = -Adr^2 - B[d\theta^2 + \sin^2\theta\, d\varphi^2] + Cdx^{0\,2}, \tag{23}$$

where A, B and C are the functions of the radial coordinate r. The general solution of the system (16) – (17) at the conditions

$$\lim_{r\to\infty} A = 1, \ \lim_{r\to\infty} (B/r^2) = 1, \ \lim_{r\to\infty} C = 1. \tag{24}$$

is of the form [10]

$$A = (f')^2(1 - \mathcal{Q}/f)^{-1}, \ B = f^2, \ C = 1 - \mathcal{Q}/f \tag{25}$$

where

$$f = (r^3 + \mathcal{K}^3)^{1/3}$$

$f' = df/dr$, $\mathcal{Q}$ and $\mathcal{K}$ are constants.

The equations of the motion of a test particle resulting from Lagrangian (10) is given by

$$\ddot{x}^\alpha + (\Gamma^\alpha_{\beta\gamma} - c^{-1}\Gamma^0_{\beta\gamma}\dot{x}^\alpha)\dot{x}^\beta\dot{x}^\gamma = 0. \tag{26}$$

In the nonrelativistic limit $\ddot{x}^r = -c^2\,\Gamma^r_{00}$, where $\Gamma^r_{00} = C'/2A = r^4C'/f^4C$. Therefore, to obtain the Newton gravity law it should be supposed that at large r the function $f \approx r$ and $\mathcal{Q} = r_g = 2GM/c^2$ is the classical Schwarzshild radius.

We can also argue that the constant $\mathcal{K} = r_g$. Indeed, consider the 00-component of the first of eq. (20). Let us set $T_{\alpha\beta} = \rho c^2 u_\alpha u_\beta$, where ρ is the matter density and u_α is the 4-velocity of matter points. At the small macroscopic velocities of the matter we can set $u_0 = 1$ and $u_i = 0$. Therefore, the equation is of the form

$$\Box\psi_{\alpha\beta} = \chi(\rho c^2 + t_{00}) \tag{27}$$

[2] It should be noted that, when we introduce it in some way, we cannot be sure apriori that the equation for $\psi_{\alpha\beta}$ yields all solutions of the equations for $B^\alpha_{\beta\gamma}$. We may introduce a potential $\psi_{\alpha\beta}$ also in another way.

where $\chi = 4\pi G/c^4$ and t_{00} is the 00-component of the tensor (22). Let us find the energy of a gravitational field of the point mass M as the following integral in the pseudo-Euclidean space-time

$$\mathcal{E} = \int t_{00} dV, \tag{28}$$

resulting from the above solution, where dV is the volume element. In the Newtonian theory this integral is divergent. In our case we have:

$$\mathcal{E} = \frac{\mathcal{Q}}{\mathcal{K}} M\,c^2 \tag{29}$$

We arrive at the conclusion that at $\mathcal{K} \neq 0$ the energy of the point mass is finite and at $\mathcal{K} = \mathcal{Q}$ the rest energy of the point particle in full is caused by its gravitational field: $\mathcal{E} = M\,c^2$

The spacial components of the vector $P_\alpha = t_{0\alpha}$ are equal to zero. Due to these facts we assume in the present paper that $\mathcal{K} = \mathcal{Q} = r_g$ and consider the solution (25) in the spherical coordinate system at the used gauge condition as a basis for the subsequent analysis.

7 Acceleration of the Universe as a Consequence of Gravitation Properties

The analysis of the recent observations data gives evidence that the deceleration parameter $q_o = -\ddot{a}(t)\ a(t)/\dot{a}(t)$ (a is the scale factor) is negative at the moment ([12] , [11]) . It means that $\ddot{a} > 0$ i.e. the expansion is accompanied with acceleration, while according to classical insights the gravity force must retard the expansion.

Equations (12) were successfully testified by the classical tests and binary pulsar PSR1913+16 [10], [9]. The motion of test particles in the spherically - symmetric gravitational field differs very little from that in general relativity at distances from the center $r \gg r_g$. However, they are completely different at $r \leq r_g$. The solution of (12) has no the event horizon and physical singularity in the center. The gravitational force (as the mass multiplied by the acceleration) affecting escape particles is repulsive from $r = 0$ up to distances of the order of r_g. The observed radius R_U of the Universe is about $10^{27} cm.$ and the observed mass M_U is $10^{56} \div 10^{57} g$, so the magnitude $2GM_U/c^2$ is of the order of R_U . For this reason, we can expect some manifestation of the repulsive force for distant objects.

Consider in flat space-time a simple model of an expanding selfgraviting homogeneous dust - ball with the sizes of the observed Universe. According to [7], the equation of the motion of a test particle in the spherically-symmetric field are given by

$$\dot{r}^2 = (c^2 C/A)[1 - (C/\overline{E})(1 + r_g^2 \overline{J}^2/B)], \tag{30}$$

$$\dot{\varphi} = c\,C\overline{J} r_g/(B\overline{E}) \tag{31}$$

where (r, φ, θ) are the spherical coordinates (θ is supposed to be equal to $\pi/2$), $\dot{r} = dr/dt$, $\dot{\varphi} = d\varphi/dt$, $\overline{E} = E/(mc^2)$, $\overline{J} = J/(amc)$, E is the particle energy, J is the angular momentum.

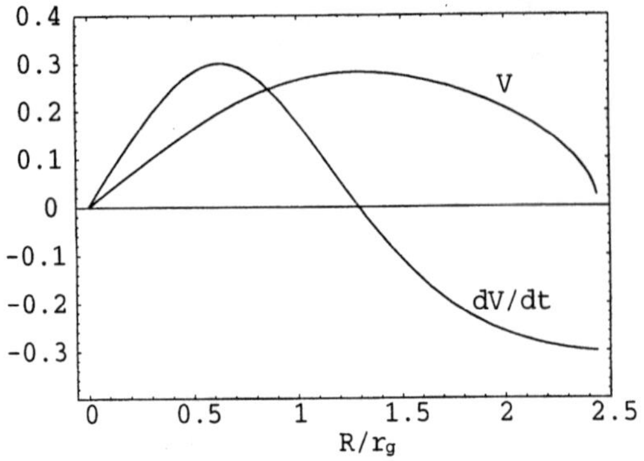

Figure 1: The radial velocity on the ball surface vs. the radius R at $\overline{E}^2 < 1$.

Consequently, the radial velocity of the specks of dust of the ball surface as a function of its radius R is given by

$$V^2 = \frac{c^2 C}{A}\left[1 - \frac{C}{\overline{E}^2}\right], \tag{32}$$

where , $V = \dot{R} = dR/dt$,

$$C = 1 - \frac{1}{\bar{f}}; \quad A = \frac{\bar{r}^4}{\bar{f}^4} C^{-1}; \quad , \bar{f} = (1 + \bar{r}^3)^{1/3} \tag{33}$$

and the dependence of the acceleration $\dot{V} = dV/dt$ on the radius R is given by

$$\dot{V} = VV' \tag{34}$$

Figs.1 and 2 show the plot of the velocity and the acceleration (arbitrary units) as the function of the radius at $\overline{E}^2 = 0.60 < 1$ and $\overline{E}^2 = 1.20 > 1$.

It follows from the figures that the acceleration is negative at f $R/r_g \gg 1$, and is positive if R/r_g is of the order of r_g or less than that. For example, if the radius $R = R_U$ and the matter density $\rho = 2 \cdot 10^{-28} g/cm^3$, the value of $R_U/r_g = 0.9 < 1$ and the acceleration is equal to $1 \cdot 10^{-8} cm/s^2$ which is half as large as this magnitude resulting from the value of $q_0 = -1$ that was found in [12].

In paper [13] the Riess et al. results were studied in detail in view of the model above. A good compliance was found.

8 Conclusion

The key reason preventing a correct inclusion of the Einstein theory of gravitation in the interactions unification is that gravity is identified with space-time curvature. It is also a cause of such unsolved problems of the theory as an operational definition of an observable quantity, the energy - momentum tensor problem and

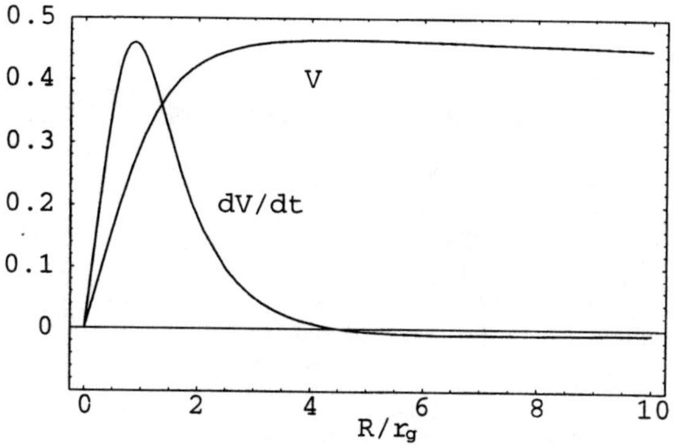

Figure 2: The radial velocity and acceleration on the ball surface vs. the radius R at $\overline{E}^2 < 1$.

gravity quantization. So, if the analysed above possibility really takes place in nature, then it will remove an isolation of the geometrical gravitational theories from the theories of other fields.

References

[1] Poincaré, H., *Dernières pensées* , Flammarion Paris (1913)

[2] Rund, H., *The differential geometry of Finsler space*, Springer (1959)

[3] Verozub, L., *Ukr. Phys.Journ.* **10**, 131, 778, 1598 (1981) (In Russian)

[4] Verozub, L., *Phys. Essays,* **8**, 518 (1995)

[5] Rodichev, V., In: *Einstein's Collection*, 286 (1974) (In Russian)

[6] Thirring, W., *Ann. Phys.* **16**, 96 (1961)

[7] Verozub, L., *Phys. Lett. A,* **156**, 404 (1991)

[8] Verozub, L., *Gravit., Cosm. and Relat. Astroph., Kharkov*, 44 (2001)

[9] Verozub, L. and Kochetov, A., *Grav and Cosmol.*, **6**, 246 (2000)

[10] Verozub, L., *Astr. Nachr.*, **317**, 107 (1996)

[11] Perlmutter, S., Aldering, G., Boyle, B., et al., 1998, In: "*Abstracts of 19th Texas Symp. on Relat. Astroph. and Cosm.*", (1998) 14.

[12] Riess, A., Filipenko, A., Challis, P., et al., *AJ*, **116** (1998) 1009.

[13] Verozub L. and Kochetov A., *Gravit, Cosm. and Relat. Astroph, Kharkov*, 90 (2001).

The Cosmological Constant Problem, The Spontaneously Broken Symmetry, And The Generalized Rest Energy

Freydoon Mansouri

Physics Department, University of Cincinnati, Cincinnati, OH 45221

Abstract. We study the consequences of spontaneously broken symmetry on the structure of space-time. In particular, we look for the modification of $E = mc^2$ when Minkowski space is replaced by de Sitter or anti-de Sitter space as a result of a non-vanishing vacuum energy.

INTRODUCTION

The notion of the spontaneously broken symmetry plays an important role in our understanding of particle physics at the fundamental level. It also plays an important role in the inflationary cosmology. We have therefore good reasons to take this concept as an important factor in the development of the universe. For whatever dynamical reasons, when a symmetry is broken spontaneously, there will be a contribution to the vacuum energy density ρ_{vac} of space-time. On the basis of Einstein's field equations, such a non-vanishing vacuum energy is equivalent to having a cosmological, Λ, given by [1, 2]

$$\Lambda = 8\pi G_N \rho_{vac} \tag{1}$$

This relation raises a number of interesting issues. One is that, according to current theories, the contribution to Λ from each phase transition, such as that in electroweak theory, is very large [1, 2]. It follows that unless a delicate cancellation mechanism such as supersymmetry has been in operation over the entire range of scales from the time of big bang to present, it would be difficult to account for the smallness of $\Lambda_{observed}$ according to standard interpretations. As will be seen below, there may be a way out of this impasse if we view the vacuum energy arising from spontaneouly broken symmetries as the primary source of Λ.

Another issue which arises from the equality given by Eq. (1) is that for finite Λ, the asymptotic symmetry of space-time is not Poincaré but de Sitter (dS) or anti-de Sitter (AdS) group. On the other hand, the well known Einstein mass-energy relation $E = mc^2$ is consequence of Poincaré symmetry and is valid only in Minkowski space-time. One of the aims of the present work is to point out how the rest energy expression is modified in dS_4 and AdS_4 spaces, with their respective radii of curvature determined by vacuum energy. As mentioned above, if we take the connection between Λ and spontaneously broken symmetry (SBS) seriously, there must have been some periods in the history of the universe, at which the value of Λ changed significanly because of phase transitions.

CP624, *Cosmology and Elementary Particle Physics*, edited by B. N. Kursunoglu et al.

In particular, at some such periods, the value of Λ must have been very large. It would be of interest to see how the kinematics of high energy reactions were modified during such periods and whether this led to any testable consequences. It will be seen below that if again we take SBS as the primary source of a non-vanishing Λ, it may be possible to test some of the consequences at the next generation of high energy colliders. To accomplish this, it will be necessary to depart from conventional wisdom and reexamine the period in the history of universe, which the current high energy experiments are testing. The main result is that in both dS and AdS spaces the ground state energies, i.e., the analogs of the energy in Einstein's mass-energy relation, depend not only on the mass but also on spin and on Λ, with the correct value of Λ to be determined such that it is appropriate to the relevant period.

The mathematical frameworks for obtaining the analogs of Einstein's mass-energy relation in AdS_4 and dS_4 spaces depend on a mixture of the representation theory of the corresponding groups and the field theoretic constructions in these spaces. Most of this already exist in the literature in one form or another. The construction of the representations for these groups relies on the structure of appropriate subgroups. For AdS_4 group, the relevant subgroup turns out to be [3, 4, 5] the compact group $SO(2) \times SO(3)$. For dS_4 group, the simplest choice for a subgroup is the maximal compact subgroup [6] $SO(4)$. On the other hand, the direct analog of subgroup for AdS_4 is the non- compact subgroup $SO(1,1) \times SO(3)$. In both spaces, to relate the eigenvalues of the quadratic Casimir operators to the physical massive states, it is necessary to introduce the concept of mass as the pole of the propagator in an appropriate quantum field theory. To lowest order this pole would be that of the Green's function for a dS_4 (or AdS_4) invariant differential equation. As is familiar from field theories in Minkowski space time, such an equation must become conformally invariant in the limit of vanishing mass [7, 8]. Combining these concepts, and assuming the existence of an invariant vacuum state [7, 9, 10], it is then possible to obtain an expression for the ground state energy. In carrying out this project, one must also carefully keep track of the dependence on the cosmological constant, Λ. In the discussions of the representation theories of AdS and dS groups mentioned above, the explicit values of the radii of curvatures of the corresponding spaces play no essential roles and are often set equal to unity. In view of Eq. (1) and the point of view emphasized in this work, to extract physical consequences it is essential that we explicitly display the dependence on this dimensional parameter or,equivalently the comological constant.

THE REPRESENTATION THEORY ASPECTS

We begin with the representation theory aspects of the problem. Here we confine ourselves to representations of the Lie algebras of AdS_4 and dS_4 groups. Whether such representations will integrate to the appropriate representaions of these groups is an interesting question that will not be addressed in this work. The algebraic description of the representations of dS and Ads spaces and groups are have many common features. So, we give the details for one of them and then indicate the changes, if any, in the corresponding experssions for the other space and algebra. For the AdS case, the construction of the representations from the physicist's point of view are more familiar [3, 4, 5]. As much as possible, we will follow the notation and conventions of reference [11].

The anti-de Sitter space in 3+1 dimensions can be viewed as a subspace of a flat 4-dimensional space with the line element

$$ds^2 = dX_A dX^A = dX_0^2 - dX_1^2 - dX_2^2 - dX_3^2 + dX_4^2 \tag{2}$$

It is determined by the constraint

$$(X_0)^2 - (X_1)^2 - (X_2)^2 - (X_3)^2 + (X_4)^2 = l^2 \tag{3}$$

where l is a real constant and is related to the cosmological constant according to $\Lambda = -l^{-2}$. The set of transformations which leave the line element invariant form the AdS_4 group $SO(3,2)$. We will actually deal with the universal cover of this group.

Similarly, the dS_4 space can be viewed as a subspace of a flat 4-dimensional space with line element

$$ds^2 = dX_A dX^A = dX_0^2 - dX_1^2 - dX_2^2 - dX_3^2 - dX_4^2 \tag{4}$$

It is determined by the constraint

$$(X_0)^2 - (X_1)^2 - (X_2)^2 - (X_3)^2 - (X_4)^2 = -l^2, \tag{5}$$

where now $\Lambda = l^{-2}$.

The AdS_4 algebra consists of the elements M_{AB} satisfying the commutation relations

$$[M_{AB}, M_{CD}] = i\left(\eta_{AD}M_{BC} + \eta_{BC}M_{AD} - \eta_{AC}M_{BD} - \eta_{BD}M_{AC}\right). \tag{6}$$

With $A = (\mu, 4)$ and $\mu = 0,1,2,3$, we can write the algebra in a more familiar form by setting

$$M_{\mu 4} = l\Pi_\mu. \tag{7}$$

With the notation

$$\varepsilon^{0123} = 1; \qquad \eta^{ab} = (1,-1,-1,-1), \tag{8}$$

we get

$$\begin{aligned} [M_{\mu\nu}, M_{\rho\sigma}] &= i\left(\eta_{\mu\sigma}M_{\nu\rho} + \eta_{\nu\rho}M_{\mu\sigma} - \eta_{\mu\rho}M_{\nu\sigma} - \eta_{\nu\sigma}M_{\mu\rho}\right) \\ [M_{\mu\nu}, \Pi_\rho] &= i\left(\eta_{\nu\rho}\Pi_\mu - \eta_{\mu\rho}\Pi_\nu\right) \\ [\Pi_\mu, \Pi_\nu] &= -il^{-2}M_{\mu\nu} = i\Lambda M_{\mu\nu}. \end{aligned} \tag{9}$$

As expressed in terms of the cosmological parameter Λ, this algebra also holds for dS_4. It can be seen that, under suitable conditions, in the limit of vanishing Λ, the AdS_4 and dS_4 algebras contract to the Poincaré algebra.

The AdS_4 and dS_4 groups have each two Casimir invariants. The quadratic ones are given, with appropriate signs for Λ, by

$$I_1 = -\frac{\Lambda}{2}M_{AB}M^{AB} = \Pi_\mu \Pi^\mu - \frac{\Lambda}{2}M_{\mu\nu}M^{\mu\nu} \tag{10}$$

As $\Lambda \to 0$, the second term vanishes, and $\Pi_\mu \to P_\mu$, so that I_1 reduces to the Casimir operator $P_\mu P^\mu$ of the Poincaré group. The important point here is that, in contrast to Poincaré algebra, for AdS$_4$ as well as dS$_4$ algebras the quantity $\Pi_\mu \Pi^\mu$ is not an invariant by itself. As a result, the analogs of Einstein mass-energy relation $E = mc^2$ will also depend on spin and the comological constant, as we shall see below.

The second Casimir invariant I_2 for either AdS$_4$ or dS$_4$ algebras can be written as

$$I_2 = \frac{\Lambda}{16} V_A V^A, \tag{11}$$

where

$$V^A = \varepsilon^{ABCDE} M_{BC} M_{DE}. \tag{12}$$

Defining the analog of the Pauli-Lubanski operator as

$$W^\mu = V^\mu = \varepsilon^{\mu\nu\rho\sigma} \Pi_\nu M_{\rho\sigma}, \tag{13}$$

we can write I_2 in four dimensional notation for both algebras:

$$I_2 = W_\mu W^\mu - \frac{\Lambda}{16} (V^4)^2 \tag{14}$$

Again, in the limit of a vanishing cosmological constant this experesion reduces to the second Casimir operator of the Poincaré group.

MASSIVE UNITARY REPRESENTATIONS

It will be recalled that for the Poincaré group, the construction of unitary representations begins by identifying an appropriate little group. For massive momenta, e.g., the little group is $SO(2) \times SO(3)$. Then the corresponding induced representation would be specified by the labels of this little group, i.e., by the (rest) energy E_0 and spin s. The quantity E_0^2 is thus the ground state eigenvalue of the quadratic Casimir operator $P_\mu P^\mu$ of the Poincaré algebra. The connection with mass comes about by considering the relativistically invariant free massive Klein Gordon equation and relating the d'Alembertian to $P_\mu P^\mu$ in the usual way:

$$P_\mu P^\mu = m^2 c^4 \tag{15}$$

This provides a group theoretic basis for obtaining the expression $E_0 = mc^2$. One can follow the same recipe for obtaining mass-energy relations for dS$_4$ and AdS$_4$ groups. So, we must first obtain the ground state eigenvalues of the corresponding quadratic Casimir operators.

The analog the above massive unitary representations for AdS$_4$ algebra has been known for sometime [4, 5]. With $SO(2) \times SO(3)$ as maximal compact subgroup, let

$$H = l\Pi_0; \quad J_\pm = M_{23} \pm iM_{31}; \quad J_3 = M_{12}. \tag{16}$$

The J's form the algebra of the $SO(3)$. They commute with H which generates the $SO(2)$ part. The quantity Π^μ has the same dimension as P^μ in the Poincaré algebra, so that it is natural to identify Π^0 as the energy operator. The generator H is then the dimensionless version of the energy operator. So, we can label our states by the eigenvalues of one or the other of these operators.

We can choose the remaining six generators of AdS_4 algebra such that they can act as raising and lowering operators for eigenvalues of Π^0. To this end, let

$$B_i^{\pm} = M_{0i} \pm il\Pi_i; \quad i = 1,2,3. \tag{17}$$

All three of the plus (minus) operators raise (lower) the eigenvalues of H :

$$[H, B_i^{\pm}] = \pm B_i^{\pm}. \tag{18}$$

In this $\{H, J, B\}$ basis, the quadratic Casimir operator will take the form

$$I_1 = \Pi_0^2 - \frac{1}{l^2}\vec{J}^2 - \frac{1}{2l^2}\{B_i^+, B_i^-\}, \tag{19}$$

where in the anticommutator a sum over the index "i " is to be taken.

Using the above preparations, we can now label [4, 5] the discrete representations of AdS_4 algebra by the labels induced by subgroup $SO(2) \times SO(3)$. Depending on whether we use eigenvalues of Π_0 or H, they are:

$$|E, j, j_3, E_\Lambda>; \quad |h, j, j_3, E_\Lambda>. \tag{20}$$

The quantity E_Λ is the energy scale associated with the value of Λ or, equivalently, with the radius of curvature l. These unitary representations have the distinct property of being bounded from below. Denoting the lowest eigenvalues by E_0 and s, respectively, this means that

$$K_i^- |E_0, s, s_3, E_\Lambda >= 0. \tag{21}$$

Applying the quadratic Casimir operator on a ground state, we get

$$I_{AdS}|E_0, s, s_3, E_\Lambda >= [E_0(E_0 - 3E_\Lambda) + E_\Lambda^2 s(s+1)]|E_0, s, s_3, E_\Lambda >, \tag{22}$$

where E_Λ is the energy scale corresponding to the radius of curvature l of the AdS_4 space. From comparing this expression with the corresponding eigenvalue for the Poincaré algebra, it is already clear that E_0 will depend on s and E_Λ.

GENERALIZED REST ENERGY RELATION

To obtain a generalized rest energy relation for a particle represented by a unitary representation of dS_4 or AdS_4 algebra, we must determine how the notion of mass gets related to the eigenvalues of our quadratic Casimir operator. In a standard field theory, the mass of a particle is identified as the pole of the propagator for the corresponding field. To lowest order, the propagator is the classical Green's function for an invariant differential equation. For a scalar field in Minkowski space, e.g., it is the Green's function for the Klein-Gordon equation

$$\left(\partial^\mu \partial_\mu + m^2\right) \phi(x) = 0, \tag{23}$$

This equation immediately generalizes to curved space-time by replacing ∂_μ with an appropriate covariant derivative ∇_μ:

$$\left(\nabla_\mu \nabla^\mu + m^2\right) \phi(x) = 0. \tag{24}$$

The massless limit of this equation is not conformally invariant. If we require on physical grounds that the massless limit of the relevant differential equation be conformally invariant, we must modify this equation by a term proportional to the scalar curvature [7, 8]:

$$\left(\nabla_\mu \nabla^\mu + \frac{R}{3} + m^2\right) \phi(x) = 0. \tag{25}$$

To be specific, we limit the discussion to AdS_4 space. For this space,

$$R = -\frac{3}{2l^2} \tag{26}$$

More generally, for a particle of any spin, we have

$$\left(\nabla_\mu \nabla^\mu - \frac{\lambda_s}{4l^2} + m^2\right) \Psi_s = 0, \tag{27}$$

where λ_s can be different for different spin.

With a suitable differential equation at hand, it is now straight forward to connect it to the quadratic Casimir operator, I_1, of the AdS_4 algebra. This can be worked out in a manner which is familiar from the relation indicated above between the d'Alembertian operator and the quadratic Casimir operator of the Poincaré (Lie) algebra. One obtains

$$\nabla_\mu \nabla^\mu - \frac{\lambda_s}{4l^2} = -I_1 + b(s). \tag{28}$$

where the spin dependent quantity $b(s)$ can be fixed using various consistency conditions [4]. It is given by

$$b(s) = +\frac{2}{l^2}(s^2 - 1) \tag{29}$$

Making use of the above equations and, as discussed above, replacing the radius of curvature l with the corresponding energy scale E_Λ, one is led to the following rest energy relation:

$$m^2c^4 = E_0(E_0 - 3E_\Lambda) + E_\Lambda^2(2 + s - s^2). \tag{30}$$

One can work out a corresponding expression for the rest energy equation for the dS_4 space. What is important for our purposes is that both for dS_4 and AdS_4 the rest energy expression is of the form

$$E_0 = E_0(m, s, E_\Lambda). \tag{31}$$

In the limit of vanishing cosmological cosmological constant, i.e., when $E_\Lambda \rightarrow 0$, this equation reduces to the familiar expression $E = mc^2$.

POTENTIAL EXPERIMENTAL CONSEQUENCES

It is clear from Eq. (30) that a non-vanishing cosmological constant leads to a rest energy relation which is significantly different from what we are familiar with. The first question that comes to mind is the potential experimental consequences of this expression. To this end, it will be recalled that the quantity E_Λ is the energy associated with the radius of curvature, l_Λ of the AdS_4 or dS_4 space. For definiteness, let us consider the AdS_4 case given by Eq. (30). As a first try, let us assume, somewhat naively, that it is the current value Λ_0 of the cosmological constant that should set the energy scale E_Λ. Then, given the current bounds [1, 2],

$$l_{\Lambda_0} = |\Lambda_0|^{-\frac{1}{2}} \sim 10^{28} cm \tag{32}$$

It follows that

$$E_{\Lambda_0} \sim 10^{-33} eV \tag{33}$$

So, if the relevant energy scale E_Λ in Eq. (30) were set by the current value of the cosmological constant, the deviation from $E = mc^2$ would be very small, and there would be no hope for the experimental detection of such deviations.

I would like to argue however that the relevant value of the cosmological constant in the generalized rest energy relations is not the value for the current epoch but the on which is related by equation (1) to the scale at which phase transitions have occured in the history of the Universe. For example, for electroweak phase transition, the correct value of Λ will be not Λ_0 but Λ_{EW} given by Eq. (1). When we do a high energy experiment at the electroweak scale, what we are doing is to recreate the processes that went on everywhere in the universe at that scale. So, assuming that the primary source of Λ at a given scale is that coming from the change in the ground state energy of the corresponding phase transition, then the value of Λ relevant to experiments carried out at the electroweak scale is that given by Eq. (1) at the electroweak scale. It might appear at first sight that this arguement is irrelevant because we are doing the experiments not in the electroweak epoch but in the present epoch at which time the appropriate value of Λ is Λ_0. However, I would like to suggest that a collider experiment at electroweak energy scale involves a local contribution to the energy-momentum tensor, so that it will affect the local behavior of Einstein's field equations. It then follows that the nearby asymptotic behavior of space-time will be altered from the large scale behavior governed by Λ_0 to the one governed by the value of Λ peculiar to electroweak scale.

The most interesting aspect of this proposal is that it is in principle testable. For example, for energies of the order of 200 GeV corresponding to the electroweak phase transition, the radius of curvature $l_{\Lambda_{EW}} \sim \frac{1}{4}$ cm. So, $E_{\Lambda_{EW}} \sim 10^{-4}$ eV. For high energy experiments of order 20 TeV, one finds $E_{\Lambda_{TeV}} \sim 1$ eV. Finally, for energies of order 1000 TeV, the value of $E_\Lambda \sim 350$ eV. For particles of small mass such as neutrinos, assuming that they are massive, such a deviation from the standard rest energy will significantly affect their kinematics at very high energies. It may be possible to test this proposal in not too distant a future.

REMARKS

One of the straight forward but important consequences of a non-vanishing cosmological constant is that the asymptotic symmetry group of space-time is not the Poincaré group but de Sitter or anti-de Sitter group. One immediate consequence of this is that the familiar Einstein rest energy expression $E_0 = mc^2$ generalizes to an expression involving mass, spin, and an energy scale determined by the cosmological constant:

$$E_0 = E_0(m, s, E_\Lambda) \tag{34}$$

One can trace this difference to the difference in the structure of the quadratic Casimir operators of these groups. The main issue to deal with is then whether these generalized rest energy expressions lead to any experimental consequences. As we have seen, if one identifies Λ with its current value, the deviation from $E_0 = mc^2$ would be too small to detect.

On the other hand, if we consider the notion of a spontaneouly broken symmetry as an important factor in shaping the universe from its early beginning to its present form, it would be natural to expect that the relevant value of Λ is correlated with the energy scale at which a certain phase transition took place. In particular, as a consequence of various phase transitions, there must have been periods in the history of the universe during which Λ and hence E_Λ were large. So, if we view a high energy experiment (and theory) as a way of recreating in the laboratory early epochs of the universe when these processes occurred naturally, then the relevant value of Λ is not its present value Λ_0 but its value appropriate to that epoch. In that case, it should be possible to test the dependence of rest energy on the cosmological constant in the future generations of high energy experiments.

Another interesting consequence of this point of view is that it provides a simple explaination for the current small value of Λ_0. All that one has to do is to take into account the size of the universe at the time when a given phase transition took place. Consider again the electroweak phase transition as an example. The total change in vacuum energy as a result of this transition is given by the vacuum energy density times the volume occupied by the universe at that time. As the universe expanded, the magnitude of this contribution to the vacuum energy remained the same, but magnitude of vacuum energy density decreased continuously. Let ρ_{Λ_0} and ρ_Λ be, respectively, the current and the initial values of the vacuum energy density corresponding to some phase transition, such as electroweak, in a previous epoch. If we regard the universe as an expanding 3-sphere with the corresponding radii R_0 and $R(t)$, respectively, then we must have

$$\frac{\Lambda_0}{\Lambda} = \frac{\rho_{\Lambda_0}}{\rho_\Lambda} = \frac{R(t)^3}{R_0^3} \tag{35}$$

For electroweak theory, we have [2]

$$\rho_{\Lambda_{EW}} \sim 3 \times 10^{47} \mathrm{ergs/cm}^3 \tag{36}$$

In this rough approximation, if we substitute the rough values for the current value of the radius of universe and its value at the time of electroweak transition, it is easy to check that the value of Λ_0, comes close to the current bounds!

Lest it might appear that this is just a pure coincidence, one can repeat the above computation for the QCD phase transition corresponding to chiral symmetry breaking. Again one finds unexpectedly a good value for Λ_0. For earlier possible phase transitions such as that at the GUT scale or the Planck scale, the results become less reliable because this rough computation is very sensitive to the values of universe radii at the time of phase transition. For very early times, the determination of the radius of universe becomes more complicated due to inflation, etc.

ACKNOWLEDGMENTS

This work was supported, in part by the Department of Energy under the contract number DOE-FGO2-84ER40153.

your information.

REFERENCES

1. S. Weinberg, Rev. Mod. Phys. **61** (1989) 1.
2. S. Carroll, [astro-ph/0004075].
3. F. Gürsey, in Group Theoretical Concepts and Methods in Elementary Particle Physics, ed. F. Gürsey, Gordon and Breach, 1962; F. Gürsey, in Relativity, Groups, and Topology, Les Houches Summer School 1963, Gordon and Breach, 1963; F. Gürsey and T.D. Lee, Pro. Natl. Acad. Sci. USA, **49** (1963) 179.
4. C. Fronsdal, Rev. Mod. Phys. **37** (1965); Phys. Rev. **D10** (1974) 589; Phys. Rev. **D12** (1975) 3819; C. Fronsdal and R.B. Haugen, Phys. Rev. **D12** (1975) 3810.
5. N.T. Evans, Jour. Math. Phys. **8** (1967) 170
6. L.H. Thomas, Ann. Math **42** (1941) 113; T.D. Newton, Ann. Math. **51** (1950) 730; J. Dixmier, Bull. Soc. Math. France **89** (1961) 9.
7. E. A. Tagirov, Ann. Phys., **76** (1973) 561.
8. S.J. Avis, C.J. Isham, D. Storey, Phys. Rev. **D 18** (1978) 3565.
9. E. Mottola, Phys. Rev. **D 31** (1985) 754.
10. B. Allen, Phys. Rev. **D 32** (1985) 3136.
11. S. Fernando and F. Mansouri, Int. Nat. Jour. Phys. (1999), [hep-th/9804].

Einstein's Elevator

Edward J. Bacinich

Alpha Omega Research Foundation Inc., www.alphaomegafoundation.org
1048 South Ocean Blvd, Palm Beach, FL 33480, USA

Abstract. Simple gravitational experiments show that gravity is the acceleration of the quantum vacuum towards matter. Two diagrams presented in this paper (Fig. 1 And Fig. 2) demonstrate that the curvature of space is merely an effect of this accelerated spacetime.

"Einstein's Elevator" has provided a good straight forward description of the principle of equivalence, a cornerstone of relativity, and certainly one of the most recognized models for explaining the central idea of general relativity among physicists today. Simply stated, the experiments involve a man dropping a ball and timing a pendulum within a closed box. An identical experiment is performed by a woman in a closed box that is deep in outer space in a zero gravity environment where she is floating. Now a rope is attached to the woman's box with an unseen force pulling the box upward with acceleration equal to gravity.

The principle of equivalence asserts that both experiments are the same and that neither person can distinguish whether his or her own box is resting on Earth or accelerating in space. It is concluded therefore that the effect of gravity is reproduced by the acceleration of a reference frame, and, conversely, acceleration is imitated by gravitation [1].

Einstein's mental experiment was taken one step forward by affixing a flashlight to the wall of each box one meter above the floor. In the woman's box in space, the beam hits the opposite wall only after the floor has risen fractionally. As a result, the beam makes contact with the opposite wall at a height slightly shorter than one meter. The results should be the same in the man's box on Earth and he too should find that the beam makes contact just less than one meter. The conclusion that physicist have made from this experiment is that the light beam is deflected or *pulled* down by gravity and that the principle of equivalence predicts that light is affected by gravity.

We believe that the conclusion just drawn has some logical and conceptual problems. The very concept that the beam is pulled down by gravity reflects Newtonian and pre-relativistic views. This is not a correct concept in the spirit of relativity containing the principle of equivalence. Physicists infer from Einstein's elevator that the effects of gravity are *reproduced* by the acceleration of a reference frame, and conversely that the effect of acceleration is *imitated* by gravitation. This is not a strict converse. We believe that the effect of gravity is produced by the

CP624, *Cosmology and Elementary Particle Physics*, edited by B. N. Kursunoglu et al.

acceleration of a reference frame, and, conversely, the effect of acceleration of a reference frame is produced by gravity. Our statement indicates that both phenomena are identical, but opposite, and analogous to action-reaction.

21st Century Einstein Elevator

Let us construct an experiment that takes place in deep space in a gravity free environment (see Figure 1.). There are two spacecraft (A&B) travelling close parallel paths. Vessel A is travelling with uniform motion and Vessel B is accelerating at one 'g' catching up to Vessel A where they momentarily reach an identical velocity side by side. A ball is thrown just before the laser is fired from Vessel A to the accelerating Vessel B. From the reference frame of observers onboard Vessel A, the ball and the beam of light appear to travel away from them in a straight parallel path right through the cage of Vessel B. Observers on Vessel B however witness the laser and the ball enter their spacecraft at the same time but inside Vessel B they curve downward so that the laser beam's exit is slightly lower than it's entrance. The slow moving ball is even more dramatic. It enters the vessel at the same time just below the laser beam but exits the vessel through the bottom of the vessel. The observations of the crews on the two vessels could not be more different. It should be noted that the crew on Vessel A are experiencing a zero gravity environment as it moves at a constant velocity. The crew on Vessel B is experiencing a gravitational-like force as they are pressed to the floor by their weight or inertial reaction to their one 'g' accelerating spacecraft.

In the woman's box as previously described by Einstein, we can see that the light only appears to be curved downward when in actual fact the light travels a straight line. If the light went a straight path or for that matter, a thrown ball from a tennis gun, then as far as the rest of the universe is concerned, neither the photons from the flashlight nor the ball experiences angular momentum along with centripetal or centrifugal forces. One may argue that the environment of the woman's elevator is that of an accelerated reference frame. This is certainly true of the woman and the box's structure, but is it true of the quantum vacuum within the box? Is the quantum vacuum within the box experiencing acceleration along with the elevator's structure? Or is it that a 'weird curved geometry' is generated in the box by the differential between the accelerated state of the box and the non-accelerated state of the local vacuum.

The exact same physical conditions of action-reaction exist when an experiment is performed in which Vessel B is in a stationary position hanging from a tall building on Earth (see Figure 2.). This experiment is particularly effective in describing how an object in free fall (Vessel A) although observed to be accelerating, is in fact in a state of uniform motion while an object (Vessel B) on Earth observed to be standing still is in actuality accelerating. How could this be? It can only be if the coordinates of the reference frame of the vacuum itself are accelerating towards mass.

FIGURE 1. - DRAWN IN INERTIAL REFERENCE FRAME OF VESSEL A

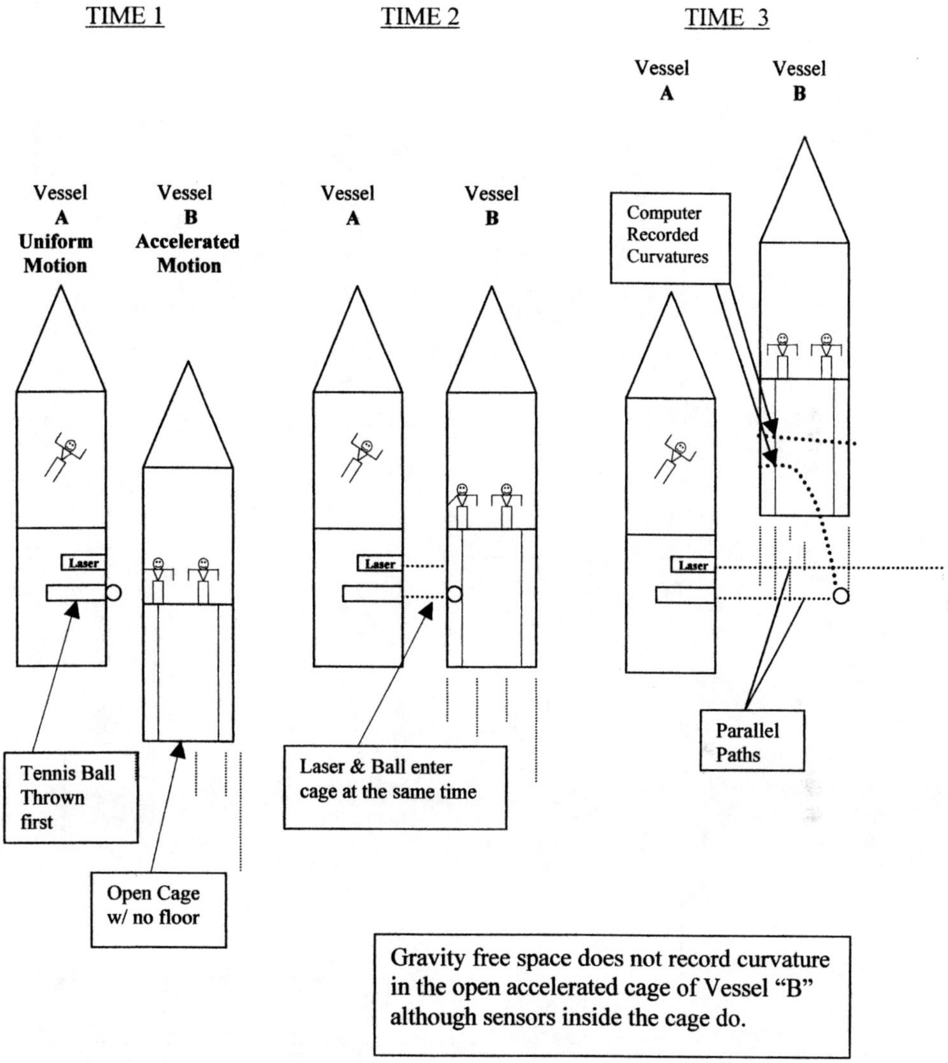

FIGURE 2. – DRAWN IN ACCLERATED FRAME OF VESSEL B

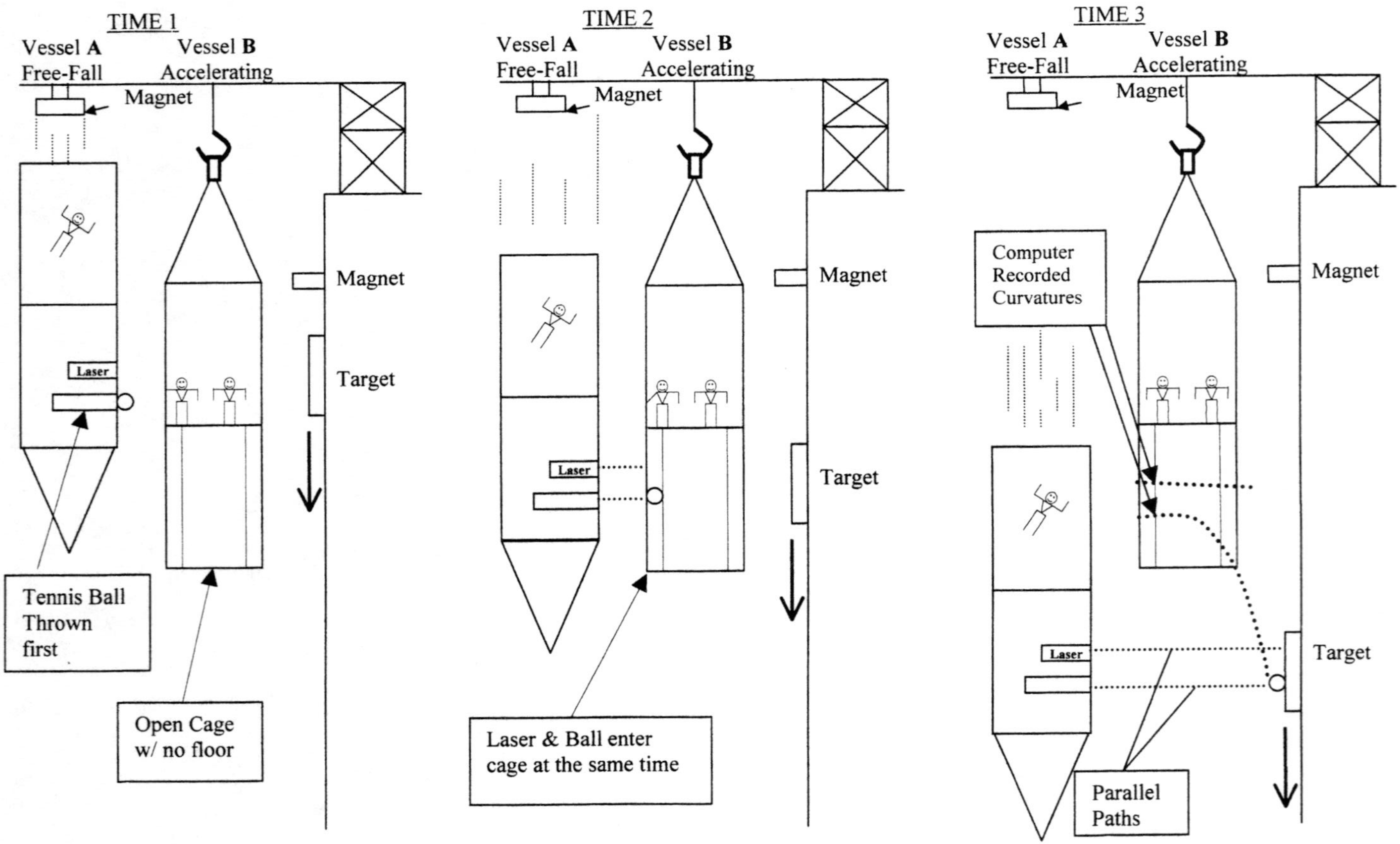

If it is universally true that every action has an equal and opposite reaction, then one should suspect that ordinary acceleration and an ordinary gravitational field (along with all inertial effects) should arise from the dichotomy of massive quantum particles and the Planck state quantum particles of the vacuum. The quantum vacuum is also a state of the background electromagnetic field and thus has physical significance.

It is *not an illusion* that the outside light or ball travels a straight path through space-time; *the illusion is* that the light or ball curves in the woman's box and in the man's box. According to the principle of equivalence, this *apparent illusion* that it curves also applies to planets travelling in elliptical paths around the sun, as opposed to travelling in straight geodesic lines. How can this be? Einstein said, "planets travel the straightest geodesic paths possible. But, according to Galilean or Einsteinian relativity, one cannot tell if one is moving at a uniform motion or standing still. Today we know that this is not strictly true. If a space ship positioned itself so that the background cosmic blackbody radiation showed no red or blue shift in any direction, then, the space ship would indeed be standing still at only *one place* in a universal reference frame.

Einstein jumped to his second conclusion, which in our opinion was an illusionary effect of his intuitive first conclusion that geometry itself was accelerating (and dismissed as a possibility). His second conclusion was: In an accelerated reference frame spacetime is curved by a massive body; furthermore, gravitational effects are regarded as arising from a curvature in spacetime which is considered somehow '*dynamical*'. In other words, this curvature produces gravity. This stands in stark contrast to Quantum Field Theory where spacetime is considered *'passive'*.

Returning to the woman's box in space in Einstein's mental experiment, we see that the light actually travels a straight path through space-time. It only *appears* to travel a curved path in the woman's box. Acceleration produces the distorted geometric environment in both the woman's and man's box. This **distorted environment in the woman's box certainly does not produce the box's acceleration.** The woman, by accelerating through space-time, observes that straight trajectories in the universe appear to curve in her reference frame. How can she tell if she actually is accelerating through space-time or whether the *real converse* is true, i.e., space-time is accelerating through her? The answer is she cannot. Both are equivalent phenomena. Accelerating at one 'g' through space or standing on the earth where space accelerates at one 'g' through you are identical phenomena, except, that one is the *action* and conversely, the other is the *reaction* respectively [2].

Quantum Physicists believe that the quantum vacuum, as a state of the electro-magnetic field with ZPF, carries super-energy density of 10 to the 120th power GeV per cubic centimeter. This quantum vacuum state can also give us the geometry of a reference frame. We think that all motion is relative to this super energetic frame, which also has the dynamic ability of accelerated motion toward matter called "the gravitational field." In our opinion, the principle of equivalence should be:

"One cannot tell if one is accelerating through the quantum vacuum or if the quantum vacuum is accelerating through one".

The curvature of space is only an effect of our equivalence principle above. General relativity uses this space curvature to explain gravity - which nevertheless is macroscopically correct. But, it does not connect conceptually to mass nor to quantum theory. The equivalence principle is exact if interpreted as action-reaction; which is compatible with quantum theory.

The understanding of general relativity and quantum theory must evolve if they are ever to unify or cohere.

REFERENCES

1. Isham, C., "Quantum Gravity" in *The New Physics*, edited by P. Davies, Cambridge University Press, Cambridge, 1989, pp. 70-93
2. Pound, R. V., and G. A. Rebka, "Effect of gravity on nuclear resonance," *Phys. Rev. Lett. 4*, A.I.P., 1960, pp.337-341

II. COSMOLOGICAL THEORIES AND THEIR EXPERIMENTAL IMPLICATIONS

Dark Energy from Strings

Paul H. Frampton

Department of Physics and Astronomy, University of North Carolina, Chapel Hill, NC 27599-3255

Abstract. A long-standing problem of theoretical physics is the exceptionally small value of the cosmological constant $\Lambda \sim 10^{-120}$ measured in natural Planckian units. Here we derive this tiny number from a toroidal string cosmology based on closed strings. In this picture the dark energy arises from the correlation between momentum and winding modes that for short distances has an exponential fall-off with increasing values of the momenta.The freeze-out by the expansion of the background universe for these transplanckian modes may be interpreted as a frozen condensate of the closed-string modes in the three non-compactified spatial dimensions.

INTRODUCTION

This talk is based on a paper with Bastero-Gil and Mersini, necessarily shortened to fit the space available, so I refer to it[1] for more detail.

In this work we will attempt to make a quantitative argument about the origin of dark energy from string theory. The transition from string theory to conventional cosmology is of importance not only to theoretical physics in general but to inflationary cosmology in particular. Corrections to short distance physics due to the nonlocal nature of strings contribute to dark energy. The possibility to detect their signature observationally is very intriguing. In Ref. [2] it was shown that a nonlinear dispersion function modifying the frequency of the transplanckian perturbation modes [3] can produce the right contribution to the dark energy of the universe [4]. The physics mechanism that gave rise to dark energy was the freeze-out of these ultralow frequency modes by the expansion of the background universe. Superstring duality [5] was invoked to justify the dispersion function. This work attempts to carry out this derivation.

In Section 2 we review some preliminaries of the Friedman-Robertson-Walker (FRW) cosmological solutions found for string theory in a D-dimensional torus [6, 7, 8, 9]. The quantum hamiltonian from closed string theory obtained in [10] by using the correspondence principle between string and quantum operators, is reviewed in Section 3. Although the background is an FRW universe, it is globally nontrivial and thus it allows two types of quantum string field configurations, twisted and untwisted fields.

Based on the equivalence between Euclidean path integral and statistical partition functions, we perform in Section 4 the calculation of a coarse-grained effective action [11, 12] for the momentum and winding modes of the system described in Section 2 for the case of 3 expanding spatial dimensions R in the T^D toroidal topology. The string scale is taken as the natural UV lattice cutoff scale of the theory. The renormalization group equations (RGE) of the coupling constants for the winding and momentum modes describe the evolution from early to late times of their entanglement. Based on T-

CP624, *Cosmology and Elementary Particle Physics,* edited by B. N. Kursunoglu et al.

duality the whole spectrum is obtained by exchanging momentum to winding modes and $R \to R^{-1}$. Their coupling is strong when the radius of the torus R is of the same order as the string scale $\sqrt{\alpha'}$, i.e. during the phase transition from a winding dominated universe to a momentum mode dominated universe. Due to the expanding background, we have a non-equilibrium dynamics and calculate the effective action by splitting our modes into the open system degrees of freedom (low energy modes, mainly momentum modes) and the environment degrees of freedom (high energy modes, mainly winding modes). The coarse-graining is performed by integrating out the environmental degrees of freedom. The scale factor $a(t) = R(t)$ serves as the collective coordinate that describes the order parameter for the environment degrees of freedom. The effective action calculated in this way contains the influence of the environment at all times in a systematic way and the coarse graining process encodes the dispersion function and corrections to short distance physics due to the correlation between the two types of modes in the system and environment. This procedure results in the RGEs for the coupling constants that offers information about their running to trivial and nontrivial fixed points at early and late times, therefore the flow of one family of lagrangians (string theory phase) to another family of lagrangians at late times (conventional 3+1 quantum theory). Results of this non-equilibrium phase transition are summarised in Section 5 with a discussion about the possibility of their observational signatures through the equation of state of the frozen short distance modes. In this section we also briefly touch upon the issue of the two field configurations in a globally nontrivial topology and the instabilities in the theory arising from their interaction. A detailed summary of the main coarse-graining formulas and procedure [11] needed in Section 4, are attached in the Appendix. In essence, the dark energy arises from the study of the UV behavior of the correlations with environmental modes.

TOROIDAL STRING COSMOLOGY.

We consider the string cosmological scenario proposed by Brandenberger and Vafa[1] [6, 8, 13]. Strings propagate in compact space, a box with D spatial dimensions and periodic boundary conditions, the T^D torus. It was argued that [6] a thermodynamic description of the strings with positive specific heat, is well defined only when all the spatial dimensions are compact.

Let us begin with the Universe placed in a T^D box with a size of the order of the string scale, that we are taking to be the Planck scale. In such a space, string states also contain winding modes, which are characteristic of having an extended object like a string, "winding" around the compact spatial dimension, besides the usual momentum modes, and oscillator modes with energy independent of the size of the box. The energy of the winding modes increases with the size of the box as wR, while the energy of the momentum modes decreases as m/R. The spectrum is symmetric under the exchange $R \leftrightarrow 1/R$ and $m \leftrightarrow w$. This symmetry known as T-duality [5] is not only a symmetry of

[1] Herein referred to as the BV model.

the spectrum but of the theory.

The BV model [6] argues that if the Universe expands adiabatically in more than 3 spatial dimensions, it would not be possible to maintain the winding modes in thermal equilibrium. As their energy density grows with the radius, their number would have to decrease, for example through annihilation processes. But typically strings do not meet in more than 3 spatial dimensions and do not interact with each other; therefore the winding modes fall out of equilibrium [14]. In summary, their growing energy density will tend to slow down the expansion of the universe and eventually stop it. But if the Universe starts to contract, the dual scenario of the momentum modes opposing contraction would take place and the Universe may oscillate between expanding/contracting eras. In what follows we use this argument of [6] to justify the assumption that only $D = 3$ dimensions of the T^D torus will expand to create an FRW universe.

Cosmological solutions for an arbitrary number of anisotropic toroidal spatial dimensions T^D were found by Mueller in [7]. He studied the cosmology of bosonic strings propagating in the background defined by a time-dependent dilaton field, $\Phi(t)$, and space-time metric

$$ds_d^2 = G_{\mu\nu}(X)dX^\mu dX^\nu = -dt^2 + \sum_{i=1}^{D} 4\pi R_i^2(t)dX_i^2\,, \tag{1}$$

The radii of the torus, $R_i(t)$, become the time-dependent scale-factors, and the spacetime dimensions is $d = 1 + D$. The equations of motion of the bosonic string in background fields are obtained from the following action[2] [15]

$$I = \frac{1}{4\pi\alpha'}\int d^2\sigma\sqrt{g}\left[g^{mn}G_{\mu\nu}(X)\partial_m X^\mu \partial_n X^\nu + \frac{1}{2}\alpha'\Phi R^{(2)}\right]\,, \tag{2}$$

where g_{mn} is the two-dimensional world-sheet metric, and $R^{(2)}$ the world-sheet scalar curvature. The background field equations are obtained by imposing the condition that the theory be free from Weyl anomalies. To lowest order in perturbation theory this leads to the equations:

$$\beta^G_{\mu\nu} = R_{\mu\nu} + \nabla_\mu\nabla_\nu\Phi = 0\,, \tag{3}$$

$$\beta^\Phi = \frac{d-26}{3\alpha'} - R + (\nabla\Phi)^2 - 2\nabla^2\Phi = 0\,, \tag{4}$$

Using the metric given in Eq. (1), they reduce to:

$$\ddot{\Phi} - \sum i\frac{\ddot{R}_i}{R_i} = 0\,, \tag{5}$$

$$\frac{\ddot{R}_i}{R_i} + \sum_{j\neq i}\frac{\dot{R}_i\dot{R}_j}{R_iR_j} - \dot{\Phi}\frac{\dot{R}_i}{R_i} = 0\,, \tag{6}$$

---

[2] The antisymmetric tensor field is taken to be zero.

$$\ddot{\Phi} - \frac{1}{2}\dot{\Phi}^2 + \sum_{i<j} \frac{\dot{R}_i \dot{R}_j}{R_i R_j} = \frac{d-26}{3\alpha'} . \quad (7)$$

When $D = 25$, the solutions obtained in [7] are:

$$e^{-\Phi(t)} \propto t^p , \quad (8)$$

$$R_i(t) \propto t^{p_i} , \quad (9)$$

with the constraints,

$$\sum_{i=1}^{D} p_i^2 = 1 , \quad \sum_{i=1}^{D} p_i = 1 - p . \quad (10)$$

Note that these solutions are found in the absence of matter sources. In general the back-reaction of the matter action of the strings in T^D alters the solutions for the background geometry[3]. It is clear that we can have an arbitrary number of compact spatial dimensions D_c with $p_i < 0$, that are decreasing with time [4], and $D - D_c$ expanding spatial dimensions with $p_i > 0$. Among the many solutions found in [7] we select the solution $D - D_C = 3$ that although it is not unique it is justified by the BV argument. The assumption that our Universe is expanding in only 3 spatial dimensions, with the remaining $D-3$ being small and compact, as well as considering a constant dilaton field[5] ($p = 0$), are consistent with Mueller's solutions Eqs. (10). The issue of stabilising the dilaton is beyond the scope of this paper, and we assume that the dilaton has acquired a mass and become stable at some fixed value. It is also assumed that the backreaction of the matter string sources on the backround geometry is small enough such that the deviations from the FRW metric, Eq. (10), can be neglected.

Due to the toroidal string cosmology, the three expanding dimensions contain both types of modes: *momentum* and *winding*, propagating in the 3+1 FRW space-time. The number of winding modes at each stage of the evolution of the Universe is determined by the dynamics of the background. In the next section, we touch base with quantum field theory through correspondence principle between string and quantum operators, in order to use coarse graining techniques for studying the influence of the winding modes on the momentum modes as the Universe expands.

QUANTUM HAMILTONIAN FROM CLOSED STRING THEORY.

Let us consider BV model [6] of a D-dimensional anisotropic torus with radius $\bar{R}_i$, by including the dynamics of both modes: momentum modes, $p_{1,i} = m/\bar{R}_i$ (where m is

---

[3] See [8, 9] and references therein for the geometry solutions in the presence of a matter action. Inclusion of matter sources alters the solutions of [7] due to the backreaction of the winding modes, such that the scale factor approaches asymptotically a constant value at late times .

[4] We do not address the concern that the time dependence of the compactified R_i endangers the constancy of the dimensionless parameters in the $D = 3$ theory.

[5] The authors of [8] argued that a constant dilaton background may not be consistent with a *high* temperature phase of strings thermodynamics.

the wavenumber), and winding modes with momenta $p_{2,i} = w\bar{R}_i/\alpha'$. The dimensionless quantity for the radius is $R_i = \bar{R}_i/\sqrt{\alpha'}$, where α' is the string scale. Based on the arguments reviewed in Section 2, we *choose* a cosmology with three toroidal radii equal and large $R \gg 1$ in units of the string or Planckian scale, with the other $(D-3)$ toroidal radii equal and small $R_C \ll 1$. Here the subscript C refers to compactified dimensions. Then, $R(t)$ becomes the scale factor for the 3+1 metric in conventional FRW (Friedman Robertson Walker) cosmology $R(t) = a(t)$, while R_C corresponds to the radius, in this factorizable metric, of the $D-3$ compact dimensions z_j that decrease with time,

$$ds_D^2 = -dt^2 + 4\pi R^2(t)dx_i^2 + 4\pi R_C^2(t)dz_j^2 = a(\eta)^2[-d\eta^2 + dy^2] + ds_{D-3}^2. \tag{11}$$

Using the string toroidal solution of [7] the time-dependence of these radii is:

$$R(t) = \alpha_U t^{p_U} \tag{12}$$

$$R_C(t) = \alpha_C t^{p_C} \tag{13}$$

The solutions in Ref. [7] show that p_U and p_C depend on the dimensionality D in an interesting way. There is a plethora of possible solutions but if we assume, for example, that the dilaton is time-independent and the compactification is isotropic we find that for $4 \leq D < \infty$, then $0.5 \leq p_U < 1/\sqrt{3} \simeq 0.577$. Let us take $D = 4$ where the scale factor behaves as a radiation-dominated universe; if, in fact, $D \geq 5$ we can assume that the $D-4$ additional dimensions have $p_C' \ll p_C$ to achieve the same result. In this case, $p_C = -0.5$. Here we do not, however, need to specialise to a particular solution.

What we have in mind for the dark energy is the correlation of momentum to *winding* string modes. The question is, given the well-known form for the kinetic energy of these strings, *e.g.* [16], how to describe best the interaction between the winding and momentum modes. Some aspects are addressed in [16] which focuses on the smallness of temperature (T/T_H). For temperature T very much below the Hagedorn or string temperature T_H we expect that only very small winding numbers $w_i = 0$ or 1 in the compact dimensions are of any significance [16]. Similar arguments apply to the momentum modes m_i for the time-reversed case.

Let us consider the small parameter $\delta(t)$, taken to be:

$$\delta = \frac{R_C}{R} \sim t^{p_C - p_U} \tag{14}$$

For the case $D = 3$ (d=4), for example $\delta \sim t^{-1} \sim (T/T_H)^2$ and is an extremely small number ($\sim 10^{-60}$) at present. The point is that in the $\delta \to 0$ limit these modes are in separate spaces and for very small δ are therefore expected to be highly restricted. The compactified dimensions can be integrated out, and we are left with the momentum and winding modes in the remaining $D = 3$ spatial dimensions.

The partition function for this system was calculated, from first principles, by summing up over their momenta in [16]:

$$Z = \sum_{\sigma} e^{-n_\sigma \varepsilon_\sigma}, \tag{15}$$

where n_σ is the number of strings in state σ with energy ε_σ

$$\varepsilon_\sigma = p_0 = \sqrt{\left(\frac{m}{R}\right)^2 + (wR)^2 + N + \tilde{N} - 2}\,, \tag{16}$$

and σ counts over (m, w), with the constraint $N - \tilde{N} = mw$ for closed strings where N and $\tilde{N}$ are the sums over the left- and right- mover string excitations, respectively. By now, in Eq.(16), we are considering only the large 3 spatial dimensions. The string state can also be described by its left and right momenta, $k_L = p_1 + p_2$, $k_R = p_1 - p_2$. The string state for left and right modes can be expanded in terms of the creation and annihilation operators α_m, $\tilde{\alpha}_n$, with higher excitation string states given by $N = \sum_{n=1}^{\infty} \alpha_{-n}\alpha_n$ (similarly for $\tilde{N}$), and string energy $L_0 + \tilde{L}_0 = p_1^2 + p_2^2 + (N + \tilde{N} - 2)/\alpha'$.

We would like to write the path integral for this configuration in terms of quantum fields[6]. The path integral is calculated from the hamiltonian density. In order to use the correspondence between the Euclidean path integral of the persistence vacuum amplitude $|\langle in|out\rangle|^2$ and the partition function Z, we need to write a hamiltonian density over the fields in configuration space in such a way that its Fourier transform in k-space corresponds to the string energy expression Eq. (16).

Thus in writing a Coarse-Grained Effective Action (CGEA), the kinetic terms are unambiguous while for the interaction terms we must appeal to simplicity and the requirement of T-duality. Closed-string field theory provides guidance, since in *e.g.* [17] truncation at a quartic coupling can be sensible, and this will lead to a CGEA which is renormalisable and satisfies useful RG equations.

Generally, closed string field theory contains couplings of all non-polynomial orders. In a semi-classical approximation we may restrict to genus $g = 0$ since the genus g contribution is proportional to$\hbar^g$ [18].

The quantum hamiltonian is in any case known for the classical string in axi-symmetric or toroidal backgrounds [10]. They explicitly calculated the quantum hamiltonian and demonstrated the correspondence principle between the string operators L_0, $\widetilde{L}_0$ and quantum field operators in the form (in the notation of [10])

$$\begin{aligned} \hat{H} &= \hat{L}_0 + \hat{\widetilde{L}}_0 = \frac{1}{2}\alpha'\left(-E^2 + p_a^2 + \frac{1}{2}(Q_+^2 + Q_-^2)\right) + N + \widetilde{N} - 2c_0 \\ &\quad -\alpha'\left[(q+\beta)Q_+ + \beta E\right] - \alpha'\left[(q-\alpha)Q_- + \alpha E\right] J_L \\ &\quad \frac{1}{2}\alpha' q\left[(q+2\beta)J_R^2 + (q-2\alpha)J_L^2 + 2(q+\beta-\alpha)J_R J_L\right] \end{aligned} \tag{17}$$

$$\hat{L}_0 - \hat{\widetilde{L}}_0 = N - \widetilde{N} - mw \tag{18}$$

where $J_{R,L}$ are bilinear quadratic operators in terms of creation and annihilation operators and the higher string oscillators $N, \tilde{N}$ contribute the string mass. Therefore the J_R^2 term would be a quartic interaction in terms of creation and annihilation operators.

---

[6] Below we use quantum string equations under the assumption that the dilaton is massive and stable.

This particular solution is for a cylindrical topology where the uncompactified x_1 and x_2 are written in polar coordinates $x_1 + ix_2 = \rho e^{i\phi}$ and x_3 is also uncompactified (but could be compactified along with additional similar coordinates), together with time and one additional compactified dimension $y \subset (0, 2\pi R)$. Although an exact solution for the hamiltonian of the string matter in a toroidal background is not yet known, a quartic potential energy was advocated and found in [9] by arguments similar to those of Eq. (17), for the classical string and the three string coupling level. We take this as an indication, in the subsequent section (if the exact solution were known to all orders), that an quantum hamiltonian analogous to Eq.(17) for closed strings on a torus, similarly containing only quartic terms as suggested by [10], exists for our present case of $(T_3) \times (T_{D-3}) \times (time)$ and focus on the uncompactified 3 spatial dimensions.

The hamiltonian depicted in Eq.(17) is for a static background,i.e a constant scale factor $R(t)$. In the next section, we base our calculation in the coarse-grained effective action (CGEA) formalism where the dynamics of an expanding background is replaced by scaling on a static background.

Thus Eq.(17) which applies to a static background (as in Eq.(16)) can be generalized to a cosmologically-expanding background as in Eq.(11) by using this technique of re-scaling, for details of which see [1]. The result is a dispersion formula characterized by a dispersed frequency with short distance modifications contained in a $\tilde{m}_0^2$ term:

$$\tilde{m_0}^2 \overset{p\to\infty}{\to} \frac{m_0^2}{2\cosh^2 p\sqrt{\alpha'}} \simeq \frac{1}{2} m_0^2 e^{-2\sqrt{\alpha'}p}. \tag{19}$$

DARK ENERGY FROM CLOSED STRING THEORY. DISCUSSION

We argue that closed strings on a toroidal cosmology lead to a plausible explanation of the dark energy phenomenon. Although bosonic strings have been used, it is expected that superstrings will lead to a similar conclusion. Certainly it is crucial that closed strings are involved because open strings do not have the same aspect of winding around the torus.

The scale factor of the universe $a(\eta)$ has been used as a collective coordinate for the environment degrees of freedom, and as the fundamental scaling parameter in the coarse-graining. The choice of a $D = 3$ expanding cosmology was chosen phenomenologically. An argument for this choice in the BV model was presented in [6], and we believe this argument does provide a possible justification. It is encouraging that inclusion of branes gives a similar result [13]. It has further been assumed that the mass gap Δ_p can be safely assumed to be slowly-varying during our coarse-graining procedure.

We would like to make the reader aware of another subtlety related to the torus topology of our background. A globally nontrivial topology like $T^3 \times R^1$ admits two types of quantum field configurations, twisted and untwisted fields, due to the periodic and anti-periodic boundary conditions imposed on the fibre bundle of the manifold. This is a long-standing problem [19] that does not have a definite remedy. The problem is the following: twisted fields can have a negative two-point function. These fields interact with each other while preserving the symmetries of the hamiltonian. Their interaction thus contributes a negative mass squared term to the effective mass of the untwisted field due to the negative two-point function of the twisted field and render the untwisted field unstable. It is often assumed that Nature simple chose to preserve the untwisted configuration only or forbids their interaction due to some, as yet unknown, symmetry [20].

String theory preserves Lorentz invariance. This symmetry has been broken for the open system of our low energy string modes due to the backreaction from the coarse grained environment. Their correlation results in our dispersion relation. If a specific frame must be chosen, it could be *e.g.* the rest frame of the CMB. The initial condition is a vacuum state conformally equivalent to the Minkowki spacetime - the so-called Bunch-Davis vacuum[21]. Finally, before summarising we should note that if there are other modes without the exponential suppression at high k, all that we need is one such mode to lead to the frozen tail comprising the dark energy.

The high wave number behaviour e^{-ak} of the dispersion relation $\omega(k)$ leads again to the correct estimate for the dark energy as a fraction $\sim 10^{-120}$ of the total energy during inflation. This dark energy is certainly completely stringy because our derivation depends on the existence of winding modes, as seen by the role of the generalised level-matching condition

$$N - \widetilde{N} = \Sigma_i m_i w_i$$

This correlation between momentum and winding modes leads to the quantum hamiltonian and hence to the interpretation of the dark energy as the weak correlation with the winding mode energy at short distances. The excitation modes of these correlations with energy less than the current expansion rate are currently frozen by the expanding background.

Within string cosmology there has always been the question of the fate of the winding modes in the uncompactified three spatial dimensions, whether they combine to a single string per horizon which wraps around the universe. Our remedy is intuitively appealing that while the momentum modes are in evidence as quarks, leptons, gauge bosons, etc. the winding modes are now condensed uniformly in the environmental background, hence with a weak correlation at short distances to the momentum modes, frozen by the expansion of the FRW universe in the form of the dark energy.

The observed small value $\Lambda \sim 10^{-120}$ in natural units bserved small value $\Lambda \sim 10^{-120}$ in natural units has an explanation in the toroidal cosmology of closed strings and thus the dark energy provides an exciting opportunity to connect string theory to precision cosmology. We may argue that numerically the size of the cosmological constant in the present approach is a combination of the string scale and the Hubble expansion rate in the sense that $\Lambda/M^4_{Planck} \simeq 10^{-120} \simeq (H_0/M_{Planck})^2$. Therefore the correct amount of dark energy obtained by this frequency dispersion function does not require any fine tuning and relies, besides a physical mechanism (such as freeze-out), only on the string scale as the parameter of the theory. However, our approach does not solve the second puzzle about the dark energy namely, the coincidence problem for the following reason: The expansion rate of the universe is determined by the *total energy density* in the universe by the relation given in the Friedman equation. As can be seen from our dispersion function which approaches conventional cosmology in the subplanckian regime ($k \leq M_{pl}$), most of the other contributions to the energy density are not frozen modes. Therefore the

Hubble rate H^2 is not always proportional to the dark energy of the frozen modes due to the contributions in H^2 from other forms of energy densitites.H^2 is dominated by frozen modes (and thus proportional to the dark energy ρ_{DE}) only at some late times $t \geq t_E$ when all other energy contributions ρ_{other} have diluted enough below ρ_{DE} due to their redshift.

The quantitative effort we have made in this work suggests that an interpretation of the dark energy in terms of string theory is more convincing than either a simple cosmological constant or the use of a slowly- varying scalar field with fine tuned parameters.

ACKNOWLEDGEMENTS

PHF acknowledges the support of the Office of High Energy, US Department of Energy under Grant No. DE-FG02-97ER41036.

REFERENCES

1. M. Bastero-Gil, P.H. Frampton and L. Mersini. hep-th/0110167
2. L. Mersini, M. Bastero-Gil and P. Kanti, Phys. Rev. **D64** (2001) 043508.
3. J. Martin and R. H. Brandenberger, Phys. Rev. **D63,** 123501 (2001); R. H. Brandenberger and J. Martin, Mod. Phys. Lett. **A16,** 999 (2001); J. C. Niemeyer, Phys. Rev. **D63**, 123502 (2001); J. C. Niemeyer and R. Parentani. astro-ph/0101451; J. Kowalski-Glikman, Phys. Lett. **B499**,1 (2001); A. Kempf, Phys. Rev. **D63**, 083514 (2001); A. Kempf and J. C. Niemeyer. astro-ph/0103225; R. Easther, B. R. Greene, W. H. Kinney and G. Shiu. hep-th/0104102; T. Tanaka. astro-ph/0012431; A. A. Starobinsky. astro-ph/0104043; M. Bastero-Gil and L. Mersini. Phys. Rev. **D** (in press). astro-ph/0107256; M. Visser, C. Barceló and S. Liberati. hep-th/0109033.
4. S. Perlmutter, M. S. Turner and M. White, Phys. Rev. Lett. **83**, 630 (1999).
5. K. Kikkawa and M. Yamasaki, Phys. Lett. **B149** (1984) 357; N. Sakai and I. Senda, Prog. Theor. Phys. 75 (1986) 692; V. P. Nair, A. Shapere, A. Strominger and F. Wilczek, Nucl. Phys. **B287** (1987) 402; P. Ginsparg and C. Vafa, Nucl. Phys. **B289** (1987) 414; B. Sathiapalan, Phys. Rev. Lett. **58** (1987) 1597; R. Dijkgraaf, E. Verlinde and H. Verlinde, Commun. Math. Phys. **115** (1988) 649; A. Shapere and F. Wilczek, Nucl. Phys. **320** (1989) 669.
6. R. Brandenberger and C. Vafa, Nucl. Phys. **B316,** 391 (1988).
7. M. Mueller, Nucl. Phys. **B337,** 37 (1990). Solutions with a linear dilaton background were studied in: R. Myers, Phys. lett. **199** (1987) 371; I. Antoniadis, C. Bachas, J. Ellis and D. Nanopoulos, Phys. Lett. **B211** (1988) 393; Nucl. Phys. **B328** (1989) 115.
8. A.A. Tseytlin and C. Vafa, Nucl. Phys. **B372,** 443 (1992).
9. A. A. Tseytlin, Class. and Quant. Grav. **12** (1995) 2365.
10. J. G. Russo and A. A. Tseytlin, Nucl. Phys. **B448** (1995) 293; Nucl. Phys. **B449** (1995) 91; A. A. Tseytlin, Class. and Quant. Grav. **12** (1995) 2365.
11. B. L. Hu and Y. Zhang, "Coarse-Graining, Scaling and Inflation", Unv. Maryland Preprint 90-186; B. L. Hu, in *Relativity and Gravitation: Classical and Quantum Proc.* SILARG VII, Cocyoc, Mexico 1990, eds. J.C. D'Olivo et al. (World Scientific, Singapore 1991); B. L. Hu, Class. Quant. Grav. **10** (1993) S93; E. A. Calzetta, B. L. Hu and F. D. Mazzitelli,*Proceedings of RG-200: Conference on Renormalization Group Theory at the Turn of the Millennium*, Taxco, 1999.
12. K. Huang, "Statistical Mechanics" (John Wiley & Sons, Singapore, 1987). P.H. Frampton, Phys. Rev. Lett. **37,** 1378 (1976); Phys. Rev. **D15,** 2922 (1977). S. Coleman and F. De Luccia, Phys. Rev. **D21,** 3305 (1980).
13. S. Alexander, R. Brandenberger and D. Easson, Phys. Rev. **D62** (2000) 103509.

14. G. Cleaver and P. Rosenthal, Nucl. Phys. **457** (1995) 621; M. Sakellariadou, Nucl. Phys. **468** (1996) 319.
15. C. Vafa and N. Warner, Phys. Lett. **218** (1989) 51; B. Greene, C. Vafa and N. Warner, Nucl. Phys. **324** (1989) 371; E. Martinec, Phys. Lett. **217** (1989) 431; E. S. Fradkin and A. A. Tseytlin, Nucl. Phys. **261** (1985) 1.
16. K. Hotta, K. Kikkawa and H. Kunitomo, Prog. Theor. Phys. **98,** 687 (1997).
17. A. Belopolsky, Nucl. Phys. **B448,** 245 (1995).
18. A. Sen and B. Zwiebach, Nucl. Phys. **B423** 580 (1994); A. Sen and B. Zwiebach, JHEP, **0003,** 002 (2000); N. Moeller, A. Sen and B. Zwiebach, JHEP, **0008** 039 (2000).
19. L.H. Ford, Phys. Rev. **D21,** 933 (1980); Phys. Rev. **D22,** 3003 (1980); D.J. Toms, Phys. Rev. **D21,** 928 (1980); Phys. Rev. **D21,** 2805 (1980).
20. L. Mersini, Mod. Phys. Lett. **A14,** 2393 (1999); Phys. Rev. **D59,** 123521 (1999); L. Randall and R. Sundrum, Phys. Rev. Lett. **83,** 3370 (1999).
21. T.S. Bunch and P.C.W. Davies, Proc. Roy. Soc. Lond. **A357,** 381 (1977).

The Big Flow

Pierre Sikivie

Department of Physics, University of Florida
Gainesville, FL 32611, USA

Abstract. The late infall of cold dark matter onto an isolated galaxy, such as our own, produces streams and caustics in its halo. The outer caustics are topological spheres whereas the inner caustics are rings. The self-similar model of galactic halo formation predicts that the caustic ring radii a_n follow the approximate law $a_n \sim 1/n$. In a study of 32 extended and well-measured external galactic rotation curves evidence was found for this law. In the case of the Milky Way, the locations of eight sharp rises in the rotation curve fit the prediction of the self-similar model at the 3% level. Moreover, a triangular feature in the IRAS map of the galactic plane is consistent with the imprint of a ring caustic upon the baryonic matter. These observations imply that the dark matter in our neighborhood is dominated by a single flow. Estimates of that flow's density and velocity vector are given.

INTRODUCTION

There are compelling reasons to believe that the dark matter of the universe is constituted in large part of non-baryonic collisionless particles with very small primordial velocity dispersion, such as axions and/or weakly interacting massive particles (WIMPs) [1]. Generically, such particles are called cold dark matter (CDM). Knowledge of the distribution of CDM in galactic halos, and in our own halo in particular, is of paramount importance to understanding galactic structure and predicting signals in experimental searches for dark matter.

One should expect this dark matter to form caustics. A caustic is a place in physical space where the density is very large because the sheet on which the dark matter particles lie in phase-space has a fold there. Caustics are commonplace in the propagation of light. An instructive example is given by the sharp luminous lines at the bottom of a swimming pool on a breezy sunny day. Two conditions must be satisfied for caustics to occur generically. First, the propagation must be collisionless. Second the flow must have low velocity dispersion. Light propagation is collisionless, and the flow of light from a point source has zero velocity dispersion. Thus caustics are common in light. Caustics in ordinary matter are very unusual because ordinary matter is not normally collisionless. But CDM is collisionless and has very small velocity dispersion. This guarantees that caustics are common in the distribution of CDM.

The primordial velocity dispersion of the cold dark matter candidates is indeed very small, of order

$$\delta v_a(t) \sim 3 \cdot 10^{-17} \left(\frac{10^{-5} eV}{m_a}\right) \left(\frac{t_0}{t}\right)^{2/3} \tag{1}$$

CP624, *Cosmology and Elementary Particle Physics*, edited by B. N. Kursunoglu et al.

for axions, and

$$\delta v_W(t) \sim 10^{-11} \left(\frac{GeV}{m_W}\right)^{1/2} \left(\frac{t_0}{t}\right)^{2/3} \tag{2}$$

for WIMPs. Here t_0 is the present age of the universe and m_a and m_W are respectively the masses of the axion and WIMP. The small velocity dispersion means that the dark matter particles lie on a thin 3-dim. sheet in 6-dim. phase-space. The thickness of the sheet is δv. The sheet cannot break and hence its evolution is constrained by topology.

Where a galaxy forms, the sheet wraps up in phase-space, turning clockwise in any two dimensional cut $(x, \dot{x})$ of that space. x is the physical space coordinate in an arbitrary direction and $\dot{x}$ its associated velocity. The outcome of this process is a discrete set of flows at any physical point in a galactic halo [2]. Two flows are associated with particles falling through the galaxy for the first time ($n = 1$), two other flows are associated with particles falling through the galaxy for the second time ($n = 2$), and so on. Scattering in the gravitational wells of inhomogeneities in the galaxy (e.g. molecular clouds and globular clusters) are ineffective in thermalizing the flows with low values of n.

Caustics appear wherever the projection of the phase-space sheet onto physical space has a fold [3, 4, 5, 6]. Generically, caustics are surfaces in physical space. On one side of the caustic surface there are two more flows than on the other. At the surface, the dark matter density is very large. It diverges there in the limit of zero velocity dispersion. There are two types of caustics in the halos of galaxies, inner and outer. The outer caustics are topological spheres surrounding the galaxy. They are located near where a given outflow reaches its furthest distance from the galactic center before falling back in. The inner caustics are rings [3]. They are located near where the particles with the most angular momentum in a given inflow reach their distance of closest approach to the galactic center before going back out. A caustic ring is a closed tube whose cross-section is a D_{-4} (also called *elliptic umbilic*) catastrophe [6]. The existence of these caustics and their topological properties are independent of any assumptions of symmetry.

As was mentioned earlier, the primordial velocity dispersion of the leading cold dark matter candidates is extremely small. However, to a coarse-grained observer, the dark matter falling onto a galaxy may have additional velocity dispersion because the phase-space sheet on which the dark matter particles lie may be wrapped up on scales which are small compared to the galaxy as a whole. This effective velocity dispersion is associated with the clumpiness of the dark matter before it falls onto the galaxy. For the caustics in a galaxy not to be washed out, the effective velocity dispersion of the infalling dark matter must be much less than the rotation velocity of the galaxy, say less than 30 km/s for our galaxy. However, an upper bound of order 50 m/s can be obtained from observation, as explained below.

Primordial peculiar velocities are expected to be the same for baryonic and dark matter particles because they are caused by gravitational forces. Later the velocities of baryons and CDM differ because baryons collide with each other whereas CDM is collisionless. However, because angular momentum is conserved, the net angular momenta of the dark matter and baryonic components of a galaxy are aligned. Since the caustic rings are located near where the particles with the most angular momentum in a given infall are at their closest approach to the galactic center, they lie close to the galactic plane.

CAUSTIC RING RADII

A specific proposal has been made for the radii a_n of caustic rings [3]:

$$\{a_n : n = 1, 2, ...\} \simeq (39,\ 19.5,\ 13,\ 10,\ 8, ...)\mathrm{kpc} \times \left(\frac{j_{\max}}{0.25}\right)\left(\frac{0.7}{h}\right)\left(\frac{v_{\mathrm{rot}}}{220\frac{\mathrm{km}}{\mathrm{s}}}\right) \quad (3)$$

where h is the present Hubble constant in units of $100\mathrm{km}/(\mathrm{s\ Mpc})$, v_{rot} is the rotation velocity of the galaxy and $j_{\max}$ is a parameter with a specific value for each halo. For large n, $a_n \sim 1/n$. Eq. 3 is predicted by the self-similar infall model [7, 8] of galactic halo formation. $j_{\max}$ is then the maximum of the dimensionless angular momentum j-distribution [8]. The self-similar model depends upon a parameter ε [7]. In CDM theories of large scale structure formation, ε is expected to be in the range 0.2 to 0.35 [8]. Eq. 3 is for $\varepsilon = 0.3$. However, in the range $0.2 < \varepsilon < 0.35$, the ratios a_n/a_1 are almost independent of ε. When $j_{\max}$ values are quoted below, $\varepsilon = 0.3$ and $h = 0.7$ will be assumed.

Since the caustic rings lie close to the galactic plane, they cause bumps in the rotation curve, at the locations of the rings. In ref. [9] a set of 32 extended well-measured rotation curves was analyzed and statistical evidence was found for bumps distributed according to Eq. 3. That study suggests that the $j_{\max}$ distribution is peaked near 0.27. The rotation curve of NGC3198, one of the best measured, by itself shows three faint bumps which are consistent with Eq. 3 and $j_{\max} = 0.28$.

A recent paper [10] gives evidence for ring caustics in our own galaxy.

RING CAUSTICS IN THE MILKY WAY

A detailed north inner galactic rotation curve was obtained [11] from the Massachusetts-Stony Brook Galactic Plane CO survey [12]. It exhibits a series of eight sharp rises in the range of (galactocentric) radii 3 to 7 kpc. For each, Table I lists the radius r_1 where the rise starts, the radius r_2 where it ends, and the increase Δv in rotation velocity. The rises are interpreted here as due to the presence of caustic rings of dark matter in the galactic plane. Each r_1 should therefore be identified with a caustic ring radius a_n, and $r_2 - r_1$ with the caustic ring width p_n [6]. The ring widths depend in a complicated way on the velocity distribution of the infalling dark matter at last turnaround [6] and are not predicted by the model. They also need not be constant along the ring. In Table I, the numbers in parentheses are for two less distinct rises between 7 kpc and our own radius $r_\odot$, taken to be 8.5 kpc.

The fourth column shows the caustic ring radii a_n^{I} of the $\varepsilon = 0.3$ self-similar infall model fitted to the eight rises between 3 and 7 kpc, assuming that these are due to caustic rings $n = 7...14$ (fit I). This is a one-parameter ($j_{\max}$) fit minimizing $rmsd \equiv [\frac{1}{8}\sum_{n=7}^{14}(1 - \frac{a_n}{r_{1n}})^2]^{\frac{1}{2}}$. It yields $j_{\max} = 0.263$ and $rmsd = 3.1\%$. The fifth column shows the radii a_n^{II} assuming that the eight rises between 3 and 7 kpc are due to caustic rings $n = 6...13$ (fit II). In this case $j_{\max} = 0.239$ and $rmsd = 2.8\%$. Fits of similar quality

TABLE 1. Radii at which rises in the Milky Way rotation curve start (r_1) and end (r_2), the corresponding increases in velocity Δv, the caustic ring radii a_n of the self-similar infall model in the two different fits (I and II) discussed in the text, and typical velocity increases $\bar{\Delta v}_n$ predicted by the model, in fit I, without amplification due to baryon accretion.

r_1 (kpc)	r_2 (kpc)	Δv (km/s)	$a_n^{\rm I}$ (kpc) n = 1 ..14	$a_n^{\rm II}$ (kpc) n = 1 .. 13	$\bar{\Delta v}_n$ (km/s) n = 1 .. 14
			41.2		26.5
			20.5	37.2	10.6
			13.9	18.6	6.8
			10.5	12.5	5.0
(8.28)	(8.38)	(12)	8.50	9.51	3.9
(7.30)	(7.42)	(8)	7.14	7.68	3.2
6.24	6.84	23	6.15	6.45	2.6
5.78	6.01	9	5.41	5.56	2.3
4.91	5.32	15	4.83	4.89	2.0
4.18	4.43	8	4.36	4.36	1.7
3.89	4.08	8	3.98	3.94	1.5
3.58	3.75	6	3.66	3.60	1.3
3.38	3.49	14	3.38	3.31	1.2
3.16	3.25	8	3.15	3.05	1.1

are obtained for the other values of ε in the range 0.20 to 0.35, or by assuming simply $a_n \sim 1/n$. On the other hand, the assumption that the eight rises between 3 and 7 kpc are due to caustic rings $n = 6+s...13+s$, where s is an integer other than 0 or 1, yields considerably worse fits. Up to this point it is unclear whether $s = 0$ or 1 is preferred. However the two less distinct rises between 7 kpc and $r_\odot$ strongly suggest $s = 1$ since their r_1 values agree at the 2.6% level with ring radii $a_5^{\rm I}$ and $a_6^{\rm I}$, but do not agree well with $a_4^{\rm II}$ and $a_5^{\rm II}$. Henceforth $s = 1$ will be assumed.

The velocity increase due to a caustic ring is given by

$$\Delta v_n = v_{\rm rot} f_n \frac{\Delta I(\zeta_n)}{\cos\delta_n(0) + \phi_n'(0)\sin\delta_n(0)} . \quad (4)$$

The f_n, defined in ref. [3], are predicted by the self-similar infall model, but $\Delta I(\zeta_n), \delta_n(0)$ and $\phi_n'(0)$, defined in ref. [6], are not. Like the p_n, the latter parameters depend in a complicated way on the velocity distribution of the dark matter at last turnaround. On the basis of the discussion in ref. [6], the ratio on the RHS of Eq. 4 is expected to be of order one, but to vary from one caustic ring to the next. The size of these fluctuations is easily a factor two, up or down. The sixth column of Table I shows Δv_n with the fluctuating ratio set equal to one, i.e. $\bar{\Delta v}_n \equiv f_n v_{\rm rot}$.

For the reasons just stated, the fact that the observed Δv fluctuate by a factor of order 2 from one rise to the next is consistent with the interpretation that the rises are due to caustic rings. However the observed Δv (column 3) are typically a factor 5 larger than the velocity increases caused by the caustic rings acting alone (column 6). To account for the discrepancy I assume that the effect of the caustic rings is amplified by baryonic matter they have accreted. First I'll argue that the gas in the disk has sufficiently high density and low velocity dispersion for such an explanation to be plausible. Second I'll give observational evidence in support of the explanation.

The equilibrium distribution of gas is:

$$d_{\rm gas}(\vec{r}) = d_{\rm gas}(\vec{r}_0)\exp[-\frac{3}{< v_{\rm gas}^2 >}(\phi(\vec{r}) - \phi(\vec{r}_0))] \, , \tag{5}$$

where d is density and ϕ gravitational potential. In the solar neighborhood, $d_{\rm gas} \simeq 3 \cdot 10^{-24} \frac{\rm gr}{\rm cm^3}$ [13], which is comparable to the density of dark matter inside the tubes of caustic rings near us. From the scale height of the gas [13] and the assumption that it is in equilibrium with itself and the other disk components, I estimate $\langle v_{\rm gas}^2 \rangle^{\frac{1}{2}} \simeq 8$ km/s. The variation in the gravitational potential due to a caustic ring over the size of the tube is of order $\Delta\phi_{\rm CR} \simeq 2 f v_{\rm rot}^2 p/a \simeq (5 \frac{\rm km}{\rm s})^2$. Because $\frac{3}{<v_{\rm gas}^2>}\Delta\phi_{\rm CR}$ is of order one, the caustic rings have a large effect on the distribution of gas in the disk. The accreted gas amplifies and can dominate the effect of the caustic rings on the rotation curve. To check whether this hypothesis is consistent with the shape of the rises would require detailed modeling, as well as detailed knowledge on how the rotation curve is measured. In the meantime, I found observational evidence in its support.

The accreted gas may reveal the location of caustic rings in maps of the sky. Looking tangentially to a ring caustic from a vantage point in the plane of the ring, one may recognize the tricusp [6] shape of the D_{-4} catastrophe. The IRAS map of the galactic disk in the direction of galactic coordinates $(l, b) = (80°, 0°)$ shows a triangular shape which is strikingly reminiscent of the cross-section of a ring caustic. The vertices of the triangle are at $(83.5°, 0.3°), (77.3°, 3.5°)$ and $(77.4°, -2.7°)$ galactic coordinates. Images can be obtained from the Skyview Virtual Observatory (http://skyview.gsfc.nasa.gov/). The shape is correctly oriented with respect to the galactic plane and the galactic center. Moreover its position is consistent with that of a rise in the rotation curve, the one between 8.28 and 8.38 kpc ($n = 5$ in fit I). The caustic ring radius implied by the image is 8.31 kpc, and its dimensions are $p = 134$ pc and $q = 200$ pc, in the directions parallel and perpendicular to the galactic plane respectively.

In principle, the feature at $(80°, 0°)$ should be matched by another in the opposite tangent direction to the nearby ring caustic, at approximately $(-80°, 0°)$. Although there is a plausible feature there, it is much less compelling than the one in the $(+80°, 0°)$ direction. There are several reasons why it may not appear as strongly. One is that the $(+80°, 0°)$ feature is in the middle of the Sagittarius spiral arm, whose stellar activity enhances the local gas emissivity, whereas the $(-80°, 0°)$ feature is not so favorably located. Another is that the ring caustic in the $(+80°, 0°)$ direction has unusually small dimensions. This may make it more visible by increasing its contrast with the background. In the $(-80°, 0°)$ direction, the nearby ring caustic may have larger transverse dimensions.

Our proximity to a ring means that the associated flows, i.e. those flows in which the caustic occurs, contribute very importantly to the local dark matter density. Using the results of refs. [3, 6, 8], and assuming axial symmetry of the caustic ring between us and the tangent point (approx. 1 kpc away from us), the densities and velocity vectors on Earth of the associated flows can be derived:

$$d^+ = 1.7\ 10^{-24}\ \frac{\text{gr}}{\text{cm}^3}\ ,\ d^- = 1.5\ 10^{-25}\ \frac{\text{gr}}{\text{cm}^3}\ ,\ \vec{v}^{\pm} = (470\ \hat{\phi} \pm 100\ \hat{r})\ \frac{\text{km}}{\text{s}}, \tag{6}$$

where $\hat{r}, \hat{\phi}$ and $\hat{z}$ are the local unit vectors in galactocentric cylindrical coordinates. $\hat{\phi}$ is in the direction of galactic rotation. The velocities are given in the (non-rotating) rest frame of the Galaxy. Because of an ambiguity, it is not presently possible to say whether $d^{\pm}$ are the densities of the flows with velocity $\vec{v}^{\pm}$ or $\vec{v}^{\mp}$. Eq. 6 has implications for dark matter searches. Previous estimates of the local dark matter density, based on isothermal halo profiles, range from 5 to 7.5 $10^{-25}\ \frac{\text{gr}}{\text{cm}^3}$. The present analysis implies that a single flow (d^+) has three times that much local density, i.e. that the total local density is four times higher than previously thought. The large size of d^+ is due to our proximity to a cusp of the nearby caustic. Assuming axial symmetry, that cusp is only 55 pc away from us. The exact size of d^+ is sensitive to our distance to the cusp but, in any case, d^+ is very large. If we are inside the tube of the fifth caustic, there are two additional flows on Earth, aside from those given in Eq. 6. A list of approximate local densities and velocity vectors for the $n \neq 5$ flows can be found in ref. [14]. An updated list is in preparation.

The sharpness of the rises in the rotation curve and of the triangular feature in the IRAS map implies an upper limit on the velocity dispersion δv_{DM} of the infalling dark matter. Caustic ring singularities are spread over a distance of order $\delta a \simeq \frac{R\ \delta v_{\text{DM}}}{v}$ where v is the velocity of the particles in the caustic, δv_{DM} is their velocity dispersion when they first fell in, and R is the turnaround radius then. The sharpness of the IRAS feature implies that its edges are spread over $\delta a \lesssim 20$pc. Assuming that the feature is due to the $n = 5$ ring caustic, $R \simeq 180$ kpc and $v \simeq 480$ km/s. Therefore $\delta v_{\text{DM}} \lesssim 53$ m/s.

There may be evidence for the accretion of baryonic matter onto the $n \leq 4$ rings as well. Binney and Dehnen studied [15] the outer rotation curve of the Milky Way and concluded that its anomalous behaviour can be explained if most of the tracers of the rotation are concentrated in a ring of radius $1.6\ r_{\odot} = 13.6$ kpc. This is very close to the expected radius (13.9 kpc) of the $n = 3$ ring. Recently, the SDSS collaboration detected [16] an overdensity of stars which appears to be lying in the galactic plane on an arc of circle at least 40° in length, and of galactocentric radius approximately 18 kpc. These stars have properties consistent with those of spheroid stars but their spatial distribution is not consistent with a power law spheroid. The observed feature may be due to the accretion of stars onto the $n = 2$ ring.

The caustic ring model may explain the puzzling persistence of galactic disk warps [17]. These may be due to outer caustic rings lying somewhat outside the galactic plane and attracting visible matter. Such disk warps would not damp and would persist on cosmological time scales.

The caustic ring model, and more specifically the prediction Eq. 6 of the locally dominant flow associated with the nearby ring, has important consequences for axion dark matter searches [18], the annual modulation [14, 19, 20] and signal anisotropy [21, 20] in WIMP searches, the search for γ-rays from dark matter annihilation [22], and the search for gravitational lensing by dark matter caustics [4]. The model allows precise predictions to be made in each of these approaches to the dark matter problem.

REFERENCES

1. E.W. Kolb and M.S. Turner, *The Early Universe*, Addison-Wesley, 1990; M. Srednicki, Editor *Particle Physics and Cosmology: Dark Matter*, Nort-Holland, 1990.
2. P. Sikivie and J.R. Ipser, Phys. Lett. **B291** (1992) 288.
3. P. Sikivie, Phys. Lett. **B432** (1998) 139.
4. C. Hogan, astro-ph/9811290, to be published in Ap.J.
5. S. Tremaine, MNRAS **307** (1999) 877.
6. P. Sikivie, Phys. Rev. **D60** (1999) 063501.
7. J.A. Filmore and P. Goldreich, Ap.J. **281** (1984) 1; E. Bertschinger, Ap. J. Suppl. **58** (1985) 39.
8. P. Sikivie, I. Tkachev and Y. Wang, Phys. Rev. Lett. **75** (1995) 2911; Phys. Rev. **D56** (1997) 1863.
9. W. Kinney and P. Sikivie, Phys. Rev. **D61** (2000) 087305.
10. P. Sikivie, astro-ph/0109296.
11. D.P. Clemens, Ap.J. **295** (1985) 422.
12. D.B. Sanders et al., Ap.J.S. **60** (1986) 1; D.P. Clemens et al., Ap.J.S. **60** (1986) 297.
13. J. Binney and S. Tremaine, *Galactic Dynamics*, Princeton U. Press, 1987.
14. P. Sikivie, in the Proc. of the Second International Workshop on *The Identification of Dark Matter*, edited by N. Spooner and V. Kudryavtsev, World Scientific 1999, p. 68.
15. J. Binney and W. Dehnen, MNRAS **287** (1997) L5.
16. H.J. Newberg, B. Yanny, et al., *The Ghost of Sagittarius and Lumps in the Halo of the Milky Way*, SDSS preprint, June 2001. I thank H. Newberg for making me aware of refs. [15, 16].
17. R.W. Nelson and S. Tremaine, MNRAS **275** (1995) 897; J. Binney, I.-G. Jiang and S. Dutta, MNRAS **297** (1998) 1237.
18. C. Hagmann et al., Phys. Rev. Lett. **80** (1998) 2043; I. Ogawa, S. Matsuki and K. Yamamoto, Phys. Rev. **D53** (1996) 1740.
19. J. Vergados, Phys. Rev. **D63** (2001) 063511; A. Green, Phys. Rev. **D63** (2001) 103003; G. Gelmini and P. Gondolo, Phys. Rev. **D64** (2001) 023504.
20. D. Stiff, L.M. Widrow and J. Frieman, astro-ph/0106048.
21. C. Copi, J. Heo and L. Krauss, Phys. Lett. **B461** (1999) 43.
22. L. Bergstrom, J. Edsjo and C. Gunnarsson, Phys. Rev. **D63** (2001) 083515; C. Hogan, astro-ph/0104106.

Theoretical Rates for Direct Detection of SUSY Dark Matter

J.D. Vergados

Physics Department, University of Ioannina, Ioannina, Gr 451 10, Greece

Abstract. Exotic dark matter together with the vacuum energy (associated with the cosmological constant) seem to dominate in the Universe. Thus its direct detection is central to particle physics and cosmology. Supersymmetry provides a natural dark matter candidate, the lightest supersymmetric particle (LSP). Furthermore from the knowledge of the density and velocity distribution of the LSP, the quark substructure of the nucleon and the nuclear structure (form factor and/or spin response function), one is able to evaluate the event rate for LSP-nucleus elastic scattering. The thus obtained event rates are, however, very low. So it is imperative to exploit the modulation effect, i.e. the dependence of the event rate on the Earth's motion and the directional signature, i.e. the dependence of the rate on the direction of the recoiling nucleus. In this paper we study such experimental signatures employing a supersymmetric model with universal boundary conditions at large $tan\beta$.

INTRODUCTION

In recent years the consideration of exotic dark matter has become necessary in order to close the Universe [1]. The COBE data [2] suggest that CDM (Cold Dark Matter) is at least 60% [3]. On the other hand evidence from two different teams, the High-z Supernova Search Team [4] and the Supernova Cosmology Project [5] , [6] suggests that the Universe may be dominated by the cosmological constant Λ. Thus the situation can be adequately described by a baryonic component $\Omega_B = 0.1$ along with the exotic components $\Omega_{CDM} = 0.3$ and $\Omega_\Lambda = 0.6$ (see next section for the definitions). In another analysis Turner [7] gives $\Omega_m = \Omega_{CDM} + \Omega_B = 0.4$. Since the non exotic component cannot exceed 40% of the CDM [1], [8], there is room for the exotic WIMP's (Weakly Interacting Massive Particles). In fact the DAMA experiment [9] has claimed the observation of one signal in direct detection of a WIMP, which with better statistics has subsequently been interpreted as a modulation signal [10].

In the most favored scenario of supersymmetry the LSP can be simply described as a Majorana fermion, a linear combination of the neutral components of the gauginos and Higgsinos [1, 11, 12, 14].

AN OVERVIEW OF DIRECT DETECTION - THE ALLOWED SUSY PARAMETER SPACE.

Since this particle is expected to be very massive, $m_\chi \geq 30 GeV$, and extremely non relativistic with average kinetic energy $T \leq 100 KeV$, it can be directly detected [15, 16]

CP624, *Cosmology and Elementary Particle Physics*, edited by B. N. Kursunoglu et al.

mainly via the recoiling of a nucleus (A,Z) in elastic scattering. In order to compute the event rate one needs the following ingredients:

1) An effective Lagrangian at the elementary particle (quark) level obtained in the framework of supersymmetry as described , e.g., in Refs. [1, 14].

2) A procedure in going from the quark to the nucleon level, i.e. a quark model for the nucleon. The results depend crucially on the content of the nucleon in quarks other than u and d. This is particularly true for the scalar couplings as well as the isoscalar axial coupling [18]–[20].

3) Compute the relevant nuclear matrix elements [22, 23] using as reliable as possible many body nuclear wave functions. The situation is a bit simpler in the case of the scalar coupling, in which case one only needs the nuclear form factor.

Since the obtained rates are very low, one would like to be able to exploit the modulation of the event rates due to the earth's revolution around the sun [24, 25]–[27]. To this end one adopts a folding procedure assuming some distribution [1, 25, 27] of velocities for the LSP. One also would like to know the directional rates, by observing the nucleus in a certain direction, which correlate with the motion of the sun around the center of the galaxy and the motion of the Earth [11, 28].

The calculation of this cross section has become pretty standard. One starts with representative input in the restricted SUSY parameter space as described in the literature [12, 14]. We will adopt a phenomelogical procedure taking universal soft SUSY breaking terms at M_{GUT}, i.e., a common mass for all scalar fields m_0, a common gaugino mass $M_{1/2}$ and a common trilinear scalar coupling A_0, which we put equal to zero (we will discuss later the influence of non-zero A_0's). Our effective theory below M_{GUT} then depends on the parameters [12]:

$$m_0,\ M_{1/2},\ \mu_0,\ \alpha_G,\ M_{GUT},\ h_t,\ ,\ h_b,\ ,\ h_\tau,\ \tan\beta\ ,$$

where $\alpha_G = g_G^2/4\pi$ (g_G being the GUT gauge coupling constant) and h_t, h_b, h_τ are respectively the top, bottom and tau Yukawa coupling constants at M_{GUT}. The values of α_G and M_{GUT} are obtained as described in Ref.[12]. For a specified value of $\tan\beta$ at M_S, we determine h_t at M_{GUT} by fixing the top quark mass at the center of its experimental range, $m_t(m_t) = 166$GeV. The value of h_τ at M_{GUT} is fixed by using the running tau lepton mass at m_Z, $m_\tau(m_Z) = 1.746$GeV. The value of h_b at M_{GUT} used is such that:

$$m_b(m_Z)^{\overline{DR}}_{SM} = 2.90 \pm 0.14 \text{ GeV}.$$

after including the SUSY threshold correction. The SUSY parameter space is subject to the following constraints:

1.) The LSP relic abundance will satisfy the cosmological constrain:

$$0.09 \leq \Omega_{LSP}h^2 \leq 0.22 \qquad (1)$$

2.) The Higgs bound obtained from recent CDF [29] and LEP2 [30], i.e. $m_h > 113\ GeV$.

3.) We will limit ourselves to LSP-nucleon cross sections for the scalar coupling, which gives detectable rates

$$4 \times 10^{-7}\ pb \leq \sigma_{scalar}^{nucleon} \leq 2 \times 10^{-5}\ pb \qquad (2)$$

We should remember that the event rate does not depend only on the nucleon cross section, but on other parameters also, mainly on the LSP mass and the nucleus used in target. The condition on the nucleon cross section imposes severe constraints on the acceptable parameter space. In particular in our model it restricts $tan\beta$ to values $tan\beta \simeq 50$. We will not elaborate further on this point, since it has already appeared [31].

EXPRESSIONS FOR THE DIFFERENTIAL CROSS SECTION .

The effective Lagrangian describing the LSP-nucleus cross section can be cast in the form [15]

$$L_{eff} = -\frac{G_F}{\sqrt{2}}\{(\bar{\chi}_1\gamma^\lambda\gamma_5\chi_1)J_\lambda + (\bar{\chi}_1\chi_1)J\} \tag{3}$$

where

$$J_\lambda = \bar{N}\gamma_\lambda(f_V^0 + f_V^1\tau_3 + f_A^0\gamma_5 + f_A^1\gamma_5\tau_3)N \ , \ \ J = \bar{N}(f_s^0 + f_s^1\tau_3)N \tag{4}$$

We have neglected the uninteresting pseudoscalar and tensor currents. Note that, due to the Majorana nature of the LSP, $\bar{\chi}_1\gamma^\lambda\chi_1 = 0$ (identically).

With the above ingredients the differential cross section can be cast in the form [11, 24, 25]

$$d\sigma(u,\upsilon) = \frac{du}{2(\mu_r b\upsilon)^2}[(\bar{\Sigma}_S + \bar{\Sigma}_V \frac{\upsilon^2}{c^2})\ F^2(u) + \bar{\Sigma}_{spin}F_{11}(u)] \tag{5}$$

In the present work we will focus on the scalar contribution and ignore the vector and spin contributions. Then

$$\bar{\Sigma}_S = \sigma_0(\frac{\mu_r(A)}{\mu_r(N)})^2\,\{A^2\,[(f_S^0 - f_S^1\frac{A-2Z}{A})^2] \simeq \sigma^S_{p,\chi^0}A^2(\frac{\mu_r(A)}{\mu_r(N)})^2 \tag{6}$$

with $\sigma^S_{p,\chi^0} = \sigma_0\ (f_S^0)^2\ (\frac{\mu_r(N)}{m_N})^2$ (scalar) , (the isovector scalar is negligible, i.e. $\sigma_p^S = \sigma_n^S$) where m_N is the nucleon mass, $\eta = m_x/m_N A$, and $\mu_r(A)$ is the LSP-nucleus reduced mass, $\mu_r(N)$ is the LSP-nucleon reduced mass and

$$\sigma_0 = \frac{1}{2\pi}(G_F m_N)^2 \simeq 0.77 \times 10^{-38} cm^2 \tag{7}$$

$$Q = Q_0 u \ , \qquad Q_0 = \frac{1}{A m_N b^2} = 4.1 \times 10^4 A^{-4/3}\ KeV \tag{8}$$

where Q is the energy transfer to the nucleus and $F(u)$ is the nuclear form factor.

In the present paper we will concentrate on the coherent mode. For a discussion of the spin contribution, expected to be important in the case of the light nuclei, has been reviewed elsewhere [31].

EXPRESSIONS FOR THE RATES.

The non-directional event rate is given by:

$$R = R_{non-dir} = \frac{dN}{dt} = \frac{\rho(0)}{m_\chi}\frac{m}{Am_N}\sigma(u,\upsilon)|\upsilon| \quad (9)$$

Where $\rho(0) = 0.3GeV/cm^3$ is the LSP density in our vicinity and m is the detector mass The differential non-directional rate can be written as

$$dR = dR_{non-dir} = \frac{\rho(0)}{m_\chi}\frac{m}{Am_N}d\sigma(u,\upsilon)|\upsilon| \quad (10)$$

where $d\sigma(u,\upsilon)$ was given above.

The directional differential rate [11],[27] in the direction $\hat{e}$ is given by :

$$dR_{dir} = \frac{\rho(0)}{m_\chi}\frac{m}{Am_N}\upsilon.\hat{e}H(\upsilon.\hat{e})\,\frac{1}{2\pi}\,d\sigma(u,\upsilon) \quad (11)$$

where H the Heaviside step function. The factor of $1/2\pi$ is introduced, since the differential cross section of the last equation is the same with that entering the non-directional rate, i.e. after an integration over the azimuthal angle around the nuclear momentum has been performed. In other words, crudely speaking, $1/(2\pi)$ is the suppression factor we expect in the directional rate compared to the usual one. The precise suppression factor depends, of course, on the direction of observation. The mean value of the non-directional event rate of Eq. (10), is obtained by convoluting the above expressions with the LSP velocity distribution $f(\upsilon,\upsilon_E)$ with respect to the Earth, i.e. is given by:

$$\left\langle\frac{dR}{du}\right\rangle = \frac{\rho(0)}{m_\chi}\frac{m}{Am_N}\int f(\upsilon,\upsilon_E)|\upsilon|\frac{d\sigma(u,\upsilon)}{du}d^3\upsilon \quad (12)$$

The above expression can be more conveniently written as

$$\left\langle\frac{dR}{du}\right\rangle = \frac{\rho(0)}{m_\chi}\frac{m}{Am_N}\sqrt{\langle\upsilon^2\rangle}\langle\frac{d\Sigma}{du}\rangle\ ,\ \langle\frac{d\Sigma}{du}\rangle = \int\frac{|\upsilon|}{\sqrt{\langle\upsilon^2\rangle}}f(\upsilon,\upsilon_E)\frac{d\sigma(u,\upsilon)}{du}d^3\upsilon \quad (13)$$

After performing the needed integrations over the velocity distribution, to first order in the Earth's velocity, and over the energy transfer u the last expression takes the form

$$R = \bar{R}\,t\,[1 + h(a,Q_{min})cos\alpha)] \quad (14)$$

where α is the phase of the Earth ($\alpha = 0$ around June 2nd) and Q_{min} is the energy transfer cutoff imposed by the detector. In the above expressions $\bar{R}$ is the rate obtained in the conventional approach [15] by neglecting the folding with the LSP velocity and the momentum transfer dependence of the differential cross section, i.e. by

$$\bar{R} = \frac{\rho(0)}{m_\chi}\frac{m}{Am_N}\sqrt{\langle v^2\rangle}[\bar{\Sigma}_S + \bar{\Sigma}_{spin} + \frac{\langle\upsilon^2\rangle}{c^2}\bar{\Sigma}_V] \quad (15)$$

where $\bar{\Sigma}_i, i = S, V, spin$ contain all the parameters of the SUSY models. The modulation is described by the parameter h .

The total directional event rates can be obtained in a similar fashion by by integrating Eq. (11) with respect to the velocity as well as the energy transfer u. We find

$$R_{dir} = \bar{R}[(t_{dir}/2\pi)[1 + (h_1 - h_2)cos\alpha) + h_3 sin\alpha] \tag{16}$$

where the quantity t_{dir} provides the un modulated amplitude, while h_1, h_2 and h_3 describe the modulation. They are functions of the angles Θ and Φ, which specify the direction of observation $\hat{e}$, as well as the parameters a and Q_{min}. The effect of folding with LSP velocity on the total rate is taken into account via the quantity t_{dir}, which depends on the LSP mass. All other SUSY parameters have been absorbed in $\bar{R}$. In the special case previously studied, i.e along the coordinate axes, we find that: a) in the direction of the sun's motion $h_2 = h_3 = 0$, b) along the radial direction (y axes) $h_3 = 0$ and c) in the vertical to the galaxy $h_2 = 0$. Instead of t_{dir} itself it is more convenient to present the reduction factor of the un modulated directional rate compared to the usual non-directional one, i.e.

$$f_{red} = \frac{R_{dir}}{R} = t_{dir}/(2\pi\, t) = \kappa/(2\pi) \tag{17}$$

It turns out that the parameter κ, being the ratio of two rates, is less dependent on these parameters. Given the functions $h_l(a, Q_{min})$, $l = 1,2,3$, one can plot the the expression in Eqs (14) and 16 as a function of the phase of the earth α.

THE SCALAR CONTRIBUTION- THE ROLE OF THE HEAVY QUARKS

The coherent scattering can be mediated via the the neutral intermediate Higgs particles (h and H), which survive as physical particles. It can also be mediated via s-quarks, via the mixing of the isodoublet and isosinlet s-quarks of the same charge. In our model we find that the Higgs contribution becomes dominant and, as a matter of fact the heavy Higgs H is more important (the Higgs particle A couples in a pseudoscalar way, which does not lead to coherence). It is well known that all quark flavors contribute [18], since the relevant couplings are proportional to the quark masses. One encounters in the nucleon not only the usual sea quarks ($u\bar{u}, d\bar{d}$ and $s\bar{s}$) but the heavier quarks c, b, t which couple to the nucleon via two gluon exchange, see e.g. Drees *et al* [19] and references therein.

As a result one obtains an effective scalar Higgs-nucleon coupling by using effective quark masses as follows

$$m_u \rightarrow f_u m_N, \;\; m_d \rightarrow f_d m_N. \;\; m_s \rightarrow f_s m_N$$

$$m_Q \rightarrow f_Q m_N, \;\; (heavy\ quarks\ c, b, t)$$

where m_N is the nucleon mass. The isovector contribution is now negligible. The parameters f_q, $q = u,d,s$ can be obtained by chiral symmetry breaking terms in relation to phase shift and dispersion analysis. Following Cheng and Cheng [20] we obtain:

$$f_u = 0.021, \quad f_d = 0.037, \quad f_s = 0.140 \qquad \text{(model B)}$$

$$f_u = 0.023, \quad f_d = 0.034, \quad f_s = 0.400 \qquad \text{(model C)}$$

We see that in both models the s-quark is dominant. Then to leading order via quark loops and gluon exchange with the nucleon one finds:

$$f_Q = 2/27(1 - \quad \Sigma_q f_q)\,, \quad \text{i.e.} \quad f_Q = 0.060\ \text{(model B)}, \quad f_Q = 0.040\ \text{(model C)}$$

There is a correction to the above parameters coming from loops involving s-quarks [19] and due to QCD effects. Thus for large $tan\beta$ we find [11]:

$$f_c = 0.060 \times 1.068 = 0.064, f_t = 0.060 \times 2.048 = 0.123, f_b = 0.060 \times 1.174 = 0.070$$
(model B)

$$f_c = 0.040 \times 1.068 = 0.043, f_t = 0.040 \times 2.048 = 0.082, f_b = 0.040 \times 1.174 = 0.047$$
(model B)

For a more detailed discussion we refer the reader to Refs [18, 19].

RESULTS AND DISCUSSION

The three basic ingredients of our calculation were the input SUSY parameters (see sect. 1), a quark model for the nucleon (see sect. 3) and the velocity distribution combined with the structure of the nuclei involved (see sect. 2). we will focus our attention on the coherent scattering and present results for the popular target ^{127}I. We have utilized two nucleon models indicated by B and C which take into account the presence of heavy quarks in the nucleon. We also considered energy cut offs imposed by the detector, by considering two typical cases $Q_{min} = 0,\ 10$ KeV. The thus obtained results for the un modulated non directional event rates $\bar{R}t$ in the case of the symmetric isothermal model for a typical SUSY parameter choice [12] are shown in Fig. 1. The two relative parameters, i.e. the quantities t and h, are shown in Fig. 2 and Figs 3,4 respectively in the case of isothermal models. The case of non isothermal models, e.g. caustic rings, is more complicated [27] and it will not be further discussed here.

It is instructive to examine the reduction factors along the three axes, i.e along $+z, -z, +y, -y, +x$ and $-x$ [26]. Since f_{red} is the ratio of two parameters, its dependence on Q_{min} and the LSP mass is mild. So we present results for $Q_{min} = 0$ and give an average as a function of the LSP mass (see Table 1). As expected the maximum rate is along the sun's direction of motion, i.e opposite to its velocity $(-z)$ in the Gaussian distribution and $+z$ in the case of caustic rings. In fact we find that $\kappa(-z)$ is around 0.5 (no asymmetry) and around 0.6 (maximum asymmetry, $\lambda = 1.0$), It is not very different from the naively expected $f_{red} = 1/(2\pi) = \kappa = 1$. The asymmetry $|R_{dir}(-) - R_{dir}(+)|/(R_{dir}(-) + R_{dir}(+))$ is quite large in the isothermal model and smaller in caustic rings. The rate in the other directions is quite a bit smaller (see Table 1).

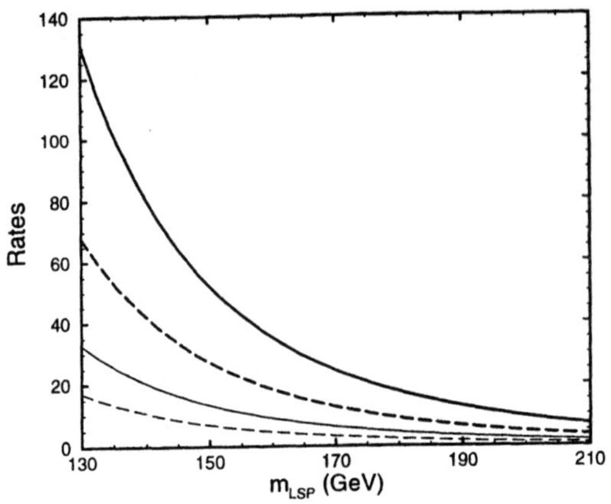

FIGURE 1. The Total detection rate per $(kg - target)yr$ vs the LSP mass in GeV for a typical solution in our parameter space in the case of ^{127}I corresponding to model B (thick line) and Model C (fine line). For the definitions see text.

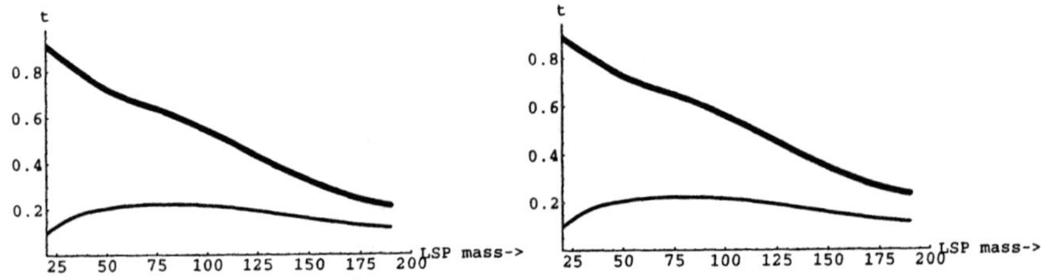

FIGURE 2. The dependence of the quantity t on the LSP mass for the symmetric case ($\lambda = 0$) on the left as well as for the maximum axial asymmetry ($\lambda = 1$) on the right in the case of the target ^{127}I. For orientation purposes two detection cutoff energies are exhibited, $Q_{min} = 0$ (thick solid line) and $Q_{min} = 10\ keV$ (thin solid line). As expected t decreases as the cutoff energy and/or the LSP mass increase. We see that the asymmetry parameter λ has little effect on the un modulated rate.

As we have seen the modulation can be described in terms of the parameters h_i, $i = 1,2,3$ (see Eq. (16)). If the observation is done in the direction opposite to the sun's direction of motion the modulation amplitude h_1 behaves in the same way as the non directional one, namely h. It is instructive to consider directions of observation in the plane perpendicular to the sun's direction of motion ($\Theta = \pi/2$)even though the un modulated rate is reduced in this direction. Along the $-y$ direction ($\Phi = (3/2)\pi$) the modulation amplitude $h_1 - h_2$ is constant, -0.20 and -0.30 for $\lambda = 0, 1$ respectively. In other words it large and leads to a maximum in December. Along the $+y$ direction the modulation is exhibited in Figs 5 and 6.

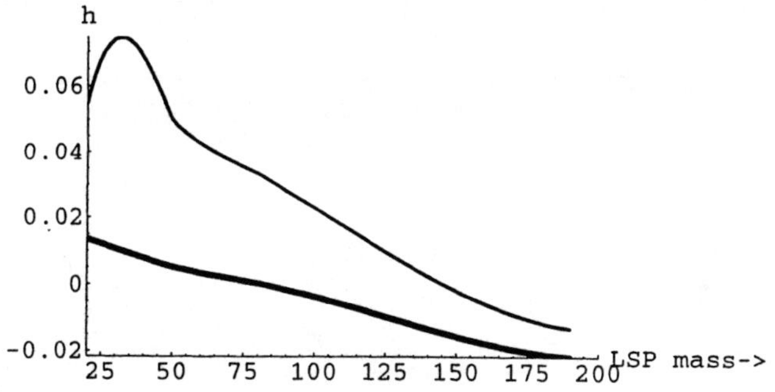

FIGURE 3. The same as in Fig. 2 for the modulation with $\lambda = 0$. We see that the modulation is small and decreases with the LSP mass. It even changes sign for large LSP mass. The introduction of a cutoff Q_{min} increases the modulation (at the expense of the total number of counts).

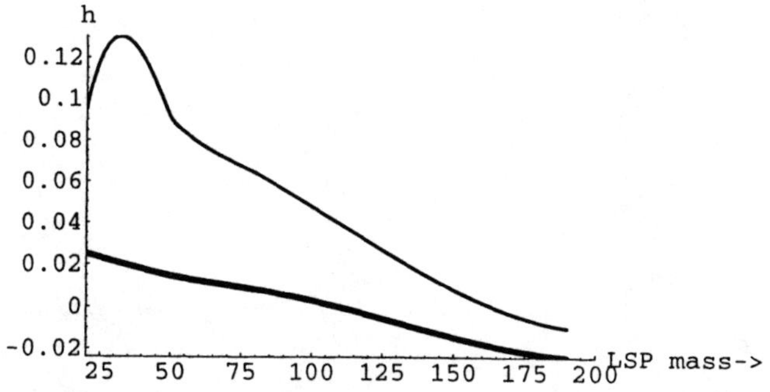

FIGURE 4. The same as in Fig. 3 for $\lambda = 1$. We see that the modulation increases with the asymmetry parameter λ.

CONCLUSIONS

In the present paper we have discussed the parameters, which describe the event rates for direct detection of SUSY dark matter. Only in a small segment of the allowed parameter space the rates are above the present experimental goals. We thus looked for characteristic experimental signatures for background reduction, i.e. a) Correlation of the event rates with the motion of the Earth (modulation effect) and b) the directional rates (their correlation both with the velocity of the sun and that of the Earth.)

A typical graph for the total un modulated rate is shown Fig. 1. The relative parameters

TABLE 1. The ratio κ of the un modulated directional rate along the three directions to the non-directional one: z is in the direction of the sun's motion, x is in the radial direction and x is perpendicular to the axis of the galaxy. The asymmetry is also given. $Q_{min} = 0$ was assumed.

		isothermal			caustic rings		
λ	dir.	+	-	asym	+	-	asym
0	z	0.02	0.50	0.92	0.75	0.25	0.50
0	y	0.16	0.16	0	0.22	0	1.00
0	x	0.16	0.16	0	0.37	0.24	0.21
1	z	0.04	0.58	0.90	-	-	-
1	y	0.12	0.12	0	-	-	-
1	x	0.17	0.17	0	-	-	-

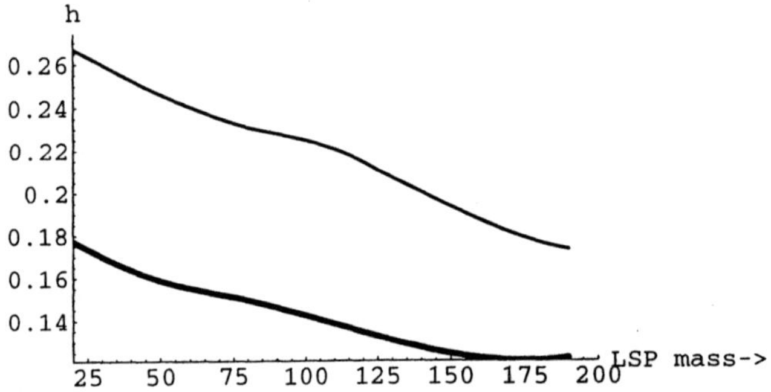

FIGURE 5. The quantity $h_1 - h_2$ in the direction $+y$ for $\lambda = 0$ (thick line) and $\lambda = 1$ thin line. In the $-y$ direction this quantity is constant and negative, -0.20 and -0.30 for $\lambda = 0$ and 1 respectively. As a result the modulation effect is opposite (minimum in June the 3nd).

t and h in the case of non directional experiments are exhibited in Fig. 2 and Figs 3 and 4. We must emphasize that these two graphs do not contain the entire dependence on the LSP mass. This is due to the fact that there is the extra factor μ_r^2 in Eq. (2.10) but mainly due to the fact that the nucleon cross section depends on the LSP mass. Fig 2 and Figs 3 and 4. were obtained for the scalar interaction, but do not expect a very different behavior in the case of the spin contribution. The overall spin contribution, however, is going to be very different, but such considerations are beyond the goals of the present paper. We should also mention that in the non directional experiments the modulation $2h_1$ is small, .i.e. for $\lambda = 0$ less than 4% for $Qmin = 0$ and 12% for $Qmin = 10\ KeV$ (at the expense of the total number of counts). For $\lambda = 1$ there in no change for $Q_{min} = 0$, but it can go as high as 24% for $Q_{min} = 10\ KeV$.

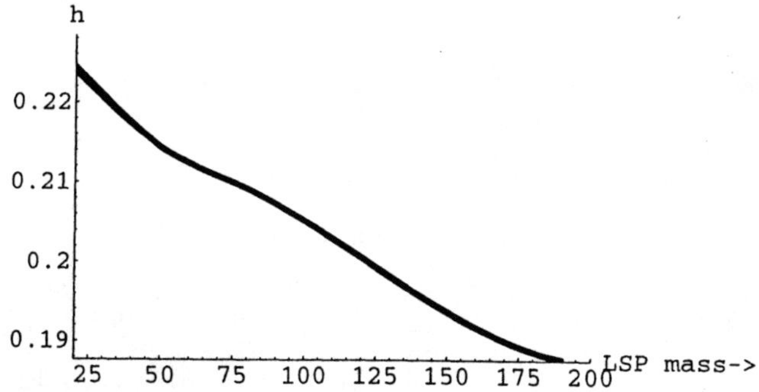

FIGURE 6. The same as in Fig. 5 for the modulation amplitude in the direction $+x$, which is essentially h_3, since$|h_1| << |h_3|$. Thus in this case the maximum occurs around September the 3nd and the minimum 6 months later. The opposite is true in the $-x$ direction.

For the directional rates It is instructive to examine the reduction factors along the three axes, i.e along $+z, -z, +y, -y, +x$ and $-x$. These depend on the nuclear parameters, the reduced mass, the energy cutoff Q_{min} and λ [26]. Since f_{red} is the ratio of two parameters, its dependence on Q_{min} and the LSP mass is mild. So we present results for $Q_{min} = 0$ and give their average as a function of the LSP mass (see Table 1). As expected the maximum rate is along the sun's direction of motion, i.e opposite to its velocity $(-z)$ in the Gaussian distribution and $+z$ in the case of caustic rings. In fact we find that $\kappa(-z)$ is around 0.5 (no asymmetry) and around 0.6 (maximum asymmetry, $\lambda = 1.0$). It is not very different from the naively expected $f_{red} = 1/(2\pi) = \kappa = 1$. The asymmetry along the sun's direction of motion, $asym = |R_{dir}(-) - R_{dir}(+)|/(R_{dir}(-) + R_{dir}(+))$ is quite characteristic, i.e. quite large in the isothermal models and smaller in caustic rings. The rate in the other directions is quite a bit smaller (see Table 1). But, as we have seen, in the plane perpendicular in the sun's velocity we have very large modulation, which may overcompensate for the large reduction factor. It is interesting to note that the dependence on the phase of the Earth α changes substantially with the direction of observation.

In conclusion in the case of directional un modulated rates we expect characteristic and pretty much unambiguous correlation with the motion on the sun, which can be explored by the experimentalists. The reduction factor in the direction of the motion of the sun is approximately only $1/(4\pi)$ relative to the non directional experiments. In the plane perpendicular to the motion of the sun we expect interesting modulation signals, but the reduction factor becomes worse. A more complete discussion will be given elsewhere [32].

This work was supported by the European Union under the contracts RTN No HPRN-CT-2000-00148 and TMR No. ERBFMRX–CT96–0090.

REFERENCES

1. For a review see e.g. Jungman, G. *et al.*,*Phys. Rep.* **267**, 195 (1996).
2. Smoot, G.F. et al., (COBE data), *Astrophys. J.* **396** (1992) L1.
3. Gawiser, E. and Silk, J.,*Science* **280**, 1405 (1988); Gross, M.A.K. Somerville, R.S., Primack, J.R., Holtzman , J. and Klypin, A.A., *Mon. Not. R. Astron. Soc.* **301**, 81 (1998).
4. Riess, A.G. *et al, Astron. J.* **116** (1998), 1009.
5. Somerville, R.S., J.R. Primack and S.M. Faber, astro-ph/9806228; *Mon. Not. R. Astron. Soc.* (in press).
6. Perlmutter, S. *et al* (1999) *Astrophys. J.* **517**,565; (1997) **483**,565 (*astro-ph*/9812133). Perlmutter, S., Turner, M.S. and White, M., *Phys. Rev. Let.* **83**, 670 (1999).
7. Turner, M.S., astro-ph/9904051; Phys. Rep. **333-334** (1990), 619.
8. Bennett, D.P. *et al.*, (MACHO collaboration), A binary lensing event toward the LMC: Observations and Dark Matter Implications, Proc. 5th Annual Maryland Conference, edited by S. Holt (1995); Alcock, C. *et al.*, (MACHO collaboration), *Phys. Rev. Lett.* **74** , 2967 (1995).
9. Bernabei, R. et al., INFN/AE-98/34, (1998); Bernabei, R. et al., it Phys. Lett. **B 389**, 757 (1996).
10. Bernabei, R. et al., *Phys. Lett.* **B 424**, 195 (1998); **B 450**, 448 (1999).
11. For more references see e.g. our previous report:
    Vergados, J.D., Supersymmetric Dark Matter Detection- The Directional Rate and the Modulation Effect, hep-ph/0010151;
12. Gómez, M.E., Vergados, J.D., hep-ph/0012020.
    Gómez, M.E., Lazarides, G. and Pallis, C., Phys. Rev. **D61** (2000) 123512 and *Phys. Lett.* **B 487**, 313 (2000).
13. Gómez, M.E. and Vergados, J.D., hep-ph/0105115.
14. Bottino, A. *et al.*, *Phys. Lett* **B 402**, 113 (1997).
    Arnowitt, R. and Nath, P., *Phys. Rev. Lett.* **74**, 4952 (1995); *Phys. Rev.* **D 54**, 2394 (1996); hep-ph/9902237;
    Bednyakov, V.A., Klapdor-Kleingrothaus, H. V. and Kovalenko, S.G., *Phys. Lett.* **B 329**, 5 (1994).
15. Vergados, J.D., *J. of Phys.* **G 22**, 253 (1996).
16. Kosmas, T.S. and Vergados, J.D., *Phys. Rev.* **D 55**, 1752 (1997).
17. Drees, M. and Nojiri, M.M., Phys. Rev. **D47** (1993) 376; (1985).
18. Drees, M. and Nojiri, M.M., *Phys. Rev.* **D 48**, 3843 (1993); *Phys. Rev.* **D 47**, 4226 (1993).
19. Djouadi, A. and Drees, M., Phys. Lett. B **484** (2000) 183; Dawson, S., *Nucl. Phys.* **B359**,283 (1991); Spira, M. it et al, *Nucl. Phys.* **B453**,17 (1995).
20. Cheng, T.P., *Phys. Rev.* **D 38** 2869 (1988); Cheng, H-Y., Phys. Lett. **B 219** 347 (1989).
21. Ressell, M.T., *et al.*, *Phys. Rev.* **D 48**, 5519 (1993);
22. Vergados, J.D. and Kosmas, T.S. *Physics of Atomic nuclei*, Vol. **61**, No 7, 1066 (1998) (from *Yadernaya Fisika*, Vol. 61, No 7, 1166 (1998).
23. Divari, P.C., Kosmas, T.S., Vergados, J.D. and Skouras, L.D., *Phys. Rev.* **C 61** (2000), 044612-1.
24. Vergados, J.D., *Phys. Rev.* **D 58**, 103001-1 (1998).
25. Vergados, J.D., *Phys. Rev. Lett* **83**, 3597 (1999).
26. Vergados, J.D., *Phys. Rev.* **D 62**, 023519 (2000).
27. Vergados, J.D., *Phys. Rev.* **D 63**, 06351 (2001).
28. Buckland, K.N., Lehner, M.J. and Masek, G.E., in Proc. *3nd Int. Conf. on Dark Matter in Astro- and part. Phys.* (Dark2000), Ed. . Klapdor-Kleingrothaus, H.V., Springer Verlag (2000).
29. *CDF Collaboration*, FERMILAB-Conf-99/263-E CDF;
    http://fnalpubs.fnal.gov/archive/1999/conf/Conf-99-263-E.html.
30. Dorman, P.J., *ALEPH Collaboration* March 2000,
    http://alephwww.cern.ch/ALPUB/seminar/lepcmar200/lepc2000.pdf.
31. Vergados, J.D., SUSY Dark Matter in Universe- Theoretical Direct Detection Rates, Proc. *NANP-01, International Conference on Non Accelerator New Physics*, Dubna, Russia, June 19-23, 2001, Editors V. Bednyakov and S. Kovalenko, hep-ph/0201014.
32. Vergados, J.D., to be published.

High-Redshift Radio Galaxies as a Cosmological Tool: Exploration of a Key Assumption and Comparison with Supernova Results

Ruth A. Daly*, Matthew P. Mory* and Erick J. Guerra†

*Department of Physics, Penn State University, Berks-Lehigh Valley College, P.O. Box 7009, Reading, PA 19610-6009, USA
†Department of Chemistry & Physics, Rowan University, Classboro, NJ 08028-1701, USA

Abstract. There are many different approaches to using observations to constrain or determine the global cosmological parameters that describe our universe. Methods that rely upon a determination of the coordinate distance to high-redshift sources are particularly useful because they do not involve assumptions about the clustering properties of matter, or the evolution of this clustering.

Two of the methods currently being used to determine the coordinate distance to high-redshift sources are the radio galaxy method and the supernova method. These methods are similar in their dependence on the coordinate distance. Here, the radio galaxy method is briefly described and results are presented. One of the underlying assumptions of the method is explored. In addition, the method is compared and contrasted to the supernova method. The constraints imposed on global cosmological parameters by radio galaxies are consistent with those imposed by supernovae.

For a universe that is spatially flat with mean mass density Ω_m in non-relativistic matter and mean mass density $1 - \Omega_m$ in quintessence, radio galaxies alone indicate at 84 % confidence that the expansion of the universe is accelerating at the current epoch. And, independent of whether or not the universe is spatially flat, radio galaxies alone indicate at 95 % confidence that Ω_m must be less than 0.6 at the current epoch.

1. INTRODUCTION

Global cosmological parameters describe the current state of the universe and indicate the future of the universe and are thus of obvious importance. One way to parameterize current values of global cosmological parameters is through the equation of state and mean mass density of the component at the present epoch. Observations of stars, galaxies, active galaxies, and other sources are then used in an attempt to constrain these parameters.

The different methods used to constrain global cosmological parameters have different assumptions and constrain the parameters in different ways. Determinations of the coordinate distance $(a_o r)$ [1] as a means to constrain global cosmological parameters is particularly useful because it allows a direct determination of global cosmological parameters. The method is independent of all aspects of density fluctuations, including the source and growth of these fluctuations. Thus, the method is particularly clean.

The properties of radio galaxies can be used to deterime the coordinate distance to sources with redshifts from zero to two. The method is outlined in section 2, and one of the primary assumptions is explored in detail in secion 2.1. General equations that

CP624, *Cosmology and Elementary Particle Physics*, edited by B. N. Kursunoglu et al.

apply to global cosmological parameters including quintessence in a spatially flat are presented in section 2.2. Supernova and radio galaxy constraints on global cosmological parameters are compared in section 3. Conclusions follow in section 4.

2. THE RADIO GALAXY METHOD AND RESULTS

The use of powerful extended FRII radio sources as a cosmological tool is presented and discussed by [2], [3], [4], [5] and [6]. The properties of the sources used for cosmology are explored in more detail by [7].

The use of very powerful FRII radio sources for cosmology is based on three assumptions. (1) Very powerful classical double radio sources are supersonic propagators, and thus the equations of strong shock physics apply. This seems justified based on the radio bridge properties of the sources, [8], [9], and [7]. (2) The average size a given source will have over its lifetime will be close to the average size of the full population of similar sources at the same redshift. This seems justified by the fact that the dispersion in source size at a given redshift is rather small. (3) The total time the AGN will produce highly collimated outflows to drive the growth of the source, t_*, is related to the beam power of the source, L_j, through the relation $t_* \propto L_j^{-\beta/3}$, where β is a parameter whose value needs to be determined.

The third assumption fits in rather nicely with currently popular models of jet production due to the electromagnetic energy extraction from a rotating black hole, and is discussed in some detail below.

2.1. Electromagnetic Energy Extraction from Rotating Holes

An AGN that produces two oppositely directed collimated outflows each with beam power L_j for a total time t_* will release a total energy $E_* = 2L_j t_*$. Thus, the relation $t_* \propto L_j^{-\beta/3}$ is equivalent to the relation $E_* \propto L_j^{1-\beta/3}$; this is equivalent to the third assumption described above. The quantities E_* and L_j for a model in which the beam power derives from the electromagnetic extraction of spin energy from a rotating black hole are described by [10]. Blandford [10] shows that the beam power and total energy available are

$$L_j = L_{EM} \sim 10^{45} (a/m)^2\, B_4^2\, M_8^2 \text{ erg s}^{-1} \propto (a/m)^2\, B^2\, M^2 \tag{1}$$

and

$$E_* = E \sim 5 \times 10^{61} (a/m)^2\, M_8 \text{ erg} \propto (a/m)^2\, M\ , \tag{2}$$

for $(a/m) << 1$, where M is the mass of the black hole, M_8 is the mass in units of $10^8 M_\odot$, a is the spin angular momentum S per unit mass M: $a = S/(Mc)$, c is the speed of light, m is the gravitational radius $m = GM/c^2$, B is the magnetic field strength, and B_4 is the magnetic field strength in units of 10^4 G [10].

An exploration of the properties of very powerful double radio galaxies indicates that $E_* \propto L_j^{1-\beta/3}$ with $\beta = 1.75 \pm 0.25$ as discussed by [4], [5], [11]and [6]. This is consistent with equations (2) and (3) when the magnetic field strength satisfies

$$B \propto M^{(2\beta-3)/2(3-\beta)} \; (a/m)^{\beta/(3-\beta)} \; . \tag{3}$$

For the case $\beta = 1.5$ simplifies beautifully to $B \propto (a/m)$. For the cases $\beta = 1.75$ and 2, the empirical determined relation is consistent with equations (3) and (4) when $B \propto (a/m)^{1.4}\, M^{0.2}$ and $B \propto (a/m)^2\, M^{1/2}$ respectively.

A comparison of the results published by [5] indicates that a when $L_j \sim 10^{45}$ erg s^{-1}, the total energy processed through large-scale jets is $\sim 5 \times 10^5 M_\odot c^2$, so the jets are active for a total lifetime of $\sim 10^7$ yr. Equations 1 and 2 show that the total source lifetime is about $t_* \sim 10^9/(M_8 B_4^2)$, indicating a total lifetime of about 10^7 yr for $M_8 B_4^2 \sim 10^2$. This is satisfied for $M_* \sim 10$, and $B_4 \sim 3$. In this case, the beam power is $\sim 10^{45}$ erg s^{-1} for $(a/m) \sim (1/30)$, and the total energy is $\sim 5 \times 10^5 M_\odot$. These values for M_8, B_4, and (a/m) seem quite reasonable. The scaling between variables required by the empirically determine relation between total energy and beam power are given above.

2.2. General Cosmological Equations for a Universe with Quintessence

Recent measurements of the cosmic microwave background anisotropy suggest that the universe has zero space curvature (de Bernardis et al. 2000, Balbi et al. 2000). Here, a universe with zero space curvature ($k = 0$)and quintessence is considered. These equations will be applied below to consider the constraints placed on quintessence by radio galaxies. These equations are well known, and can be found in [1], [12], and [14], to name a few. Quintessence is introduced and discussed by [13],[14], and [15].

The deceleration parameter q_o is defined to be $q_o = -(\ddot{a}a)/\dot{a}^2$ evaluated at the present epoch z=0, where a is the cosmic scale factor. This can be re-written $q_o = -\ddot{a}_o/(a_o H_o^2)$; quantities evaluated at z=0 have a subscript 'o', and $H_o = (\dot{a}_o/a_o)$. It is is shown in [12] that

$$(\ddot{a}/a) = -\frac{4\pi G}{3} \sum(\rho_i + 3p_i) = -\frac{4\pi G}{3} \sum \rho_i(1+3w_i) \tag{4}$$

where p_i is the pressure, ρ_i is the mean mass-energy density, w_i is the equation of state of ith component, $w_i = p_i/\rho_i$, with the quintessence term is included in the summation.

Mass-energy conservation of each component implies that

$$\dot{\rho}_i = -3(\rho_i + p_i)(\dot{a}/a) \tag{5}$$

[12]. By definition, the equation of state is $w_i = p_i/\rho_i$, so the eq. 5 implies $(\dot{\rho}_i/\rho_i) = -3(1+w_i)(\dot{a}/a)$. When the equation of state w_i does not change with time, the solution to this equation is $\rho_i = \rho_{i,o}(1+z)^{3(1+w_i)}$, where $(1+z) = a_o/a$. Thus, a component with equation of state w_i and present mean mass-energy density ρ_o will have a mean mass-energy density at redshift z of $\rho = \rho_o(1+z)^{n_i}$, where $n_i = 3(1+w_i)$.

When $k = 0$,

$$(\dot{a}/a)^2 = \frac{8\pi G}{3} \sum \rho_i \,, \tag{6}$$

where $\rho_i = \rho_{i,o}(1+z)^n$. It follows that $H_o^2 = (\dot{a}_o/a_o)^2 = (\frac{8\pi G}{3}) \sum \rho_{i,o}$. For $k = 0$, $\sum \rho_{i,o} = \rho_{c,o}$ where $\rho_{c,o}$ is the critical density at redshift zero. So,

$$H_o^2 = \frac{8\pi G}{3} \rho_{c,o} \tag{7}$$

for $k = 0$. The deceleration parameter $q_o = -\ddot{a}_o/(a_o H_o^2)$ then becomes

$$q_o = (1/2) \sum \Omega_i (1 + 3w_i) \,, \tag{8}$$

where $\Omega_i = \rho_{i,o}/\rho_{c,o}$ and equations 4 and 7 have been used.

The universe is accelerating when $q_o < 0$. This can only occur if $(1+3w_i) < 0$, or $w_i < -1/3$, which is a necessary but not sufficient condition to have an accelerating universe at the present epoch. When there are only two types of mass-energy controlling the expansion rate of the universe at the the current epoch, quintessence and non-relativistic matter, then the deceleration parameter is $q_o = \Omega_m/2 + \Omega_Q(1+3w)/2$. The universe will be accelerating in its expansion when

$$1 + 3w(1 - \Omega_m) < 0 \,, \tag{9}$$

which follows since $\Omega_Q = 1 - \Omega_m$; similar equations are presented by [14]. Thus, there are two regions on the $\Omega_m - w$ plane; points in one region represent solutions for which the universe is currently accelerating, while those in the other region represent solutions for which the universe is currently decelerating (e.g. see Figure 1). The curve separating these regions is indicated on Figure 1, which illustrates the constraints obtained using radio galaxies [6]. Note that radio galaxies alone place interesting constraints on Ω_m and w, and these results are consistent with those obtained using other methods (e.g Wang et al. 2000).

The coordinate distance to a source at redshift z follows from the equation

$$\int dr/\sqrt{1-kr^2} = \int dt/a(t) = (1/a_o) \int (\dot{a}/a)^{-1} dz \tag{10}$$

[1]. For a spatially flat universe, the left hand side of the equation reduces to the coordinate distance r. Equations 6 and 7 imply that $(\dot{a}/a)^2 = (H_o^2/\rho_{c,o}) \sum \rho_{i,o}(1+z)^{n_i}$ or

$$(\dot{a}/a)^2 = H_o^2 \sum \Omega_i (1+z)^{n_i} \,. \tag{11}$$

Following [12], $(\dot{a}/a) = H_o E(z)$, where

$$E(z) = \sqrt{\sum \Omega_i (1+z)^{n_i}} \,. \tag{12}$$

The coordinate distance to a source at redshift z in a spatially flat universe is (see equation 10)

$$(a_o r) = H_o^{-1} \int_0^z dz/E(z) \,. \tag{13}$$

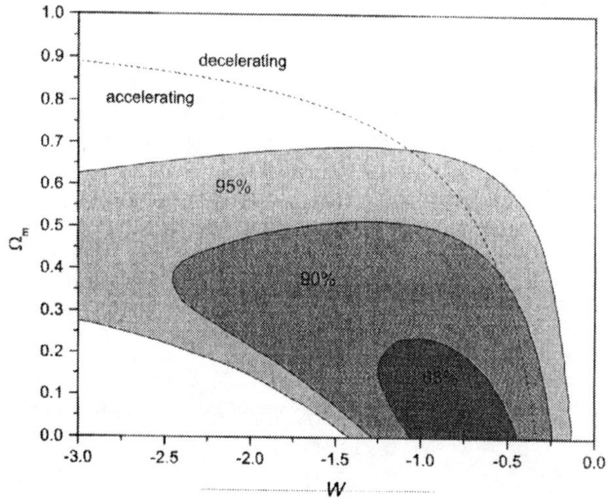

FIGURE 1. One-dimensional confidence contours obtain using radio galaxies, assuming a spatially flat universe. The dotted curve separates regions for which solutions represent an accelerating or decelerating universe. Conference participants were interested in values of w less than -1.0, so the region shown has been extended to w=-3.0. A similar figure is published by [6].

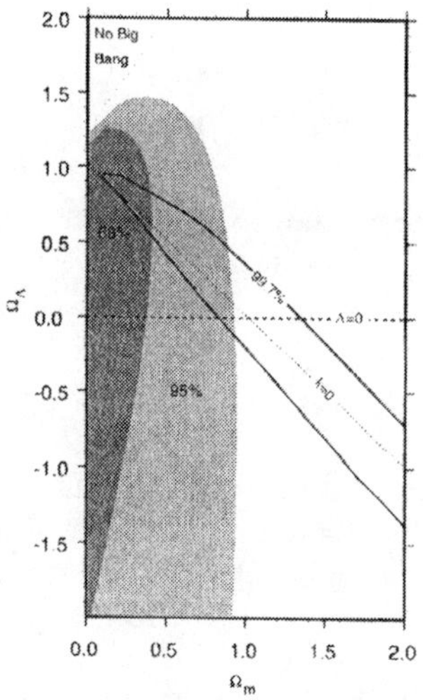

FIGURE 2. Constraints from the cosmic microwave background from [16] are indicated by the solid lines, while the two-dimensional constraints from radio galaxies are indicated by the shaded regions. Clearly, these two sets of observations imply a flat, low-density universe.

TABLE 1. Comparison Between Supernova and Radio Galaxy Methods.

Supenovae	Radio Galaxies
Type SNIa	Type FRIIb
$\propto (a_o r)^{2.0}$	$\propto (a_o r)^{1.6}$
$0 < z < 1$	$0 < z < 2$
$\sim$ 100 sources	20 sources (70 in parent pop.)
modified standard candle	modified standard yardstick
light curve $\Longrightarrow$ peak luminosity	radio bridge $\Longrightarrow$ average length
empirical relation	physical relation
written in terms of observables	written in terms of physical variables
normalized at $z = 0$	not normalized at $z = 0$
dependent on local distance scale	independent of local distance scale
universe is accelerating	Ω_m is low
	universe is acclelerating if k=0
some theoretical understanding	good theoretical understanding
well tested empirically	needs more empirical testing

A spatially flat universe has $\sum\Omega_i = 1$. The components that must be included in equation 12 are those that contribute from redshift zero out to the redshift to which the coordinate distance is being determined. Here, two components are considered: non-relativistic matter with normalized mean mass-energy density at $z = 0$ of Ω_m, and quintessence with normalized mean mass-energy density at $z = 0$ of Ω_Q. Thus, $\Omega_m + \Omega_Q = 1$, and equation 12 becomes $E(z) = \sqrt{\Omega_m(1+z)^3 + (1-\Omega_m)(1+z)^n}$. The coordinate distance is obtained by substituting this into eq. 13 and solving for $(a_o r)$.

3. A COMPARISON OF SUPERNOVA AND RADIO GALAXY RESULTS

The supernova and radio galaxy methods are compared in detail by [6]. These results are summarized in Table 1 and Figures 3 and 4.

4. CONCLUSIONS

Only a few methods are available at present to constrain global cosmological parameters through the determination of the coordinate distance to high-redshift sources. In this paper the results presented by [6], and [17] are presented and summarized.

ACKNOWLEDGMENTS

It is a pleasure to thank the organizers of the Coral Gables conference for such an interesting and interactive meeting. In particular I would like to thank Behram Kursunoglu, Sydney Meshkov, Arnold Perlmutter, and Ina Sarcevic. I would like to acknowledge numerous interesting and helpful discussions with conference participants

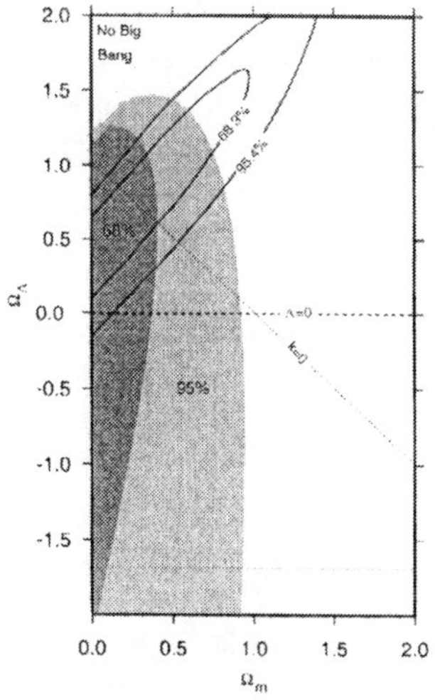

FIGURE 3. Constraints from the high-redshift supernova team [20] are indicated by the solid lines, while the radio galaxy results are indicated by the shaded regions; both are two-dimensional constraints.

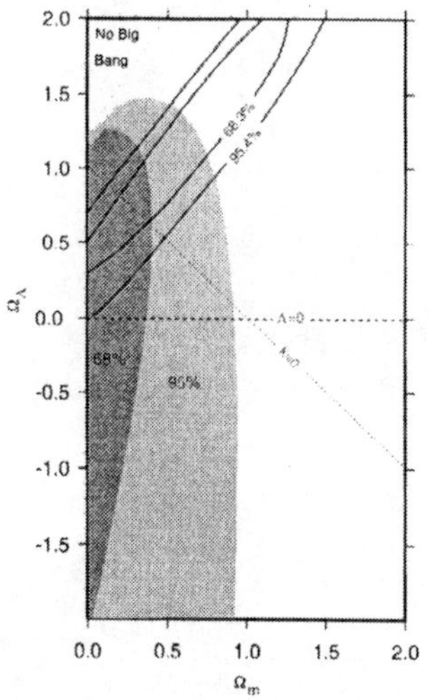

FIGURE 4. Constraints from the supernovae cosmology team [19] are indicated by the solid lines, while the radio galaxy results are indicated by the shaded regions; both are two-dimensional constraints.

Philip Mannheim, Paul Frampton, Ina Sarcevic, and Lynn Cominsky, which were particularly valuable. I would also like to thank Chris O'Dea for numerous helpful discussion related to radio sources. This work was supported in part by a National Young Investigator Award Grant number AST-0096077 from the US National Science Foundation, the Berks-Lehigh Valley College of Penn State University. Research at Rowan University was supported in part by teh College of Liberal Arts and Sciences and National Science Foundation grant AST-9905652.

REFERENCES

1. Weinberg, S. 1972, Gravitation and Cosmology, John Wiley & Sons
2. Daly, R. A. 1994, ApJ, 426, 38
3. Daly, R. A. 1995, ApJ, 454, 580
4. Guerra, E. J., & Daly, R. A. 1998, ApJ, 493, 536 (GD98)
5. Guerra, E. J., Daly, R. A., & Wan, L. 2000, ApJ, 544, 659 (GDW00)
6. Daly, R. A., & Guerra, E. J. 2001, AJ, in press
7. Daly, R. A. 2001, in Lifecycles of Radio Galaxies, ed. J. Biretta, C. O'Dea, A. Koekemder, & E. Perlman (Amsterdam: Elsevier Science), in press
8. Leahy, J. P., & Williams, A. G. 1984, MNRAS, 210, 929
9. Leahy, J. P., Muxlow, T. W., & Stephens, P. W. 1989, MNRAS, 239, 401
10. Blandford, R. D. 1990, in Active Galactic Nuclei, ed. T. J. L. Courvoisier & M. Mayor, Springer-Verlag, 161
11. Daly, R. A., & Guerra, E. J. 2001, IAU Symposium No. 201, New Cosmological Data and the Values of the Fundamental Parameters, ed. A. Lasenby, A. W. Jones, & A. Wilkinson (San Francisco: ASP Conference Series), in press
12. Peebles, P. J. E. 1993, Principles of Physical Cosmology, Princeton University Press
13. Steinhardt, P. J. 1996, Nature, 382, 768
14. Turner, M.S. & White, M. 1997, Phys. Rev. D, 56, R4439
15. Caldwell, R. R.,Dave, R. & Steinhardt, P. J. 1998, Phys. Rev. Lett., 80 1582
16. Bludman, S., & Roos, M. 2001, ApJ, 547, 77
17. Daly, R. A., & Guerra, E. J. 2001, in IAU Symposium No. 201: New Cosmological Data and the Values of the Fundamental Parameters, ed. A. Lasenby, A. W. Jones, & A. Wilkinson (San Francisco: ASP Conference Series), in press
18. Liu, R., Poole, G., & Riley, J. M. 1992, MNRAS, 257, 545
19. Perlmutter et al. 1999, ApJ, 517, 565
20. Riess et al. 1998, AJ, 116, 1009

Gravitational Microlensing and Blazar Variability: The Case of AO 0235+164

James R. Webb and Emily S. Howard

Dept. of Physics
Florida International University
and
The SARA Observatory

Abstract. We review the basics of gravitational lens phenomena in astrophysics, concentrating on its applicability to Blazar variability. A brief discussion of gravitational lenses in the context of AGN is presented, followed by a more detailed model to explain the high amplitude outburst observed in Blazar AO 0235+164. We find some preliminary model parameters for the 1997-98 outburst of AO 0235+164.

I. BRIEF HISTORY OF GRAVITATIONAL LENSES

Gravitational lenses have now become widely accepted phenomena in astrophysics. A detailed history of gravitational lenses can be found in Schneider, Ehlers, and Falco (1992). Numerous gravitational lens systems have already been detected and many more are suspected to exist, thus opening up a new and interesting field in astrophysics. S. Refsdal (1964) examined realistic scenarios for using gravitational lenses and speculated about the effects of gravitational lenses on the study of quasi-stellar objects and distant galaxies. Walsh, Carswell and Weymann (1979) subsequently detected the first gravitational lens, the quasar PKS 0957+56. Finding and analyzing faint galaxies and possible gravitational lenses is becoming easier every day with instruments such as the Hubble Space Telescope, the WYIN, the Keck, and many other large telescopes.

We distinguish between a lens system that forms multiple separated images of a distant object like PKS 0957+56, and a lens system that merely amplifies the image of the distant object. The primary difference between the two cases is the angular separation of the images. If the angular separation of the images is below the angular resolution of your instrument, then we say the object is a "microlens". In this case we see a brightening of the original image instead of a multiplicity of images.

II. BLAZARS

Quasi-Stellar Objects (QSO) were discovered in the early 1960's, and identified as a type of Active Galactic Nuclei (AGN). QSO's exhibited highly

CP624, *Cosmology and Elementary Particle Physics*, edited by B. N. Kursunoglu et al.

redshifted emission and absorptions lines, indicating large distances from us, and huge intrinsic luminosities. The classification of AGN depends on three properties: the presence of spectral lines, the strength of radio emission, and the degree of optical variability. Radio Quiet Quasi-stellar objects (RQQSO) emit very weakly in the radio but are still highly luminous galactic cores. The luminosities of RQQSOs were large, but the emission was fairly stable and the light curves over time did not show much activity. Quasars are radio loud sources and represent about 10% of all QSO's and they exhibit significant optical and radio variability. The variability allowed constraints to be placed on the size of the emitting regions and showed that the continuum sources of objects were extremely small, perhaps 10^{14} cm, indicating a massive collapsed object as the power source. A small percentage of Quasars show Optically Violent Variations (OVV) in their optical light curves. These variations can be as much as several magnitudes increase in brightness over timescales as short as a few days or a few months.

BL Lac sources are similar to quasars but do not show broad emission lines, only narrow absorption lines. The term Blazar was coined as the class of sources made up of OVV Quasars and BL Lac objects which have similar optical properties. Table I. shows the relationship and characteristics of the many classes of AGNs

Table 1. Types of Active Galaxies

	Quasi-stellar	**Objects**		**Active**	**Galaxies**
	RQQSO	**Quasars**	**BL Lacs**	**N-Gal**	**Seyfert Gal.**
Radio emission	**RQ**	**RL ($> 10^{41}$)**	**RL**	**RL**	**RL**
Optical Continuum	**PL**	**PL**	**PL**	**PL +TH**	**PL + TH**
Lines	**BEL**	**BEL**	**Abs only**	**BEL**	**NEL +BEL**
Optical Variability	**V**	**V-OVV**	**OVV**	**V**	**V**
X-ray emission	**10^{46}**	**10^{47}**	**10^{47}**	**10^{40}**	**10^{40}**
γ-ray emission	**No?**	**V**	**Barely**	**??**	**Yes**

RQ =Radio quiet **BEL = Broad Emission Lines**
RL = Radio Loud **NEL = Narrow Emission Lines**
PL = Power Law **TH = Thermal**
V = Variable **OVV = Optically Violent Variables**

Blazars are a subset of Table I and contain objects found in the classes known as *OVV quasars* and *BL Lac objects*. Thus they show OVV in the optical, may or may not have strong broad emission lines, and the continuum spectra are highly polarized. The synchrotron character of the emission spectrum and the amounts of energy necessary to explain the observations lead to models incorporating a massive black hole ($10^{7\text{-}8}$ Solar masses or more) surrounded by an accretion disk and large-scale magnetic fields. The accretion disk feeds the hole by funneling large amounts of matter and magnetic fields onto it, much of which is ejected parallel to the holes rotational axis forming relativistic jets. The relativistic jets are powered mainly by the either accretion or energy extraction from the rotation of the black hole. Relativistic electrons accelerated by shocks in the jet interact with the magnetic field and emit a synchrotron spectrum.

The "Unified model" of AGN suggests that when these jets are pointing toward us, we see the strongly beamed radiation and classify these objects as either BL Lac sources or OVV quasars, depending on the absence or presence of broad emission lines. Although we have a fair understanding of most of the broadband spectrum, we do not understand the mechanisms responsible for the variability. Suggestions for possible variability mechanisms can be classified into two distinct categories: intrinsic and extrinsic. Intrinsic mechanisms include instabilities in the accretion disk, shocks and density variations in the relativistic jet, and interactions between photons and particles in the jet and disk (Compton scattering processes). Extrinsic causes for variability include variable extinction or magnification by gravitational lenses. This last effect, apparent variability caused by gravitational lens, is the subject of this paper.

III. GRAVITATIONAL LENSES

Some blazar outbursts may be interpreted in terms of gravitational microlensing. We define the blazar as the "source", the intervening object as the "Mass", and we are the "Observer". Figure I. illustrates the geometry of the system. In Figure 1, ξ is the impact parameter, D_s is the distance from observer to source, D_{ds} is the distance from deflector mass to source, and D_d is the distance from Observer to lens. The Einstein Angle (α) is equal to $4GM/c^2\ \xi = 2R_s/\ \xi$, and the Einstein Radius is given by $r_E = [\ (4Gm/c^2)\ (D_sD_{ds})D_d^{-1}]^{1/2}$. The distant source in our case is the quasar and a nearby lens is an intervening galaxy or objects in the intervening galaxy. Here we actually consider only collapsed objects in the intervening galaxy, not the galaxy as a whole. If the entire galaxy were responsible for the lens, then the images would be widely separated and we would see non-overlapping multiple images, but here we consider a galaxy whose center is off axis from the background quasar, but individual massive object in the galaxy may occasionally pass with small enough impact angle to micro-lens the quasar. For a macro-lens, one expects two or more images, arcs, or circles, depending on the alignment of the lens with the background source. For a micro-lens, the image deflection is small and the images cannot be resolved resulting in one image, but with amplification.

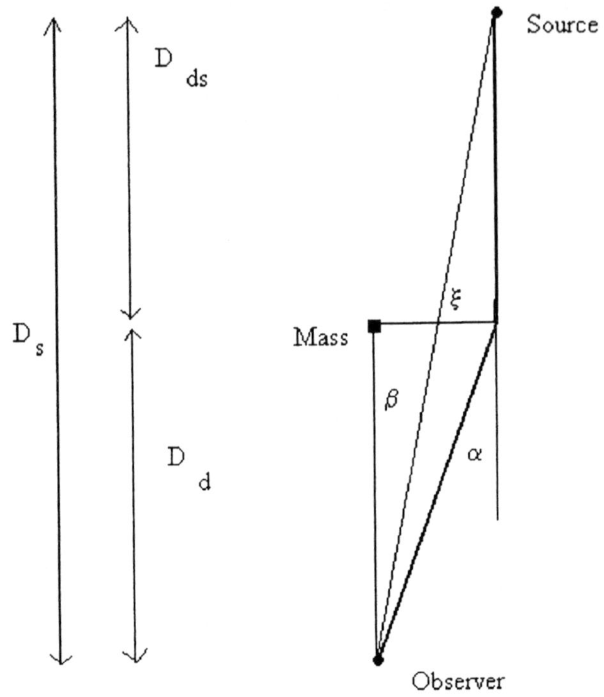

FIGURE 1. Geometry for a gravitational lens system.

If the lens is in motion relative to the source or observer and image resolution is below angular separation, we see amplitude or brightness changes and it appears to us the source brightens. This is called a High Amplitude Event or HAE. One expects the HAE to exhibit symmetric variations, thus outbursts in Blazars should show this symmetry if microvariability the cause. The outburst should also be frequency independent across spectrum, thus providing a signature that can be used to discriminate between intrinsic models and microvariability. Intrinsic models predict frequency dependsnt outbursts. The gravitational lens model is the only simple model that predicts frequency independent outbursts over many decades in frequency.

The duration of an outburst is related to the speed of the lens, specifically the angular speed across the line of sight to the source. The amplitude of the outburst is dependant on the mass, size, and impact angle of the lens with respect to the source. Pronk (1999) defines a dimensionless impact parameter $u = r/r_E$ and gives the Einstein radius in terms of this parameter as:

$$\theta_{+/-} = 0.5(-u+/-(u^2+4)^{1/2})\ \theta_E$$

The characteristic timescale t_o for an outburst depends on the velocity of the lens, $d\theta/dt = V/D_d$ and is given by: $t_0 = \theta_E/d\theta/dt$. The brightness changes is a function of the minimum distance,

$$u = [u^2_{min} + ((t-t_0)/t_0)^2]^{1/2}.$$

$$Amp(i) = u_i^2+2/(u_i(u_i^2+4)^{1/2}).$$

The above scheme was used to analyze light curves assuming the intervening lens was in motion relative to the Earth and source. A fitting program was written in IDL to model light curves using an automated *Levenberg-Marquardt least-squares minimization* routine. The program allows upper and lower bounding constraints to be placed on each parameter, or the parameter can be held fixed as other parameters vary in any combination.

III. THE CASE OF AO 0235+164

There have been several suggestions that optical outbursts in this object could be due to microlensing (Schneider and Weiss 1987, Kayser 1988, Grieger, Kayser and Refsdal 1988). AO 0235 has a redshift of 0.94, indicating a distance on the order of 4,377 Mpc (q_0=.5, H_o=75 km/sec/Mpc). An underlying galactic component has been imaged and the corresponding redshift agrees with the blazar emission and absorption redshifts. There is an intervening galaxy with a redshift of z = 0.524 less than 2 arc seconds south of AO. This intervening galaxy could theoretically provide objects responsible for microlensing the light from AO 0235+164.

We have two accepted observing programs at the SARA observatory, located at Kitt Peak National Observatory. One involves regular monitoring of selected Blazars and the other is a Target of Opportunity (TOO) program designed to observe Blazars in outburst. During a routine monitoring run, we detected Blazar AO 0235+164 going into an outburst and activated the TOO program to follow its optical activity throughout the entire event. Figure 2 shows the long-term optical light curve of AO 0235+164 in the B-band illustrating the many energetic flares seen in this source. The more detailed light curve of the 1997-1998 outburst is shown in Figure 3. Other observers were alerted so they could follow the outburst in the radio, mm, and IR portions of the spectrum (Webb, et al. 2000). Given the sparseness of the sampling, the 1997-1998 outburst was consistent with being symmetric, although the decline appeared to be slightly steeper than the rise. The most exciting results from the outburst were the similarities in the 850 mm flux, the 14.5 Ghz flux and the optical flux. All three widely separated frequency bands showed similar outburst profiles and amplitudes. This behavior is contrary to the predictions of most intrinsic variability models, but agrees well with the idea of gravitational microlensing. The intrinsic variability models generally predict strong wavelength dependence and time delays between widely separated frequency bands.

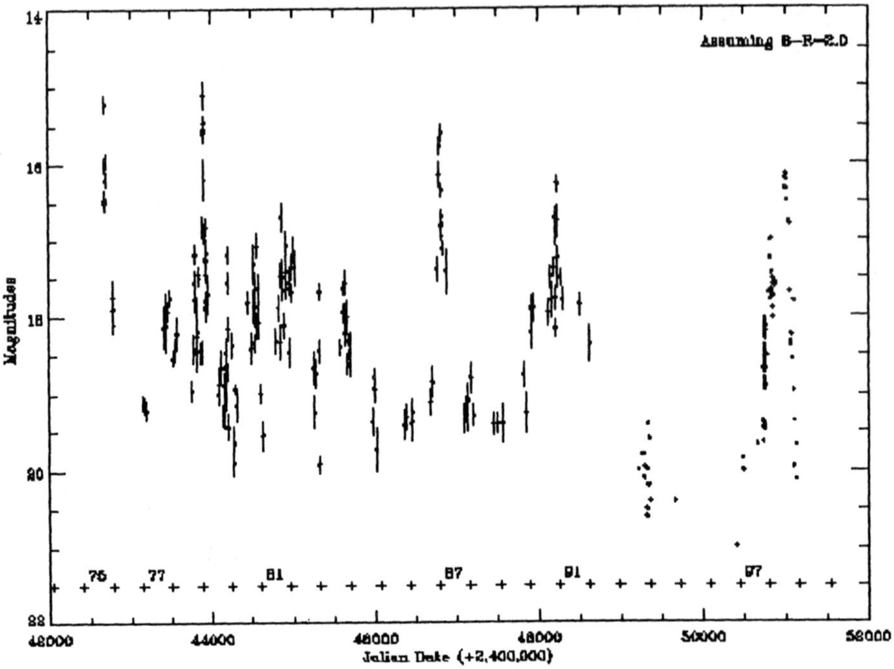

FIGURE 2. The Long-term light curve of AO 0235+164 in B-magnitudes.

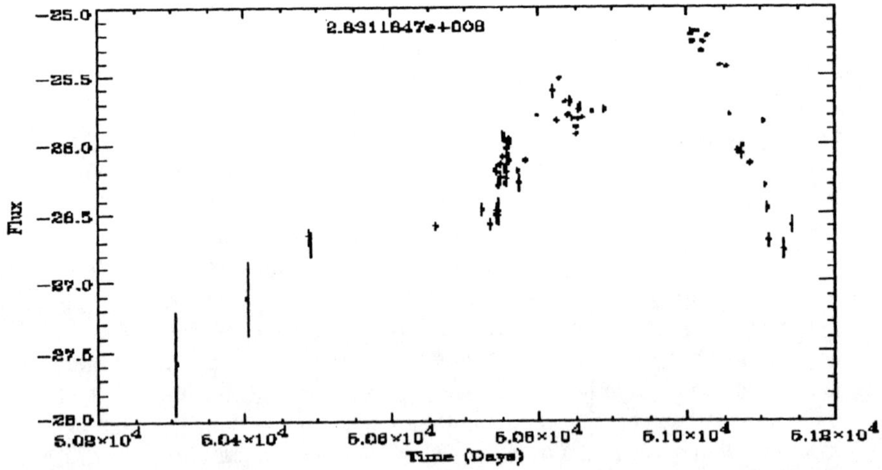

FIGURE 3. The 1997-1998 Outburst of AO 0235+164 in Log Flux(ergs/cm^2/sec/Hz).

If the 1997-1998 outburst of AO 0235+164 was due to microlensing, the model developed in the previous section can be applied to the data. We first converted the magnitudes into flux units before fitting. Figure 3 shows the light curve in units of Log F ($ergs/cm^2s^{-1}Hz^{-1}$). The computer program written in IDL requires a number of input parameters. The source redshift, the lens redshift, a value for Hubble's constant, and the value for q_0 are all initial requirements. The program computes the corresponding luminosity distances and sets up the fitting routine. The data file is also input which contains the light curve data to be fit. The data must be converted to flux (not log flux) before fitting. An initial mass estimate, linear velocity of the lens, and impact parameter is input. These values will be adjusted to obtain the best fit to the data. The program first finds the maximum in the data and uses that point as the center of the outburst. It then assumes an impact parameter of .1 and uses the input mass and lens velocity to calculate a preliminary model. The model parameters are subsequently adjusted in order to minimize the chi-square of the fit. As of the date of the Global conference are just beginning to fit models to the data. A big problem is constraining the parameters properly so the fits make sense astrophysically. A preliminary fit to the optical data yielded the following parameters:

- The mass of the Lens in Solar Masses: 3.9274166
- The impact parameter:.......................... 0.15737571
- The Einstein Radius: 1.0159770e-011
- The velocity of the Lens:.......................136618.52
- The center of the peak is:.......................51012.494
- The Chi-Square of the fit is:.....................3.6601667e+009

The very large Chi-Square is due to the fact that the outburst is poorly sampled before the maximum and the program is trying to fit a much narrower outburst. The lens mass is high, but reasonable for galactic halo objects. The lens velocity, again due to the narrow peak implied by the sampling is much too high for a halo object. Figure 4 shows the model fit.

IV. CONCLUSIONS

As we further understand the program and include the other data, we will be able to determine whether lens models are able to account for the outburst in 1997 – 1998. If so, we can then extend our analysis to historical outbursts and determine the feasibility of this model for those outbursts as well. Preliminary fits do not look promising, but much work remains to be done in properly constraining the adjustable parameters before anything concrete can be concluded.

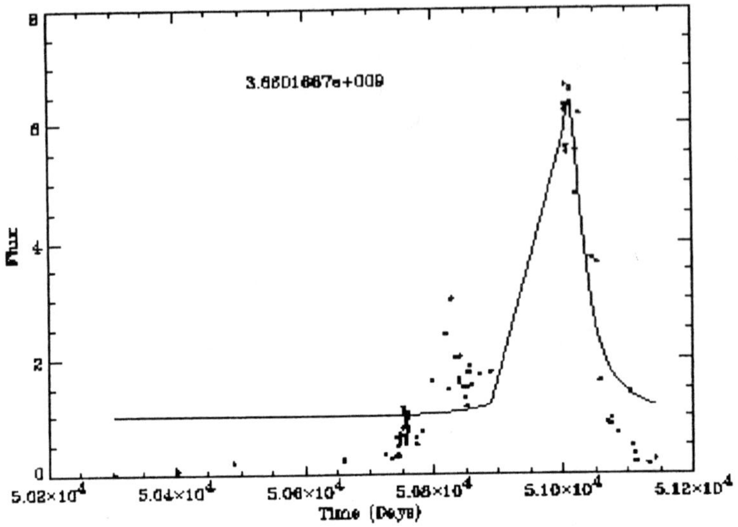

FIGURE 4. The 1997-1998 outburst and gravitational lens model. The flux is in Janskys and the solid line represents the best fit lens model.

ACKNOWLEDGMENTS

J.R.W. would like to acknowledge the 2001 Provosts grant for funding this research project.

REFERENCES

1. Schneider, Ehlers and Falco, 1992, *Gravitational lenses*, Springer-Verlag New York,Berlin, Heidelberg.
2. Refsdal, S., 1964, *M.N.R.A.S.*, **128,** 295.
3. Walsh, Carswell and Weymann, 1979, Nature, **279**, 281.
4. Schneider, P., and Weiss, A., 1987, *Astron. Astrophys.*, **171**,49.
5. Kayser, R., 1988, Astron. Astrophys., **206**, L8.
6. Greiger, B., Kayser, R., and Refsdal, S., 1988, *Astron. Astrophys.* **194**, 54.
7. Webb, J. R., Howard, E., Benetiz, E., Balonek, T., McGrath, E., Shrader, C., Robson, I., and Jenkins, P., 2000, *Astronomical Journal,* **120**, 41.
8. Pronk, D., 1999, now defunct website.

Cosmic Rays with 50 Joules of Energy: What, How, and from Where?

Thomas J. Weiler

Department of Physics & Astronomy, Vanderbilt University, Nashville, TN 37235
email: tom.weiler@vanderbilt.edu.

Abstract. The observation of cosmic ray air showers above 10^{20} eV poses three serious challenges for particle astrophysics: (i) how the primary particles are accelerated to these fantastic multi-Joule energies, (ii) how the primaries evade energy loss while crossing a Universe filled with radiation, and (iii) why events appear to cluster in direction. The data is briefly reviewed. Then some novel solutions to these problems, generally involving new physics, are explored. Emphasis is placed on the signatures by which cosmic ray experiments will discriminate among proposed solutions, and the role neutrinos may play in resolving puzzles.

THE DATA AND THE EXPERIMENTS

An unsolved astrophysical mystery, now almost forty years old, is the origin, propagation, and composition of the extreme energy cosmic ray primaries (EECRs) responsible for the observed events at highest energies. About twenty events at and above 10^{20} eV have been observed by five different experiments.[1] The origin of these events is a mystery, for there are no visible source candidates within 50 Mpc of earth, except possibly M87, a radio-loud AGN about 18 Mpc away, and Cen-A (NGC5128), a radio galaxy at 3.4 Mpc, and neither of these radio-galaxies is in the direction of any of the observed events [1]. Since the observed events display a large-scale isotropy, many sources rather than one source seem to be required. Furthermore, as emphasized long ago by M. Hillas [2], all known candidate sources (AGNs, GRBs, etc.) seem incapable of accelerating protons to the highest energy observed, 3×10^{20} eV (= 50 J, a truly macroscopic value!). The nature of the primary particle is also mysterious, because interactions with the cosmic microwave background (CMB) renders the Universe opaque to nucleons and iron nuclei (Fe) [3] at and above $E_{GZK} = 5 \times 10^{19}$ eV, and double pair production on the cosmic radio background renders the Universe opaque to photons at this and lower energies. The theoretical prediction of the end of transparency for nucleons at E_{GZK} is the famous "GZK (Greisen-Zatsepin-Kuzmin) cutoff" [4]. Figure 1 shows a recent compilation of the Japanese AGASA data set, clearly extending beyond E_{GZK}.

[1] At this past summer's ICRC in Hamburg, Germany, HiRes retracted seven of nine preliminary events at 10^{20} eV, but AGASA added six more from their data at zenith angles between 45 and 60 degrees. The two experiments disagree on the EECR flux, but not on the existence of super-GZK events. The discrepancy must lie in the different systematics of Cerenkov vs. fluorescence detectors. The AUGER experiment will have both detector types, and soon settle the flux controversy.

CP624, *Cosmology and Elementary Particle Physics,* edited by B. N. Kursunoglu et al.

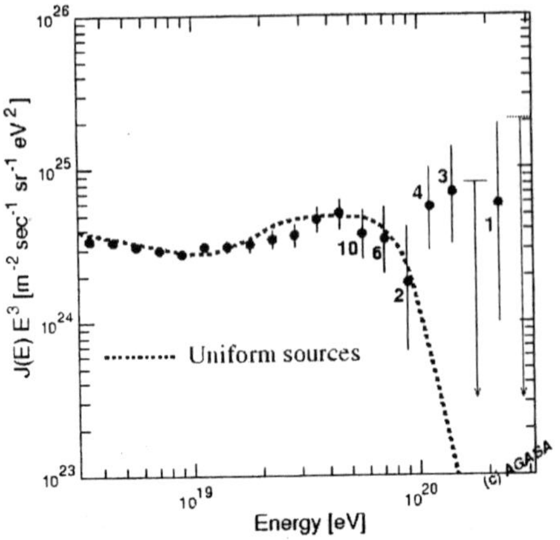

FIGURE 1. Extreme-energy cosmic ray spectrum as observed by AGASA. Error bars correspond to 68% C.L. and the numbers count the events per energy bin. The dashed line revealing the GZK cutoff is the spectrum expected from uniformly distributed astrophysical sources (from the AGASA website).

The EECR field is data-driven, with the ground-based detectors AGASA, HiRes, and Auger now, and ICECUBE in the near future, expected to deliver significant clues over the next several years, and the much larger space-based detectors EUSO (on the ISS) and OWL (free-flying tandem satellites) offering the potential to deliver definitive answers. At 10^{20} eV, AGASA currently provides about an event per year, and HiRes about an event per month. Auger will see two such events per week, while EUSO and ultimately OWL have the potential to collect such an event every few hours.

PROPOSED MODELS

Many models for the origin of the super-GZK events have been proposed. The cms energies are as large as a PeV (10^{15} eV), orders of magnitude beyond the Fermilab Tevatron and the LHC. Consequently, there is ample room in EECR data for new particle physics, and such new physics can only be studied with EECRs. Models proposed to solve the EECR mysteries fall into four basic categories. These are:
(A) nearby accelerators, which includes Galactic supershocks, young neutron stars, magnetars (highly magnetized pulsars), decaying super-massive particles (SMPs) with GUT masses $\sim 10^{15}$ GeV or with inflation-motivated masses $\sim 10^{12}$ to 10^{14} GeV, topological defects (TDs), M87 and Cen-A, rare nearby GRBs, now-dormant rare AGNs, and primordial black holes (PBHs);
(B) exotic primaries, such as light supersymmetric baryons made from a light gluino plus the usual quarks and gluons [5], or magnetic monopoles [6];
(C) exotic physical law, such as broken Lorentz Invariance or a breakdown of conventional gravity (general relativity) above some high scale; and
(D) neutrino primaries.

A fifth possibility, distant Zevatrons emitting nuclei, nucleons, and photons, is disfavored by the propagation arguments and the observed directional-clustering of events on small-scales. The classes of models are briefly discussed below, and reviewed in [7]. Each of these solutions impacts on particle physics, with (B), (C), and (D) going well beyond the SM.

No single model appears successful in explaining all signatures of existing data without stretching some of our theoretical preconceptions. Among nearby source models, magnetars require iron nuclei primaries to ease the acceleration requirement and to isotropize the flux in our galactic magnetic field. For SMPs and TDs it is necessary to tune the lifetime to be longer but not too much longer than the age of the Universe in order to maintain an appreciable secondary particle emission rate today. Sources mainly in the Galactic plane predict an unobserved planar anisotropy in the data. Sources clustered in the Galactic halo predict a dipole enhancement in the direction of the Galactic center, which may be tested by the Auger Observatory in a few years. Local GRBs and AGNs are few to none, and strong magnetic fields must be postulated to isotropize and/or confine their emissions. For PBHs, a sensible initial mass spectrum is unable to provide a sufficient number in the final stage of decay today. If the Southern Hemisphere object Cen-A is the source, then its proximity allows super-GZK neutrons to travel to Earth without decay. These line-of-sight neutrons present a detectable flux for AUGER [8]. Line-of-sight neutrino and photon fluxes are also expected; these are presently under investigation.

Exotic primary particles evade the GZK cutoff in interesting ways. With the susy baryon, the solution is kinematic – the cutoff energy varies as the square of the mass of the first excited resonant state. However, such a scenario invoking new, heavy matter renders the source-energetics issue even more challenging. The hypothesized relativistic magnetic monopoles have a negligible interaction (Thomson cross-section) with the CMB. Monopoles are produced in the early Universe with an abundance of one per correlated volume at the time of a phase-transition breaking a nonabelian gauge symmetry. Given the magnitude and coherence length data for the cosmic magnetic fields, monopoles with Dirac charge are naturally accelerated to energies $\sim 10^{20}$ to 10^{23} eV. The resulting monopole mass, number density and flux today are not incompatible with the observed EECR rate.

The remarkable proposals positing broken Lorentz invariance [9] or broken gravity are made somewhat plausible by the energy window available to EECRs ($E_{\rm LAB} \sim 10^{20}\,\rm eV \sim 10^{-8} M_P$, and $\sqrt{s} \sim$ PeV), orders of magnitude beyond that of the Tevatron or LHC. Some popular theoretical ideas, e.g. compactification of a high-dimensional spacetime, spacetime fluctuations (quantum foam), and a low string scale M_S, lend themselves to speculation at the end of the EECR spectrum. Models such as these have no photo-pion production above $E_{\rm GZK}$, and so no proton pile-up below $E_{\rm GZK}$ and no cosmogenic neutrino flux. They may contain undeflected pointing of the primary back to its source (due to a suppression of neutron decay). Even more striking signatures have been proposed, for example particle velocities that depend on the species (γ vs. p vs. ν), or an energy-dependent speed of light.

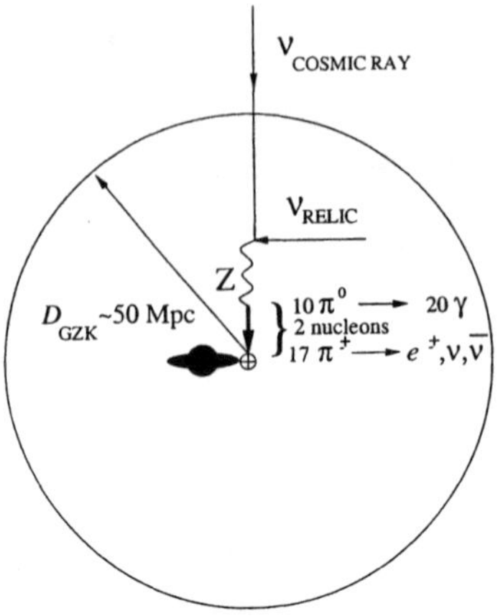

FIGURE 2. Z-burst production form resonant annihilation of a cosmic-ray neutrino on a relic (anti)neutrino. If the Z-burst occurs within the GZK zone ($\sim$ 50 to 100 Mpc) and is directed toward earth, then photons and nucleons may arrive at earth and initiate super-GZK air showers.

Neutrino Primaries

Turning to the possibility that the primaries may be neutrinos, one encounters an immediate obstacle: the Standard Model (SM) neutrino cross-section is down from that of an electromagnetic or hadronic interaction by six orders of magnitude. The two solutions to the small cross-section problem for primary neutrinos have been proposed. They are neutrino annihilation to Z-bursts, and a neutrino cross-section turning strong at highest energies. In the Z-burst model [10], the primary particles which propagate across cosmic distances above the GZK cutoff energy are neutrinos, which then annihilate with the cosmic neutrino background within the GZK zone ($D < D_{\rm GZK} \sim 50$ Mpc) to create a "local" flux of nucleons and photons above $E_{\rm GZK}$. The emchanism is displayed in Figure 2. For m_ν in the range 0.1 to 2 eV suggested by neutrino experiments, the energy in this resonant "Z-burst" is fortuitously situated sufficiently above $E_{\rm GZK}$ at $E_\nu^R = M_Z^2/2m_\nu = 4\,({\rm eV}/m_\nu) \times 10^{21}{\rm eV}$ so as to produce on average 2 baryons and 20 photons with energies right to explain the super-GZK data. This hypothesis has received considerable attention, for it is an economical EECR solution in that it does not rely on particle physics beyond the SM. The probability for each neutrino flavor at its resonant energy to annihilate within the halo of our galactic cluster or supercluster is likely within an order of magnitude of 1%, with the exact value depending on unknown aspects of neutrino mixing and relic neutrino clustering. The absolute lower bound in a hot Big Bang universe for the probability to annihilate within 50 Mpc (roughly a nucleon propagation distance) of earth is 0.04%. For a sufficiently large cosmic neutrino flux, this hypothesis successfully explains the observed air–showers above $E_{\rm GZK}$ as shown in detailed simulations [11]. Among the reasons to hope that Nature obliges is that resonant neutrino annihilation provides the best hope at present to actually measure the relic neutrino density liberated but a second after the Big-Bang.

Any nucleon flux beyond E_{GZK} is degraded in energy, photo-producing pions which in turn decay to produce cosmogenic neutrinos. The strongly-interacting neutrino model is motivated by the fact that the number of neutrinos per pion-decay somewhat compensates for their lesser energy, with the result that the neutrino flux matches well to the observed super-GZK flux. One may entertain the notion that the cosmogenic neutrinos **are** the super-GZK primaries, and that these neutrinos acquire a strong cross-section at 10^{20} eV [12]. Several recent ideas relate well to the 10^{20} eV scale. One idea is that grand unification occurs precociously at $\sqrt{s} \sim$ TeV, making the neutrino strongly-interacting. Another is that the exchange of towers of Kaluza-Klein (KK) modes from extra compactified-dimensions leads to a strong neutrino cross-section. A third is that extra-dimensional gravity provides a larger Schwarzschild radius $R_S(s)$ and therefore a strong cross-section $\sim \pi R_S^2(s)$ at high-energies. Strongly-interacting neutrinos mitigate the origin issue (the neutrinos are made as strongly as hadrons at their common sources), evade the propagation issue, and allow pointing and pairing of events on the sky, as observed.

Dispersion relations relate the hypothesized high-energy cross-section to the lower-energy neutrino-nucleon elastic amplitude [13]. Remarkably, the real forward amplitude becomes anomalous seven orders of magnitude lower in energy than does the total cross-section. In progress is a comparison of HERA $e^- + p \to \nu_e + n$ quasi-elastic charge-exchange data to SM predictions. This will yield a greatly improved bound for a large class of models having an anomalous *charged-current* interaction at high energies. The class includes precocious unification models, and possibly TeV-scale black-hole production models. The real virtue of the dispersion-relation approach is that it truly bounds $\sigma_{\nu N}(E)$, whereas direct bounds from Fly's Eye, AMANDA, Goldstone Lunar Experiment, and most recently, AGASA, bound the product $F_\nu \times \sigma_{\nu N}$ in a non-separable way.

DISCRIMINATING AMONG MODELS

There are several signatures to be investigated to discriminate among the competing models. One revealing signature, **small-scale anisotropies and pointing**, is already evident in the existing data sample. Of the 59 AGASA events above 4×10^{19} eV, 13 are contained in five doublets and one triplet with separation angle less than the angular resolution of 2.5 deg. The probability for random clustering from an a priori isotropic distribution is 0.07% [14]. A useful analytic expression for the dependence of random cluster-counts on the angular bin size is available in [15]. Work examining dynamical clustering from some of the models presented above has already appeared [16]. A recent analysis claims evidence for pointing toward distant blazars [17]. Data from AUGER will settle this issue. If the pairing does turn out to be dynamical, I would argue that neutrinos are a favorite candidate for the primary particles. This is because photons have such a short ($\sim$ 10 Mpc) absorption length below 10^{21} eV, neutrons should decay, and protons and nuclei are bent by cosmic magnetic fields during their extragalactic journey. If neutrinos are the primaries, they should point back to their sources, thereby enabling point-source astronomy to peer deeper into the Universe and deeper into the sources than is possible with light.

Other noteworthy signatures include:

Large-scale Anisotropies: Here one investigates possible associations of the CR directions with the Galactic halo (to be revealed by a dipole anisotropy favoring the direction of the Galactic center, or by a quadrupole asymmetry for a non-spherical halo), with the local galactic magnetic field, with matter distributions in nearby galactic or supergalactic clusters such as Virgo, or with possible large coherent galactic or extragalactic magnetic fields. Exposure to both Northern and Southern Hemispheres will prove invaluable. For example, M87, Cen-A, and the Galactic center are located in the Southern Hemisphere. AUGER together with AGASA and HiRes (and eventually, the space-based experiments EUSO and OWL) will provide this coverage.

Energy-Direction-Time Correlations: Because bending of charged particle trajectories by intervening magnetic fields increases as particle energy decreases, one may learn about the strength and geometry of extragalactic fields from relative time delays and angular correlations of particles from a common source. One may also learn whether the source is constant (e.g. AGN hot-spots), repeating (e.g. halo SMPs or TDs if subclustering exists on small-angular scales), or bursting (e.g. GRBs).

Composition of the Primaries: Another signature to be sought is a statistical identification of the nature of the primaries as a function of their energies. All models wherein the primaries arise from QCD jets (e.g. *Z*-bursts, and decaying SMPs and TDs) produce many more pions than baryons, and so many more neutrinos and photons than nucleons. It is interesting that at and above about 10^{22} eV, the attenuation length for photons becomes quite large, 100 Mpc or more. Thus, if there is a photon flux at 10^{22} eV, another window to point-source astronomy may become available.

Possible $\mathbf{E_{max}}$ energy cutoff: The prediction of a cutoff at $E_{\rm GZK}$ is wrong. Does Nature provide an alternative cutoff within our reach, or do the data continue beyond our reach? MHD experts claim that it is difficult for conventional shock-acceleration mechanisms to produce ZeV proton energies. Reconnection of magnetic fields generated by supermassive black holes in the center of galaxies may provide a higher energy cutoff [18]. Decaying SMP and TD models also have a natural cutoff, at half of their mass, which could be as high as $E_{\rm max} \sim 10^{24}$ eV for a long-lived GUT-mass particle. In the Z-burst model there is a natural cutoff, at most 7×10^{22} eV, related to the tiny mass of neutrinos.

CR flux above vs. below $\mathbf{E_{GZK}}$: A "smoothness" variable such as $R_j = F_j(E > E_{\rm GZK})/F_j(E < E_{\rm GZK})$ for each primary species j=nucleon, photon, iron nuclei, neutrino, etc., may be revealing [19]. For primaries with a GZK cutoff, $F(E < E_{\rm GZK})$ samples sources from the whole volume of the Universe and may include cascade products from $F(E > E_{\rm GZK})$, whereas $F(E > E_{\rm GZK})$ samples just the GZK volume; for primaries without a GZK cutoff, both $F(E < E_{\rm GZK})$ and $F(E > E_{\rm GZK})$ sample the whole Universe. Lumps, bumps, and gaps in the spectrum near $E_{\rm GZK}$ are a consequence of some models. For example, hadron and neutrino pile-ups just below $E_{\rm GZK}$ are expected from the photo-pion production process which occurs above $E_{\rm GZK}$. An analogous example for photons is the diffuse gamma-ray pile-up in the GeV range which results from electomagnetic showering of extreme-energy photons on the cosmic radio, microwave, infrared, and magnetic field backgrounds. A limit published by the EGRET experiment [20] (to be vastly improved by the next generation gamma-ray observatory, GLAST)

disfavors models with hard spectra hadronizing at cosmic distances.
Spectral index above $\mathbf{E_{GZK}}$: A continuation of the pbserved sub-GZK power law to super-GZK energies would argue against the dynamics predicting the GZK effect. Alternatively, one means of achieving more events above $E_{\rm GZK}$ is to postulate a flattening of the primary proton spectrum at highest energies.
Measurable neutrino flux above $\mathbf{E_{GZK}}$: If there is a new source of primary neutrinos above $E_{\rm GZK}$, or if the neutrino cross-section becomes strong at super-GZK energies, then there is the possibility that the primary neutrino flux at $\gtrsim 10^{20}$ eV can be measured in teraton-sized detectors. The boon for astronomy if extreme-energy neutrinos are measured is clear. Through this non-electromagnetic window one may peer back in time and out in distance far beyond what is available with photon or proton primaries which attenuate on the CMB; protons also deflect in cosmic magnetic fields. Also in contrast to photons and protons, weakly-interacting incident neutrinos may have originated deep in the internal engines of their sources, possibly bearing privy information about the source dynamics. The gain for particle physics if extreme-energy neutrinos are measured is also quite important. As shown very recently in [21], future detectors of EECR neutrinos will be able to measure the neutrino-nucleon cross section at energies as high as 10^{20} eV or higher, by comparing the horizontal and up-going air shower rates. Knowledge of $\sigma_{\nu N}$ at 10^{20} eV will provide a considerable test of subtleties in asymptotic QCD, such as unitarization of parton amplitudes, small-x gluon saturation (anti-splitting), the nature of the Pomeron, higher twist operators, etc.

SUMMARY

It seems that the ultimate explanation for the puzzles in EECRs will provide a surprise at a minimum, and possibly radically new physics at a maximum. The extreme energies of events already observed cannot be approached by terrestrial accelerators. Thus, there is ample motivation to pay attention to the strong suggestion of new physics in the EECR data.

The increased statistics expected from the next generation of air-shower and air-fluorescence detectors will readily discriminate among the many exciting models proposed to date. A "sight-bite" of the discriminatory power of selected signatures is offered in the following "Truth-Table."

ACKNOWLEDGMENTS

My own work mentioned in this contributed paper has been supported by DOE grant DE-FG05-85ER40226, and has beneffited from collaboration with L. Anchordoqui, P. Biermann, H. Goldberg, T. Kephart, A. Kusenko, and S. Wick.

TABLE 1. "Truth-table" of predicted signatures for various proposed models of EECRs

	Local Zevatrons	**SMPs & TDs**	**Z-bursts**	**SI-neutrinos**	**Q-gravity & LI-violation**	**Magnetic Monopoles**	**SUSY-baryons**
small-scale anisotropy/pointing	none	possible (sub-clustering)	yes	definitely	yes (stable neutrons)	little	none
large-scale anisotropy	planar, dipole, quadrupole	dipole, quadrupole	Gal. & SuperG halos	none	none	none	spiral arms; cosmic $\vec{B}$
energy-direction-time correlations	little	little	yes	none	some	yes	little
primary composition	protons or Fe	γ's, nucleons	messenger ν's; then γ's, protons	neutrinos	nucleons	exotic	exotic
E_{max} cutoff	10^{21} eV	$\lesssim M_{GUT}$	10^{22} eV	10^{21} eV	depends	10^{22} eV	10^{23} eV
flux above/below E_{GZK}	smooth	smooth	depends on ν-clustering	depends on scale M_S	smooth	smooth	smooth
EE neutrino flux	none	considerable	huge	large	none	negligible	none
EE spectral index	>2	hard	hard	>2	>2	>2	hard

REFERENCES

1. J. Elbert and P. Sommers, Astrophys. J. 441, 151 (1995).
2. A.M. Hillas, Ann. Rev. Astron. Astrophys. 22, 425 (1984); more recent work is C.A. Norman, D.B. Melrose, and A. Achterberg, Astrophys. J. 454, 60 (1995); R.D. Blandford, Physica Scripta T85, 191 (2000).
3. For a more romantic view of iron, see F.W. Stecker and M.H. Salamon, Astroph. J. 512, 521 (1999).
4. Named after the pioneering work of Greisen, Kuzmin, and Zatsepin in the 1960's; recent detailed explorations of the GZK cutoff include S. Lee, Phys. Rev. D58, 043004 (1998); A. Achterberg et al., astro-ph/9907060; T. Stanev et al., Phys. Rev. D62 093005 (2000) [astro-ph/0003484].
5. G.R.Farrar, Phys. Rev. Lett. 76, 4111 (1996); D.J.H.Chung, G.R.Farrar and E.W.Kolb, Phys. Rev. D57, 4606 (1998); I.F.M. Albuquerque, G.R. Farrar and E.W. Kolb, Phys. Rev. D59, 015021 (1999).
6. T. W. Kephart and T. J. Weiler, Astropart. Phys. 4, 271 (1996); S. D. Wick, T. W. Kephart, T. J. Weiler, P. L. Biermann, [astro-ph/0001233], to appear in Astropart. Phys..
7. Recent reviews of EECR data, puzzles, and models include: P. Biermann, J. Phys. G23, 1 (1997); P. Bhattacharjee and G. Sigl, Phys. Rept. 327, 109 (2000) [astro-ph/9811011]; A.V. Olinto, astro-ph/0102077; X. Bertou, M. Boratov, and A. Letessier-Selvon, Int. J. Mod. Phys. A15, 2181 (2000); M. Nagano and A.A. Watson, Rev. Mod. Phys. 72, 689 (2000); G. Sigl, Science 291, 73 (2001); F.W. Stecker, astro-ph/0101072; T. J. Weiler, Proc. RADHEP2000, Nov. 16-18, 2000, UCLA, ed. P. Gorham and D. Saltzberg, [hep-ph/0103023].
8. L. Anchordoqui, H. Goldberg, and T. J. Weiler, Phys. Rev. Lett. 87, 081101 (2001) [astro-ph/0103043].
9. Mestres-Gonzales, Proc. 25th ICRC, 1997, Durban, So. Africa, (World Sci., Singapore); S. Coleman and S.L. Glashow, hep-ph/9808446, and Phys. Rev. D59, 116008; F.W.Stecker and S.L.Glashow, Astropart. Phys. 16, 97 (2001).
10. T.J. Weiler, Astropart. Phys. 11, 303 (1999), and ibid. 12, 379E (2000) [for corrected receipt date]; building on much earlier work in T.J. Weiler, Phys. Rev. Lett. 49, 234 (1982); Astrophys. J. 285, 495 (1984); some related work is D. Fargion, B. Mele and A. Salis, Astrophys. J. 517, 725 (1999).
11. S. Yoshida, G. Sigl, and S. Lee, Phys. Rev. Lett. 81, 5505 (1998); G. Gelmini, hep-ph/0005263; Z. Fodor, S. Katz and A. Ringwald, hep-ph/0105064 (2001); O. Kalashev, V. Kuzmin, D. Semikoz, and G. Sigl, hep-ph/0112351 (2001).
12. S. Nussinov and R. Shrock, Phys. Rev. D59, 105002 (1999); G. Domokos and S. Kovesi-Domokos, Phys. Rev. Lett. 82, 1366 (1999); P. Jain, D.W. McKay, S. Panda and J.P. Ralston, Phys. Lett. B484, 267 (2000).
13. H. Goldberg and T. J. Weiler, Phys. Rev. D59, 113005 (1999).
14. The latest update of directional clusters is M. Takeda et al. (AGASA Collaboration), Proc. ICRC 2001, Hamburg, Germany.
15. H. Goldberg and T. J. Weiler, Phys. Rev. D64, 056008 (2001).
16. P.G. Tinyakov and I. I. Tkachev, JETP Lett. 74, 1 (2001) [astro-ph/0102101]; Z. Fodor and S.D. Katz, Phys. Rev. D63, 023002 (2001); S. Dubovsky, P. Tinyakov, and I.I. Tkachev, Phys. Rev. Lett. 85, 1154 (2000).
17. P.G.Tinyakov and I.I. Tkachev, "BL Lacertae are sources of the observed ultrahigh energy cosmic rays", JETP Lett., to appear [astro-ph/0102476] (2001).
18. S. Colgate and H. Li, Astrophys. Space Sci., 264, 357 (1999); S. Colgate, H. Li, and V. Pariev, Physics of Plasmas, in press (2001).
19. G. Farrar and T. Piran, astro-ph/0010370.
20. P. Sreekumar et al., Astrophys. J. 494, 523 (1998).
21. A. Kusenko and T.J. Weiler, hep-ph/0106071, submitted to Phys. Rev. Lett.

How X-ray Experiments See Black Holes: Past, Present and Future

Lynn Cominsky

Department of Physics and Astronomy
Sonoma State University, Rohnert Park, CA 94928

Abstract. This paper presents a review of the mass determinations for galactic black hole candidates that have been derived from X-ray observations with previous generations of orbiting satellites. Recent observations illustrating physical properties determined for black holes in the Milky Way and in other galaxies are also summarized. Finally, new missions are described that will further test our understanding of black holes, extending our view to the event horizons.

OVERVIEW

X-ray emission from a black hole candidate in an X-ray binary was discovered in 1972 [1], with the identification of the optical companion to the X-ray emitter Cygnus X-1. In the past 30 years, the field has matured, and satellite-based X-ray observations of black hole binaries and black holes in the cores of galaxies are conducted by scientists from many different nations, and with many different instruments. Recently, the increased collecting area and high time resolution of satellites such as NASA's Rossi X-ray Timing Explorer and the unprecedented imaging and spectroscopic capabilities of NASA's Chandra X-ray Observatory and ESA's XMM-Newton Observatory have produced data that have broken new ground in allowing high-energy astrophysicists to make physical measurements of black hole candidates. Unobtainable in earth-bound laboratories, X-ray satellite observations remain the best way to study the effects of strong field gravity and to test the predictions of General Relativity in this regime.

In this paper, I will review the basic physics of accretion, gravitationally-driven mass transfer which produces X-ray emission from black holes, and discuss the measurement of mass in black hole X-ray binaries. Recent evidence for black hole spin, disk interactions, the Lense-Thirring effect and a proposed new class of black holes of intermediate mass will be summarized. Finally, I will close with a brief look at two proposed new X-ray missions which will be able to study relativistic signatures all the way to the black hole event horizons.

ACCRETION IN BLACK HOLE BINARIES

In accreting black hole binaries, the X-ray luminosity originates in $\sim 10^7 - 10^8$ K plasma comprising the inner accretion disk around the compact object. The accretion luminosity

CP624, *Cosmology and Elementary Particle Physics*, edited by B. N. Kursunoglu et al.

observed is given by:

$$L_{accr} = \frac{\eta\, G\dot{M}M}{R} \sim 2.5 \times 10^{38}\, \eta \text{ ergs s}^{-1} \tag{1}$$

assuming that $M = 3\ M_\odot$, $R = 10$ km, $\dot{M} = 10^{-8}\ M_\odot\ yr^{-1}$, and where η is an efficiency factor for the conversion of gravitational potential energy into radiated luminosity. The efficiency differs for different assumptions about the General Relativistic metric for the black holes. Typical accretion luminosities observed in black hole binary systems vary from $10^{36} - 10^{38}$ ergs s^{-1}.

In studying this inner accretion flow, one must take into account the very rapid dynamical timescales for matter to move through the emitting region:

$$\tau_{dyn} = (\frac{r^3}{GM})^{1/2}. \tag{2}$$

This is about 0.1 ms for matter at the event horizon, $r = 30$ km, of a 10 $M_\odot$ black hole. Hence, the typical orbital period of circulating matter,

$$P_{orb} = 2\pi\tau_{dyn} \sim 1 \text{ ms}. \tag{3}$$

To study processes occurring near the event horizon, it is therefore necessary to observe X-ray timing signatures in black hole systems at frequencies of $\sim$ 1000 Hz.

Mass Determination in Black Hole Binaries

The determination of the orbital elements in binary systems with no detectable pulsations or no X-ray eclipses requires the use of optical data from the companion star. Radial velocity curves are obtained, with amplitude $K \sin i$ km s^{-1}, and orbital period P_{orb} (see Figure 1.) These data measure the effect of the unseen compact object on its companion star, and can be related through Kepler's laws to the mass function (see Equation 4), establishing a lower limit on the mass of the unseen X-ray emitting companion:

$$f(M) = \frac{P_{orb}\, K^3}{2\pi G} < M_x. \tag{4}$$

The lower limit to the compact object's mass M_x stems from the fact that without eclipses, it is not possible to precisely know the system's inclination angle. The angle can be estimated, however, from studies of the ellipsoidal light variations during the binary orbit. However, assuming that the companion star has a mass of zero, and that the inclination angle is 90 degrees, yields a mass function equal to the mass of the compact object. Since these are extreme values, realistic values for the mass of the companion and the inclination angle will result in a black hole mass greater than the derived mass function.

Black hole candidates are those objects for which the best estimates for the compact object's mass (after taking into account the mass of the optical companion and the inclination angle to the system) are greater than 3 $M_\odot$, the Rhoades-Ruffini upper limit

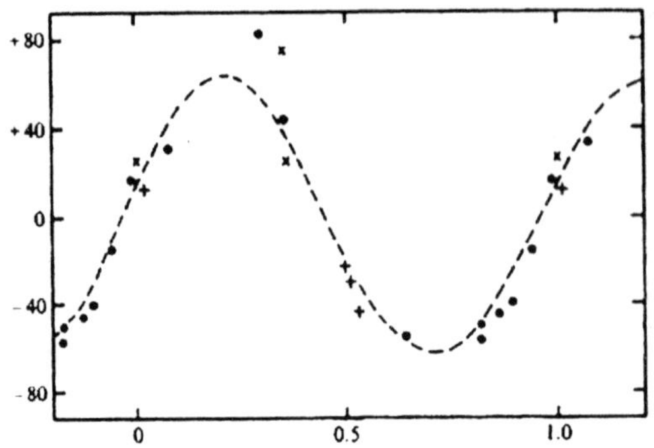

Source: Webster & Murdin 1972

FIGURE 1. Optical radial velocity curve for the companion to the black hole candidate Cygnus X-1. Radial velocities are measured in km s^{-1}, and are shown as a function of phase in the 5.6 d binary orbit

to the mass of a neutron star [2]. Other common observational characteristics shared by some of the black hole candidates include two-component energy spectra, in which one component is very soft, and highly variable, transient X-ray outbursts. The best black hole candidates are those in which the mass function itself is greater than $3M_{\odot}$. For example, in V404 Cygni [3] the mass function is $6.08 \pm 0.06\ M_{\odot}$, and the most likely mass is 12 $M_{\odot}$.

It should be noted however, that the Rhoades-Ruffini neutron star mass limit of 3 $M_{\odot}$, although independent of assumptions about the specific form of the neutron star equation of state, does make certain assumptions that can be questioned. For example, Rhoades and Ruffini assume that General Relativity provides the correct description of gravity in the strong field regime, and that gravity provides the stabilizing force which overcomes repulsive pressures in the star's core. If instead hadronic forces in the core can stabilize the star, it may be possible to increase the mass of a neutron star far above this limit [4][5].

RECENT OBSERVATIONS

In this section, I review recent, sometimes controversial evidence and ideas regarding the direct observation of relativistic effects in black hole systems, as well as evidence for a new class of "intermediate" mass black holes.

Microquasars and Quasiperiodic Oscillations

There are now about a half-dozen proposed galactic black hole candidate binaries which are known as "microquasars." The first two systems to be studied, GRS1915+105 and GRO J1655-40, exhibit superluminal radio jets (as do extragalactic quasars), and transient, highly variable X-ray emission [7][8]. Because the masses of these two objects are only $\sim 10\ M_{\odot}$ rather than $\sim 10^6\ M_{\odot}$, the radii of their accretion disks and radio lobes are also about one million times smaller. The temperature of the radiation from the inner accretion disk is correspondingly higher, with the bulk of the observable emission in the X-ray bandpass. And of course, unlike their galaxy-gobbling extragalactic counterparts, the microquasars are being fueled by accreting matter from a binary companion star.

Optical observations and careful analysis of ellipsoidal light variations from the companions to several microquasar sources have resulted in mass determinations for the compact objects that clearly establish their black hole nature (e.g., $7.02 \pm 0.22\ M_{\odot}$ for GRO J1755-40 [9], $8.73 - 11.70\ M_{\odot}$ for SAX J1819.3-2525 [10] and $9.68 - 11.58\ M_{\odot}$ for XTE J1550-564 [11].) Several of these "microquasar" black hole candidates show high-frequency quasi-periodic oscillations (HFQPOs, [12][13][14][15]). Observations of these HFQPOs may offer the best hope of a direct determination of the black hole spin [15][16], although the interpretation of both the high- and low- frequency QPOS is very complex [17]. Alternative origins for the HFQPOs seen in some microquasars include "diskoseismic" perturbations of the microquasar's relativistic accretion disk [18][19] and resonance effects [20].

Evidence For Event Horizons In Black Holes

Narayan and his colleagues [21][22][23] have suggested that accreting black hole binaries, when in quiescence (*i.e.*, not during a transient X-ray outburst) will have very different X-ray spectra than quiescent neutron star transients. Their advection dominated accretion flow (ADAF) models give good results for fits of the energy spectra over a wide bandpass, from optical through X-ray wavelengths. In these models, the presence of the event horizon in a black hole binary results in much lower accretion efficiency, as the advected material simply disappears over the horizon, without radiating much of the thermal energy carried in the infalling plasma. (The accretion efficiency, η, in these systems is estimated at 0.05%.)

An additional prediction of the ADAF models is that black hole transients should have a wider dynamic range of minimum to maximum luminosities, since their quiescent states should be ~ 500 times less luminous. Although hard to prove due to limited observations of non-emitting X-ray transients, there is some evidence that appears to support this prediction [22].

Evidence For Black Hole Spin

Zhang, Cui and Chen [24] have interpreted the often seen soft-component in X-ray energy spectra from black hole candidate objects as blackbody emission from the inner accretion disk at the marginally stable orbit, r_{ms}. They have used this idea to calculate blackbody spectra for black holes with different rotation rates. Distinct spectral differences are then found in the three different spin classes: maximally prograde Kerr, Schwarzschild (slowly or non-spinning) and maximally retrograde Kerr. The accretion efficiencies η differ for these three classes of objects from 30% to 6% to 3%, as r_{ms} changes from 1 to 6 to 9 times (GM/c^2), respectively.

In this scenario, the microquasar sources contain maximally prograde Kerr black holes. The extremely high accretion efficiency ($\eta = 30\%$) coupled with the small r_{ms} raises the temperature of the inner part of the accretion disk producing a relatively hard spectral component, in agreement with the observations.

The traditional galactic black hole candidates in binaries with a pronounced soft-component, are identified with slowly or non-rotating (Schwarzschild) black holes. And the black hole candidates which seem similar to these sources, but are lacking the spectrally-soft component are suggested to be maximally retrograde Kerr black holes. In this case, the inefficient accretion ($\eta = 3\%$) coupled with the relatively large r_{ms} combine to lower the temperature of the soft component until it falls below the standard X-ray bandpass, and is undetectable by present instruments.

Lense-Thirring Effect

Stella and Vietri [25] were the first to suggest that QPO observed in the range 20 - 40 Hz from neutron star binaries could result from the Lense-Thirring effect. Extending Stella and Vietri's suggestion to include black holes, Cui, Zhang and Chen [26] applied the Lense-Thirring formalism to the QPO seen in the two microquasar sources as well as several other candidate systems. For the black hole candidates, the weak field approximation used by Stella and Vietri was not appropriate, so Cui et al. modified the equations to apply to strong field systems. Assuming a mass for the candidate object and interpreting the observed QPO as the Lense-Thirring precession frequencies yielded spin rates which were found to be consistent with those in their earlier work [24]. In particular, the microquasar GRO J1655-40, which has a well-determined [9] black hole mass of $\sim 7\ M_{\odot}$, is found to be rotating at 95% of maximum, while the more typical black hole candidates, such as Cyg X-1 are only rotating at $\sim$ 50% of maximum. Further analysis, however, of GRO J1655-40 has complicated the picture as three other QPO frequencies have been found [13] including a HFQPO at 450 Hz [15].

Interactions between Disks and Rotating Black Holes

Observations using the Rossi X-ray Timing Explorer by Eikenberry et al. [27], combine ground-based infrared data with the X-ray data to show the first direct evidence for a connection between instabilities in the inner accretion disk around a black hole and the formation of jets emanating from outside its event horizon. As can be seen in Figure 2, the X-ray and infrared emissions are correlated at the beginnings of the flares, although with a $\sim$ 310 second lag at infrared wavelengths. As the emissions seen in the two bandpasses decouple, the X-ray luminosity begins to oscillate wildly (with a quasi-period of $\sim$10 s), until it finally cuts off. The infrared emission persists and decays slowly, following the disappearance of the X-rays. The cycle then repeats every $\sim$30 minutes.

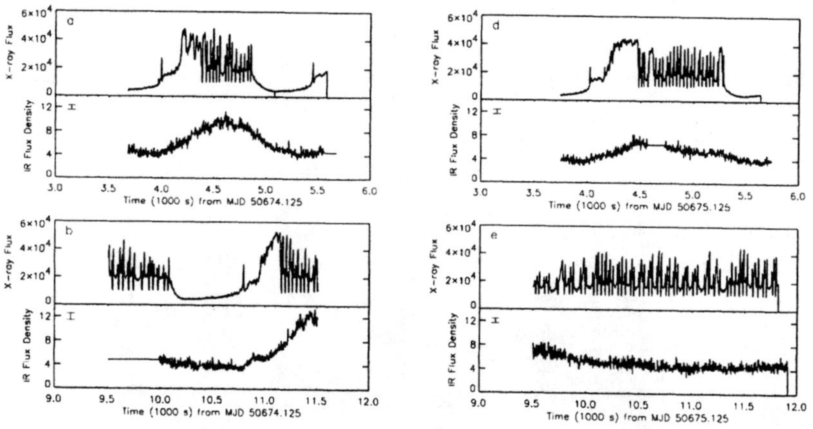

Source: Eikenberry et al. 1998

FIGURE 2. Coordinated infrared and X-ray observations of GRS 1915+105, a microquasar.

Eikenberry and his colleagues interpret these striking results in the following manner: the initial linkage between the X-ray and infrared flares indicates that the emissions in both bandpasses are triggered by the same event. The rapidly oscillating intense X-ray luminosity is associated with activity in the inner disk. Some of the disk material is then ejected to form the longer-lasting non-thermal infrared [28] (and radio [29]) emission. These flares are similar to, but not as large as the previously observed [7] superluminal events from the same object. The subsequent disappearance of the X-ray emission may indicate that the material in the inner disk has left the system: either by falling over the black hole's event horizon or in the form of the jet material.

Very recent observations using ESA's XMM-Newton satellite by Wilms et al. [30] of the detailed profile of the ~ 6 keV iron K-α line in MCG-6-30-15 have provided what may be the first direct evidence for the extraction of rotational energy from a spinning black hole by magnetic fields connecting the black hole to the disk. Although the magnetic field lines are not required by the XMM data to pierce the event horizon, the most likely interpretation supports models in which the magnetic field lines connect the rotating event horizon to either the accretion disk or its corona.

Evidence for Intermediate Mass Black Holes

The first claim of a black hole having intermediate mass (i.e., between 100 - 1000 $M_\odot$) was by Ptak and Griffiths [31]. They used ASCA observations to measure rapid changes in the hard X-ray emission from M82, and found that the low X-ray luminosity from the object implied a lower limit to its mass of $\sim 460 M_\odot$. Further support for this novel measurement of the mass of the X-ray source in M82 was provided by the location of the X-ray object inside a superbubble [32] and by precise positional information in Chandra data [33]. The Chandra image showed that the variable, hard X-ray source in

M82 (previously observed with ASCA) was located 9 arc seconds (about 140 pc) from the galaxy center, and thus could not be supermassive. Colbert and Mushotzky [34] came to a similar conclusion, based on a ROSAT study of 39 spiral and elliptical galaxies. They found that the likely mass of the black hole X-ray sources in the spiral galaxies is $10^2 - 10^4 M_{\odot}$. In very recent work, Colbert et al. [35] have conducted a similar study, using Chandra data, of 40 galaxies, and have found similar results.

An alternative explanation, which may account for some of these results, and seems likely to explain the paucity of X-rays from known supermassive black holes [36] is the ADAF model for accretion onto black holes[21][22][23] as discussed above.

Source: `http://constellation.gsfc.nasa.gov`

FIGURE 3. Mission concept for Constellation X.

FUTURE OBSERVATIONS

In this section, I discuss plans for two future X-ray Observatories: Constellation X and MAXIM (Micro Arcsecond X-ray Imaging Mission). Both missions are described in the current Cosmic Journeys roadmap for NASA's Structure and Evolution of the Universe theme area [37]. If expeditiously funded, Constellation X could be in orbit by 2008, and MAXIM could be in orbit by 2020. Both missions offer the opportunity to directly measure properties of black hole event horizons.

Constellation X: Black Hole High Resolution X-ray Spectroscopy

Constellation X is a proposed fleet of satellites, each of which contains an imaging X-ray telescope and a high resolution X-ray spectrometer. Constellation X will make high precision spectral measurements of the Kα fluorescence line of iron in nearby active galaxies. This will allow a direct probe of the physics of the accretion flow as well as the spacetime geometry within 10 Schwarzschild radii of the black hole. As shown in

Figure 4, Constellation X will be able to distinguish between spinning (Kerr) and non-rotating (Schwarzschild) black holes. From the observed distortion of the line profile, Constellation X will be able to measure the mass and spin of the black hole and also to test general relativity.

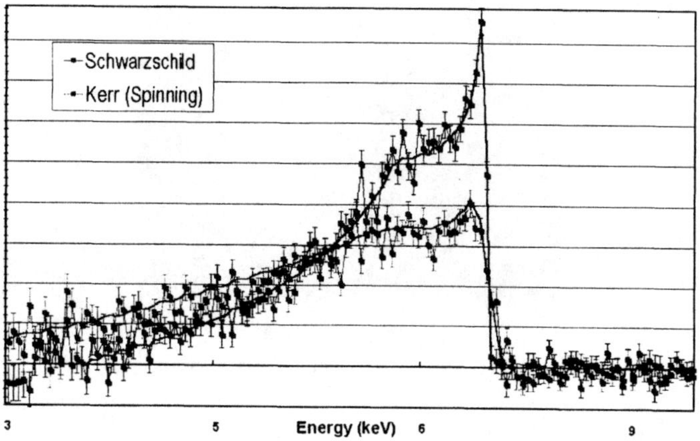

Source: `http://constellation.gsfc.nasa.gov`

FIGURE 4. Simulated iron Kα line profiles for spinning (Kerr) and non-rotating (Schwarzschild) black holes as seen by Constellation X.

MAXIM: Direct Imaging of the Event Horizon

In order to directly image the event horizon of black holes in nearby active galaxies, it is necessary to have resolution that is a million times better than that presently obtainable with the exquisite mirrors on the Chandra Observatory. In order to reach sub-micro-arcsecond resolution, Webster Cash [39] has pioneered a new experimental technique - X-ray interferometry. His laboratory experiments have achieved 0.1 arc second resolution, using two pairs of flat mirrors. Sub-microarcsecond resolution may therefore be achievable with the much larger baselines available in orbit.

A development plan for MAXIM (the Micro Arcsecond X-ray Imaging Mission) is in progress [40]. A Pathfinder mission would be flown initially, in order to demonstrate the feasibility of X-ray interferometry in space. The MAXIM Pathfinder mission would consist of a single spacecraft, with the mirrors separated by about one meter; it would be capable of 100 microarcsecond resolution. After a successful Pathfinder mission in $\sim$2010, the MAXIM mission could be in orbit by $\sim$2020. MAXIM will consist of up to 33 optics spacecraft, flying in formation with with a precision of 20 nanometers, plus a detector spacecraft 500 kilometers behind the mirrors. MAXIM would be able to image the event horizon of M87 and to map out the accretion disk around the black hole in the center of our Milky Way galaxy.

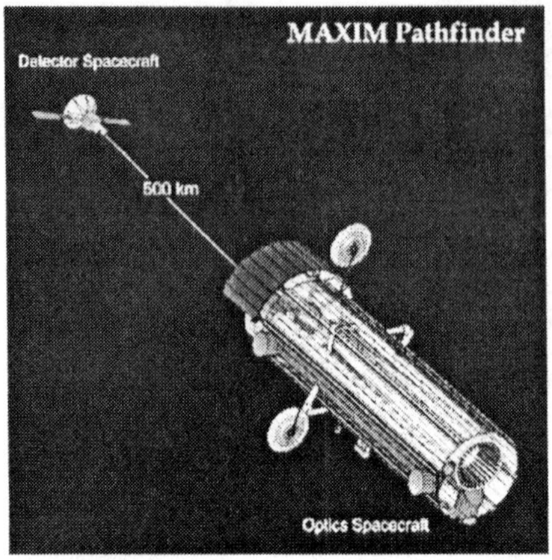

Source: `http://maxim.gsfc.nasa.gov`

FIGURE 5. MAXIM Pathfinder mission concept.

SUMMARY

X-ray observations of black holes have progressed greatly since their initial discovery thirty years ago. Observations of black holes from orbiting X-ray satellites such ASCA, ROSAT and the Rossi X-ray Timing Explorer are powerful tools for studying gravity in the strong field regime. New information about the physical properties and new populations of black holes is also being obtained with NASA's Chandra Observatory, and ESA's XMM-Newton Observatory. Future missions such as Constellation X and MAXIM offer the hope that predictions of General Relativity in this regime will be thoroughly examined.

ACKNOWLEDGMENTS

This paper has greatly benefited from discussions with Ron Remillard and J. Garrett Jernigan.

REFERENCES

1. Webster, B. L. & Murdin, P. 1972, Nature, 235, 37
2. Rhoades, C. E. & Ruffini, R. 1974, Phys. Rev. Lett., 32, 324
3. Casares, J. & Charles, P. A. 1994, MNRAS, 271, L5
4. Bahcall, S., Lynn, B. W. & Selepsky, S. B. 1989, Nucl. Phys. B, 325, 606
5. Miller, J. C., Shahbaz, T. & Nolan, L. A. 1998, MNRAS, 294, L25
6. Mirabel, I. F. & Rodriguez, L. F. 1998, Nature, 392, 673
7. Mirabel, I. F. & Rodriguez, L. F. 1994, Nature, 371, 46
8. Tingay, S. J. et al. 1995, Nature, 374, 141
9. Orosz, J. A. & Bailyn, C. D. 1997, ApJ, 477, 876

10. Orosz, J. A. et al. 2001, ApJ, 555, 489
11. Orosz, J. A. et al. 2002, ApJ in press, also astro-ph/0112101
12. Morgan, E. H., Remillard, R. A. & Greiner, J. 1997 ApJ, 482, 993
13. Remillard, R. A. et al. 1999, ApJ, 522, 397
14. Miller, J. M. et al. 2001, ApJ, 563, 928
15. Strohmayer, T. E. 2001, ApJ, 552, L49
16. Remillard, R. A. et al. 1999, ApJ, 517, L127
17. Remillard, R. A. et al. 2002, astro-ph/0202305
18. Wagoner, R. V. 1999, Physics Reports, 311, 259
19. Nowak, M. A. et al. 1997, ApJ, 477, L91
20. Abramowicz, M. A. & Kluzniak, W. 2001, A&A, 374, L19
21. Narayan, R. & Yi, I. 1994, ApJ, 428, L13
22. Narayan, R., Garcia, M. R. & McClintock,J. E. 1997 ApJ, 478, L79
23. Narayan, R., Barret, D. & McClintock, J. E. 1997, ApJ, 482, 448
24. Zhang, S. N., Cui, W. & Chen,W. 1997, ApJ, 482, L155
25. Stella, L. & Vietri, M. 1998, ApJ, 492, L59
26. Cui, W., Zhang, S. N. & Chen, W. 1998, ApJ, 492, L53
27. Eikenberry, S. S. et al. 1998, ApJ, 494, L61
28. Fender, R. P. et al. 1997, MNRAS, 290, L65
29. Rodriguez, L. F. & Mirabel, I. F. 1997, ApJ, 474, L123
30. Wilms, J. et al. 2002, MNRAS in press, also astro-ph/0110520
31. Ptak, A. & Griffiths, R. 1999, ApJ, 517, L85
32. Matsushita, S. et al. 2000, ApJ, 545, L107
33. Matsumoto, M. et al. 2001, ApJ, 547, L25
34. Colbert, E. & Mushotzky, R. 1999, ApJ, 519, 89
35. Colbert, E. et al. 2001, AAS 199, #141.05
36. Allen, S. W., Di Matteo, T. & Fabian, A. C. 2000, MNRAS, 311, 493
37. `http://universe.gsfc.nasa.gov`
38. `http://constellation.gsfc.nasa.gov`
39. Cash, W. et al. 2000, Nature, 407, 160
40. `http://maxim.gsfc.nasa.gov`

Evidence for the Black Hole Event Horizon

Ramesh Narayan*† and Jeremy S. Heyl†

*Institute for Advanced Study, Princeton, NJ 08540
†Harvard-Smithsonian Center for Astrophysics, Cambridge, MA 02138

Abstract. Roughly a dozen X-ray binaries are presently known in which the compact accreting primary stars are too massive to be neutron stars. These primaries are identified as black holes, though there is as yet no definite proof that any of the candidate black holes actually possesses an event horizon. We discuss how Type I X-ray bursts may be used to verify the presence of the event horizon in these objects. Type I bursts are caused by thermonuclear explosions when gas accretes onto a compact star. The bursts are commonly seen in many neutron star X-ray binaries, but they have never been seen in any black hole X-ray binary. Our model calculations indicate that black hole candidates ought to burst frequently if they have surfaces. Based on this, we argue that the lack of bursts constitutes strong evidence for the presence of event horizons in these objects.

INTRODUCTION

X-ray binaries [1] are double star systems in which mass is transferred from a secondary star onto a compact primary star. The gas radiates X-rays as it accretes onto the primary, which is either a neutron star (NS) or a black hole (BH). Of the over 200 X-ray binaries known in the Galaxy, about 30 are transient sources, known as "X-ray novae" or "soft X-ray transients" [2, 3]. These sources are characterized by episodic outbursts at X-ray, optical and radio frequencies, separated by long intervals (years to decades) of quiescence. X-ray novae have played a major role in the hunt for BHs.

In the quiescent state of an X-ray nova, the absorption-line velocities of the secondary star can be measured precisely because the non-stellar light from the accretion flow is modest compared to the light of the secondary. From a measurement of the velocity of the secondary as a function of the binary orbital phase, one may directly determine the mass function,

$$f(M) \equiv \frac{M_1^3 \sin^3 i}{(M_1 + M_2)^2}, \tag{1}$$

where M_1 and M_2 are the masses of the primary and secondary star, and i is the orbital inclination angle. The mass function provides a strict lower limit on the mass of the primary.

In nearly a dozen X-ray novae with reliable measurements of the mass function, it has been found that $f(M) \gtrsim 3M_\odot$ (see [4] for a list of X-ray novae with large mass functions as of mid-2001, and [5] for a newly confirmed system). Assuming that general relativity applies, the maximum mass of a NS can be calculated to be no more than about $2 - 3M_\odot$ [6–8]. Therefore, there is strong reason to think that the compact primaries in these X-ray novae are BHs.

CP624, *Cosmology and Elementary Particle Physics*, edited by B. N. Kursunoglu et al.

While the mass criterion is an excellent technique for finding BH candidates, one would like to have more direct evidence that the candidates truly are BHs. Ideally, one would like to be able to show that a candidate BH has an *event horizon*. This is the topic of the present article.

The outburst in an X-ray nova is caused by a sudden increase in the mass accretion rate onto the compact primary. The X-ray flux rises on a time scale of days, and subsequently declines on a time scale of weeks or months. Over the course of the outburst, the mass accretion rate varies by orders of magnitude, which makes these systems particularly useful for studying the nature of the central star. Basically, one can monitor the response of the star to a wide range of accretion rates, and thereby obtain a more complete understanding of the properties of the star than one could with a steadily accreting star.

Roughly half the known X-ray novae contain BH candidates and the rest contain NSs. X-ray novae with NSs have X-ray luminosities in quiescence of the order of 10^{-5} – $10^{-6}L_{Edd}$, where L_{Edd} is the Eddington luminosity. X-ray novae with BH candidates are much dimmer, with quiescent luminosities in some cases below $10^{-8}L_{Edd}$ [9]. The anomalous dimness of quiescent BH candidates has been used to argue for the presence of event horizons in these objects [9–11, see 4 for a review]. The argument is in essence that black holes are significantly *blacker* than neutron stars and must therefore have event horizons.

We discuss in this paper a different method of testing for the presence of event horizons in BH X-ray binaries. This method again makes use of the fact that the accreting primary in an X-ray nova experiences a wide range of mass accretion rates. However, instead of focusing on the luminosity of the object in quiescence, we study the presence or absence of Type I X-ray bursts.

When gas accretes onto a compact star such as a NS, it is compressed and heated as it accumulates on the surface, leading to thermonuclear reactions. In many NS systems (both X-ray novae and other types of NS X-ray binaries), the reactions occur unsteadily and cause sudden brief bursts in the X-ray flux [12]. The flux rises almost to the Eddington level within about a second and decays exponentially over a few seconds or tens of seconds. These thermonuclear bursts are called Type I bursts (to distinguish them from Type II bursts, which are not thermonuclear and are observationally distinct [13]). Type I bursts have been observed in a large number of NS systems [see 14, for a review], and the theory of the bursts is relatively well understood [14–25].

No Type I burst has been seen in any BH X-ray binary, even though, as we show below, BH candidates ought to produce bursts as efficiently as NSs if they possess surfaces. We argue that the lack of bursts represents strong evidence for the presence of event horizons.

A SIMPLE MODEL OF TYPE I BURSTS

We have developed a simple model to investigate the stability of gas accreting on the surface of a compact star. (The discussion in this section and the next closely follows ref. [26].) We consider a compact spherical star of mass M and radius R, accreting gas steadily at a rate $\dot{\Sigma}$ ($\mathrm{g\,cm^{-2}\,s^{-1}}$). In the local frame, the gravitational acceleration is $g = GM(1+z)/R^2$, where the redshift z is given by $1+z = (1-R_S/R)^{-1/2}$, and $R_S = 2GM/c^2$ is the Schwarzschild radius. We assume that the accreting material has mass fractions X_0, Y_0 and $Z_0 = 1 - X_0 - Y_0$, of hydrogen, helium and heavier elements (mostly CNO). The underlying star, as well as the fully burnt material sitting on it, is taken to have a composition $X = Y = 0, Z = 1$.

We consider a layer of accreted material of surface density Σ_{max} sitting on top of a substrate of fully burnt material. Since the physical thickness of the layer is much less than the radius, we work in plane parallel geometry and take g to be independent of depth. We solve for the density ρ, the temperature T, the outgoing flux F, and the hydrogen, helium and heavy element fractions, $X, Y, Z = 1 - X - Y$, as functions of the column density Σ.

Hydrostatic equilibrium, combined with the equation of state $P = P(\rho, T)$ of the gas, gives

$$\frac{\partial P}{\partial \Sigma} = \frac{\partial P}{\partial \rho}\frac{\partial \rho}{\partial \Sigma} + \frac{\partial P}{\partial T}\frac{\partial T}{\partial \Sigma} = g. \tag{2}$$

For the equation of state we use expressions given in [20] for the gas, radiation and degeneracy pressure, modified as needed when the degenerate electrons are relativistic.

H- and He-burning are described by two differential equations:

$$\frac{dX}{dt} = -\frac{\varepsilon_H}{E_H^*}, \quad \frac{dY}{dt} = -\frac{dX}{dt} - \frac{\varepsilon_{He}}{E_{He}^*}, \quad \frac{d}{dt} \equiv \frac{\partial}{\partial t} + \dot{\Sigma}\frac{\partial}{\partial \Sigma}. \tag{3}$$

Here, ε_H, ε_{He} are the nuclear energy generation rates for H and He burning, and E_H^*, E_{He}^* are the corresponding energy release per unit mass of H and He burned [20]. For ε_H, we include the pp chain and the CNO cycle, including fast-CNO burning, saturated CNO burning, and electron capture reactions, as described in [27, 28]. Since we are not concerned with modelling the bursts themselves, and since our stability criterion does not depend on the detailed treatment of the deep crust, we do not include proton captures onto heavier nuclei (see [29] for a discussion of some of the consequences of the *rp*-process burning on accreting NSs). For He-burning, we include the triple-α reaction, but not pycnonuclear reactions [30]; the latter are important only at greater pressures than we consider. We do not correct the reaction rates to include screening since we are concerned only with determining whether nuclear burning of H and He can proceed stably under given conditions; stable burning of H and He utilizes almost exclusively the reactions included in our model.

Radiative transfer gives another differential equation:

$$\frac{\partial T}{\partial \Sigma} = \frac{3\kappa F}{4acT^3}, \qquad \frac{1}{\kappa} = \frac{1}{\kappa_{rad}} + \frac{1}{\kappa_{cond}}. \tag{4}$$

We employ the fitting functions in [31] for the radiative opacity κ_{rad}, and an analytical formula from [32], suitably modified for relativistic electrons, for the conductive opacity κ_{cond}; the formula for the latter agrees well with more modern treatments [e.g. 33].

Finally, the energy equation gives

$$\rho T\frac{ds}{dt} = \rho(\varepsilon_H + \varepsilon_{He}) + \rho\frac{\partial F}{\partial \Sigma}, \tag{5}$$

where s is the entropy per unit mass. The above five differential equations (2)–(5) form a closed set for the five variables, ρ, T, F, X, Y.

We need five boundary conditions to solve the equations. We apply four boundary conditions at the surface of the star ($\Sigma = 0$) and one at the base of the accretion layer ($\Sigma = \Sigma_{max}$). Two of the four surface conditions are obvious: $X = X_0$, $Y = Y_0$. We obtain the third boundary condition by equating the accretion luminosity of the infalling gas, $L_{acc} = 4\pi R^2 \dot{\Sigma} c^2 z/(1+z)$, to blackbody emission from the surface: $4\pi R^2 \sigma T_{out}^4 = L_{acc}$. This fixes the surface temperature T_{out}. Then, using T_{out} and an assumed value of F_{out}, we solve for the surface density profile $\rho(\Sigma)$ from the radiative transfer equation, thus obtaining the fourth boundary condition.

At the base of the accretion layer we have an inner boundary condition, which plays an important role in the problem. For the calculations presented here, we assume that the base temperature, $T_{in} = T(\Sigma_{max})$, is fixed. We examine several values of T_{in} for layers with $\Sigma_{max} = 10^9, 10^{10}$ and 10^{11} g cm^{-2}. Applying the boundary condition at Σ_{max} rather than deeper in the substrate is an approximation, but the error due to this is not serious. For the high surface densities we consider, the heat transfer is dominated by conduction [33], so the temperature gradient for $\Sigma > \Sigma_{max}$ is small. Moreover, we examine models for several values of T_{in} which further mitigates any error.

The calculations proceed in two stages. First, we solve for the steady state profile of the accretion layer by setting $\partial/\partial t = 0$ and replacing the operator $\partial/\partial\Sigma$ in equations (2)–(5) by $d/d\Sigma$. With this substitution, we have five ordinary differential equations with four outer boundary conditions and one inner boundary condition. The solution to these equations gives the profiles of the basic fluid quantities: $\rho(\Sigma)$, $T(\Sigma)$, $F(\Sigma)$, $X(\Sigma)$, $Y(\Sigma)$.

Having calculated the steady state structure of the accretion layer, we next check its stability. Various local stability criteria have been discussed in the literature [25], in which one considers the properties of the gas at a single depth. While this approach is very useful for physical insight, it is clearly inadequate for quantitative work. Occasionally, some authors have considered a global criterion involving an integral over the entire layer [23]. Even this approach is not very satisfactory since it usually involves a restriction on the perturbations (e.g., constant temperature perturbation in [23]).

We have carried out a full linear stability analysis of the accretion layer. To our knowledge, this is the first time that this has been attempted in the field of Type I bursts. We start with the steady state solution described above and assume that it is slightly perturbed, $Q(\Sigma) \to Q(\Sigma) + Q'(\Sigma)\exp(\gamma t)$, where Q corresponds to each of our five variables, and the perturbations $Q'(\Sigma)$ are taken to be small. We linearize the five partial differential equations given in equations (2)–(5) (three of which involve time derivatives), apply the boundary conditions, and solve for the eigenvalue γ. We obtain a large number of solutions for γ, many of which have both a real and an imaginary part. We consider the accretion layer to be unstable if any eigenvalue has a real part (growth rate) greater than the characteristic accretion rate $\gamma_{acc} = \dot{\Sigma}/\Sigma_{max}$.

If the steady-state model is unstable according to the linear stability analysis, accretion cannot proceed stably with the particular Σ_{max} and $\dot{\Sigma}$. Whether this instability manifests itself as a Type I burst depends on how the burning flame, once ignited at some random point, envelopes the surface of the star. This is as yet an unsolved problem, although ref. [34] discusses a possible solution. In the following, we assume that the instability will

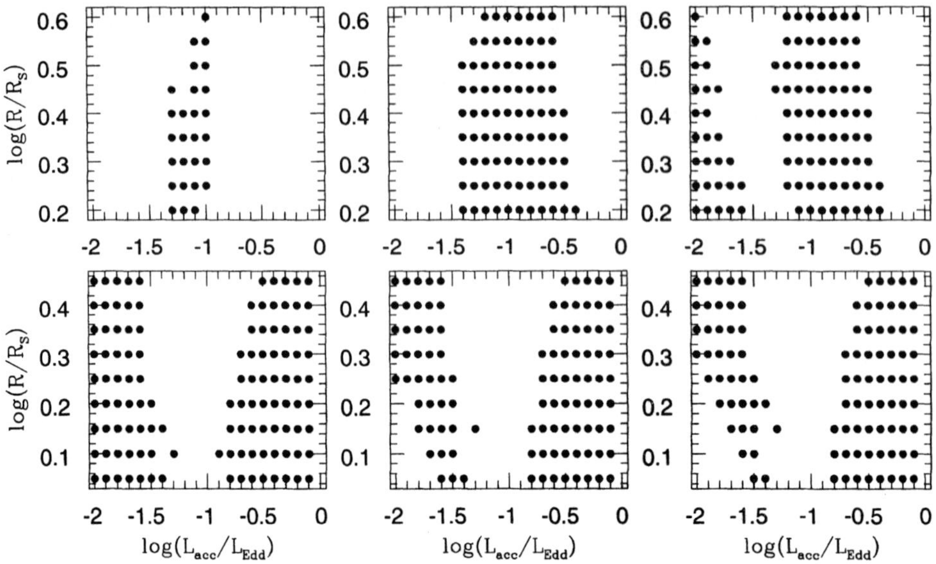

FIGURE 1. Regions of instability, shown by dots, as a function of accretion luminosity and stellar radius. Top left: $1.4M_{\odot}$ NS with a base temperature $T_{in} = 10^{8.5}$ K. Top center: $T_{in} = 10^{8}$K. Top right: $T_{in} = 10^{7.5}$ K. Bottom left: $10M_{\odot}$ BH candidate with a surface, and a base temperature $T_{in} = 10^{7.5}$ K. Bottom center: $T_{in} = 10^{7}$ K. Bottom right: $T_{in} = 10^{6.5}$ K.

grow rapidly to the nonlinear regime and that the system will exhibit Type I bursts.

RESULTS

Type I Bursts on Neutron Stars

The top three panels of Fig. 1 show results for solar composition material ($X_0 = 0.7$, $Y_0 = 0.27$, $Z_0 = 0.03$) accreting on a NS of mass $1.4M_\odot$. We consider a range of accretion rates, parameterized by the ratio L_{acc}/L_{Edd}, where we take the Eddington luminosity to be $L_{Edd} = 4\pi GMc/\kappa_{es}$ with $\kappa_{es} = 0.4\ \mathrm{cm^2\,g^{-1}}$. We also consider a range of radii for the NS: $\log(R/R_S) = 0.2 - 0.6$ (corresponding to $R = 6.5 - 16$ km). For each choice of L_{acc}/L_{Edd} and R/R_S, we try three values of the surface mass density of the accretion layer: $\Sigma_{max} = 10^9,\ 10^{10},\ 10^{11}\ \mathrm{g\,cm^{-2}}$. If any of the three cases is unstable, i.e., if it has any eigenvalue with $\mathrm{Re}(\gamma) > \gamma_{acc}$, we say that the system cannot achieve a stable steady state and that it will exhibit Type I bursts.

The results presented in Fig. 1 correspond to three choices of the temperature at the base of the accretion layer: $T_{in} = 10^{8.5},\ 10^8,\ 10^{7.5}$ K (from left to right). In the present work, we are primarily interested in X-ray novae. Since these sources have very low luminosities in quiescence ($L_X < 10^{33}\ \mathrm{erg\,s^{-1}}$ for NS X-ray novae), the core temperatures of the NSs are likely to be $T_{in} \lesssim 10^8$ K [35].

The calculations shown in Fig. 1 indicate that NSs are unstable to bursts for a wide range of T_{in}, but that the width of the instability strip (in L_{acc}) is less for higher values of T_{in}. The reason for the latter is clear from the analysis in [20] where it is shown that as the flux escaping from the stellar core into the accretion layer increases (which happens when T_{in} increases), bursting behavior is restricted to a smaller range of $\dot{\Sigma}$.

Assuming that $T_{in} \lesssim 10^8$ K, we see from Fig. 1 that NSs should be unstable to bursts for accretion luminosities up to $L_{acc} \approx 0.3L_{Edd}$. The prediction is generally consistent with observations: the only NSs that are known not to burst are bright Z sources with $L_{acc} \to L_{Edd}$ [36], and X-ray pulsars. Although X-ray pulsars are significantly less luminous than Eddington, they accrete effectively at close to the Eddington rate since the accreting matter is channeled onto a small area on the NS surface by strong magnetic fields (see [37] for a detailed discussion of this argument).

Immediately below $L_{acc} \sim 0.3L_{Edd}$, the instability is of a mixed type in which a He-burning instability triggers a burst in which He and H both burn explosively. At lower luminosities, nearly all the H is burned steadily, and the instability corresponds to a pure He burst. These results are consistent with previous work [25]. Our calculations indicate that bursting behavior cuts off below an accretion luminosity $L_{acc} \sim 10^{-1.5}L_{Edd}$. The cutoff is the result of the restriction $\Sigma_{max} \le 10^{11}\ \mathrm{g\,cm^{-2}}$ in the model. Systems with luminosities below the cutoff are still unstable to He bursts, but only at extremely high column density Σ_{max}. Correspondingly, the recurrence times t_{rec} of bursts are extremely long — years in some cases according to our calculations — which makes these systems less useful for the present study, since accretion outbursts of X-ray novae last no more than a few months.

For $T_{in} = 10^{7.5}$ K, there is a second instability strip at low luminosities, around $L_{acc} \sim 10^{-2}L_{Edd}$. This strip corresponds to pure H bursts (the possibility of such bursts was first noted in [22]). An interesting difference between the two instability strips in this panel is that the strip on the right generally has complex eigenvalues γ for the unstable modes while that on the left has real eigenvalues.

Type I Bursts on Black Hole Candidates

The bottom three panels in Fig. 1 show results for a $10M_\odot$ BH candidate with a surface. The three panels correspond to different choices of the base temperature: $T_{in} = 10^{7.5}$, 10^7, $10^{6.5}$ K (from left to right). The particular choices of T_{in} are motivated by the extraordinarily low quiescent luminosities of many BH X-ray novae ($L_X < 10^{31}\,\mathrm{erg\,s^{-1}}$, see [9]; a few BH novae are brighter than this, but even these are not likely to have $T_{in} > 10^{7.5}$ K). We consider stellar radii in the range $\log(R/R_S) = 0.05 - 0.45$, corresponding to R between $(9/8)R_S$ and $3R_S$. The former limit is the smallest radius in general relativity for an object whose density either decreases or remains level with increasing radius [30].

The calculations indicate that BH candidates with surfaces are similar to NSs in their bursting behavior. Except for a modest rightward shift of the positions of the instability zones, the results in the lower three panels are quite similar to those for a NS with a similarly low value of T_{in} (upper right panel). As in the case of NSs, bursts are expected from BH candidates even for larger values of T_{in} than we have considered here, except that the instability strip becomes narrower when T_{in} exceeds 10^8 K. (As already discussed, such high core temperatures are ruled out.) We conclude that BH candidates are as prone to Type I bursts as NSs are, provided they have surfaces.

Because our model focuses only on the most important physical effects, and neglects some details, the exact positions of the instability strips in Fig. 1 may be uncertain at the level of say a factor of two in accretion luminosity. We believe, however, that the overall pattern of instability is fairly robust. If at all, our computations are likely to be conservative in the sense that bursting behavior may be even more widespread than we predict.

DISCUSSION

It is clear from the theory of bursts ([25] and references therein) that bursting behavior is largely determined by the surface gravity g, the mass accretion rate, and the composition of the accreting material. It is also clear that the accreting object must have a surface in order to burst. If there is no surface, material cannot accumulate, and therefore cannot become hot or dense enough to trigger a thermonuclear explosion. Indeed, using this argument, astronomers routinely identify any X-ray binary that has a Type I burst as a NS system.

We suggest that the argument could equally well be applied in reverse, at least in the case of X-ray novae: If an X-ray nova with a BH candidate (identified independently via a mass measurement) does not exhibit Type I bursts at any time during its accretion outburst, then we argue that the primary star cannot have a surface — the object must be a BH with an event horizon. To our knowledge, this simple argument has not been explored so far in the literature [but see 38].

While the argument is simple, it is not obvious that it is correct. The reason is that not all systems with surfaces necessarily have bursts; for instance, many NS X-ray binaries, e.g., the brighter Z sources and X-ray pulsars, do not burst. Clearly, it is necessary to understand through physical modeling under what conditions a compact star with a surface will or will not burst. This is the motivation behind the calculations presented here. Our results are shown in Fig. 1, with the top three panels corresponding to NSs and the bottom three panels to $10M_{\odot}$ BH candidates with surfaces. In brief, we find that BH candidates with surfaces ought to burst as prolifically as NSs.

A notable feature of our results is that we find bursting behavior in BH candidates (with surfaces) for quite a wide range of L_{acc}/L_{Edd}, regardless of the radius of the accreting star. As discussed in the Introduction, the accretion rate in a typical X-ray nova spans a wide range during an accretion outburst. At maximum, the sources often approach $L_{acc} \sim L_{Edd}$, at the right edge of a given panel in Fig. 1, while in quiescence, the sources are extremely underluminous, well off the left edge of the panels. Thus, during the rise to outburst and subsequent decline, each BH X-ray nova enters the plot on the left, moves horizontally all the way across, and then moves horizontally back, to disappear off the left edge. The journey typically takes a few months. During this entire journey, it is an amazing observational fact that not one of the BH candidates has ever had a Type I burst. Our calculations indicate that this is impossible if the sources have surfaces. Regardless of the radius of the BH candidate's surface (which is of course unknown), the trajectory that the star traverses during an accretion outburst would always intersect wide stretches of instability. Moreover, according to our estimates, the burst recurrence time is of the order of a day in many regions of the diagram, which means that each source should exhibit many bursts during its months-long journey. The absence of bursts in BH X-ray novae is thus highly significant and argues strongly for the lack of surfaces in these systems.

Before we can claim that this "proves" the reality of the event horizon, more work is needed.

First, we need to explain why some NSs do not have bursts, and we should provide a convincing argument why the same explanation cannot be applied to BH candidates. We have discussed in this paper two categories of NSs that do not burst. (i) Most Z sources do not burst, and we explained this as the result of their luminosities being too close to L_{Edd}. Since BH X-ray novae span a wide range of L_{acc}, the explanation obviously does not apply to them. (ii) X-ray pulsars generally do not burst, and in this case the standard explanation is that magnetic funneling causes the "effective" accretion rate to approach the Eddington level locally. Since BH X-ray novae do not show coherent pulsations, it is extremely unlikely that they have the kind of magnetic funneling present in X-ray pulsars; therefore, once again, the explanation cannot be invoked for BH X-ray novae. We are not aware of other classes of NSs that do not burst, but if any are identified, it is necessary to explain the lack of bursts through physical modeling. It is also necessary to demonstrate that the same explanation does not apply to BH X-ray novae.

Second, we need to show that the model is able to reproduce the main features of Type I bursts as observed in NS X-ray binaries. For instance, the statistics of burst durations and recurrence times [39] ought to appear naturally in a realistic model.

Third, the role of the inner boundary condition needs to be explored in detail. In Fig. 1, we see that different choices of the base temperature for a BH candidate give similar results. We have tried other boundary conditions (e.g., we have applied a boundary condition on the flux [20] instead of the temperature), and also tried changing the composition of the accreting gas. In all cases we find that the accumulating layer is unstable to bursts for a wide range of L_{acc}/L_{Edd}, both for $1.4M_{\odot}$ NSs and $10M_{\odot}$ BH candidates with surfaces.

Fourth, the difficult issue of flame propagation over the surface of the star once the instability has been triggered needs to be addressed (e.g., [34]). In this context, we note the empirical fact that the flame clearly propagates with no trouble on NSs, as indicated by the fact that these objects do exhibit bursts. The key question is: could the propagation physics be so delicate that the burning front stops propagating when the mass of the star changes from $1.4M_{\odot}$ to $10M_{\odot}$? We do not know the answer to this question.

We should caution, morever, that the analysis presented here assumes that the accumulating gas on the surface of a BH candidate behaves like normal matter, with nucleons and electrons. Obviously, our arguments would become invalid if the properties of the gas changed drastically, e.g., if the nuclei disappeared and were replaced by quarks. Whether such extreme changes are plausible on the surface of a strange star remains to be seen. The density and pressure at the base of the bursting layer do not go above few $\times\, 10^8\,\mathrm{g\,cm^{-3}}$ and $10^{26}\,\mathrm{erg\,cm^{-3}}$, respectively, even in the most extreme cases we have considered. It is hard to imagine exotic physics being important under these conditions [40].

On the observational front, we should check whether some NSs that lie within the unstable regions of Fig. 1 are stable to bursts. Any obvious large-scale disagreement between the observed burst behavior of NSs and the results presented here would indicate that the model is missing important physics. We are not aware of serious discrepancies of this nature, but the matter clearly deserves careful study.

In the case of BH X-ray novae, we should use observational data to derive strict quantitative limits on bursting activity. The current data are already compelling — no burst has been reported from any of the dozen or so BH X-ray novae with dynamically measured primary masses. Nevertheless, a careful study of archival data on past X-ray novae as well as data to be collected on future novae is needed to rule out the possibility that bursts might have been missed. This is well worth the effort since a firm demonstration that BH transients do not have Type I bursts at any point in their light curves would, as we have shown, provide a strong argument for the presence of event horizons in these systems.

ACKNOWLEDGMENTS

We thank Andrew Cumming, Alex Ene, Kristen Menou, Bohdan Paczynski and Greg Ushomirsky for useful discussions. RN was supported in part by the W. M. Keck Foundation as a Keck Distinguished Visiting Professor. RN's research was supported by NSF grant AST-9820686 and NASA grant NAG5-10780. JSH was supported by the Chandra Postdoctoral Fellowship Award # PF0-10015 issued by the Chandra X-ray Observatory Center, which is operated by the Smithsonian Astrophysical Observatory for and on behalf of NASA under contract NAS8-39073.

REFERENCES

1. Lewin, W. H. G., van Paradijs, J., and van den Heuvel, E. P. J., *X-ray binaries*, Cambridge Univ. Press, Cambridge, 1995.
2. Tanaka, Y., and Shibazaki, N., *Ann. Rev. Astron. Astrophys.*, **34**, 607 (1996).
3. van Paradijs, J., and McClintock, J. E., "Optical and Ultraviolet Observations of X-ray Binaries," in *X-ray Binaries*, edited by W. Lewin, J. van Paradijs, and E. van den Heuvel, Cambridge Univ. Press, 1995, p. 58.
4. Narayan, R., Garcia, M. R., and McClintock, J. E., "X-ray Novae and the Evidence for Black Hole Event Horizons," in *Proc. IX Marcel Grossmann Meeting*, edited by V. Gurzadyan, R. Jantzen, and R. Ruffini, World Scientific, Singapore, 2002 (astro-ph/0107387).
5. Greiner, J., Cuby, J. G., and McCaughrean, M. J., *Nature*, **414**, 522 (2001).
6. Rhoades, C. E., and Ruffini, R., *Phys. Rev. Lett.*, **32**, 324 (1974).
7. Cook, G. B., Shapiro, S. L., and Teukolsky, S. A., *Astrophys. J.*, **424**, 823 (1994).
8. Kalogera, V., and Baym, G., *Astrophys. J.*, **470**, L61 (1996).
9. Garcia, M. R., McClintock, J. E., Narayan, R., Callanan, P., Barret, D., and Murray, S. S., *Astrophys. J.*, **553**, L47 (2001).
10. Narayan, R., Garcia, M. R., and McClintock, J. E., *Astrophys. J.*, **478**, L79 (1997).
11. Menou, K., Esin, A. A., Narayan, R., Garcia, M. R., Lasota, J. P., and McClintock, J. E., *Astrophys. J.*, **520**, 276 (1999).
12. Grindlay, J., Gursky, H., Schnopper, H., Parsignault, D. R., Heise, J., Brinkman, A. C., and Schrijver, J., *Astrophys. J.*, **205**, L127 (1976).
13. Hoffman, J. A., Marshall, H. L., and Lewin, W. H. G., *Nature*, **271**, 630 (1978).
14. Lewin, W. H. G., van Paradijs, J., and Taam, R. E., *Space Sci. Rev.*, **62**, 223 (1993).
15. Hansen, C. J., and van Horn, H. M., *Astrophys. J.*, **195**, 735 (1975).
16. Woosley, S. E., and Taam, R. E., *Nature*, **263**, 101 (1976).
17. Joss, P. C., *Nature*, **270**, 310 (1977).
18. Taam, R. E., and Picklum, R. E., *Astrophys. J.*, **224**, 210 (1978).
19. Fujimoto, M. Y., Hanawa, T., and Miyaji, S., *Astrophys. J.*, **246**, 267 (1981).
20. Paczynski, B., *Astrophys. J.*, **264**, 282 (1982).
21. Fujimoto, M. Y., Hanawa, T., Iben, I., and Richardson, M. B., *Astrophys. J.*, **278**, 813 (1984).
22. Fujimoto, M. Y., Hanawa, T., Iben, I., and Richardson, M. B., *Astrophys. J.*, **315**, 198 (1987).
23. Fushiki, I., and Lamb, D. Q., *Astrophys. J.*, **323**, L55 (1987).
24. Taam, R. E., Woosley, S. E., and Lamb, D. Q., *Astrophys. J.*, **459**, 271 (1996).
25. Bildsten, L., "Thermonuclear Burning on Rapidly Accreting Neutron Stars," in *The Many Faces of Neutron Stars*, edited by J. Buccheri, J. van Paradijs, and M. A. Alpar, Kluwer, Dordrecht, 1998, p. 419.
26. Narayan, R., and Heyl, J. S., *Astrophys. J.,* submitted (2002, astro-ph/0203089).
27. Matthews, G. J., and Dietrich, F. S., *Astrophys. J.*, **287**, 969 (1984).
28. Bildsten, L., and Cumming, A., *Astrophys. J.*, **506**, 842 (1998).
29. Schatz, H., Bildsten, L., Cumming, A., and Wiescher, M., *Astrophys. J.*, **524**, 1014 (1999).
30. Shapiro, S. L., and Teukolsky, S. A., *Black Holes, White Dwarfs and Neutron Stars*, Wiley-Interscience, New York, 1983.
31. Iben, I., *Astrophys. J.*, **196**, 525 (1975).
32. Clayton, D. D., *Principles of Stellar Evolution and Nucleosynthesis*, McGraw-Hill, New York, 1968.
33. Heyl, J. S., and Hernquist, L., *Astrophys. J.*, **324**, 292 (2001).
34. Spitkovsky, A., Levin, Y., and Ushomirsky, G., *Astrophys. J.*, **566**, 1018 (2002).
35. Brown, E. F., Bildsten, L., and Chang, P., *Astrophys. J.,* in press (2002, astro-ph/0204102).
36. Matsuba, E., Dotani, T., Mitsuda, K., Asai, K., Lewin, W. H. G., van Paradijs, J., and van der Klis, M., *Publ. Astron. Soc. Japan*, **47**, 575 (1995).
37. Lamb, D. Q., *Astrophys. J. Suppl.*, **127**, 395 (2000).
38. Menou, K., "Black Hole Accretion in Transient X-Ray Binaries," in *Proc. 2nd KIAS Astrophysics Workshop*, 2002 (astro-ph/0111469).
39. van Paradijs, J., Penninx, W., and Lewin, W. H. G., *Mon. Not. Roy. astron. Soc.*, **233**, 437 (1988).
40. Glendenning, N. K., *Compact Stars, Nuclear Physics, Particle Physics, and General Relativity*, Springer, Berlin, 1997.

Microscopic Black Holes as a Source of Ultrahigh Energy γ-rays

Roberto Casadio*, Benjamin Harms† and Octavian Micu†

*Dipartimento di Fisica, Università di Bologna and I.N.F.N., Sezione di Bologna, via Irnerio 46, 40126 Bologna, Italy
†Department of Physics and Astronomy, The University of Alabama, Box 870324, Tuscaloosa, AL 35487-0324, USA

Abstract. We investigate the idea that ultrahigh energy γ-rays ($E > 10\,\text{TeV}$) can be produced when charged particles are accelerated by microscopic black holes. We begin by showing that microscopic black holes may exist as remnants of primordial black holes or as products of the collisions in the large extra dimensions scenario of high energy cosmic rays with atmospheric particles. We then solve Maxwell's equations on curved spacetime backgrounds in 4, 5 and 6 dimensions and use the solutions to calculate the energy distributions. From the latter we obtain the black hole parameters needed to produce the energies of the observed γ-rays.

INTRODUCTION

The existence of cosmological black holes with masses of $10 - 10^8$ solar masses is now an accepted fact. The existence of microscopic black holes with masses of $1 - 10$ Planck masses has not been established, but their existence is of great theoretical interest. If they do exist, there is the possibility that they could be probed experimentally to obtain information about quantum gravity. One possible source of such microscopic black holes is primordial black holes, which have reached Planck-size masses during the present epoch through emission of Hawking radiation. Another possible source is microscopic black hole production. The recent proposal of the existence of large extra dimensions [1] and the consequent lowering of the fundamental energy scale to 1 TeV implies that microscopic black holes can be created in accelerators [2] whose center of mass energies are above the fundamental energy scale. Microscopic black holes would also be produced in the collisions of ultrahigh energy cosmic rays with the Earth's atmosphere [3]. In this talk we describe one of the signatures of black holes impinging upon the Earth's atmosphere: ultrahigh energy γ-rays. As of yet there is no firm evidence that such emissions are occurring in our atmosphere, but there is some evidence for γ-rays with energies greater than 100 TeV at the 1.6 σ-level from unknown sources within the galatic plane[4].

A charged particle being accelerated by a black hole can produce γ-rays with energies in the multi-TeV range before the particle passes beyond the horizon radius provided that the curvature gradient of the space around the black hole is large enough. Such curvature gradients occur in quantum black holes, black holes whose masses are of the order the Planck mass. A calculation taking into account special relativity (but not

CP624, *Cosmology and Elementary Particle Physics*, edited by B. N. Kursunoglu et al.

general relativity) shows us that to produce γ-ray energies in the 10 TeV range a single electronic charge would have to be accelerated by a black hole with a mass equal to five times that of the Planck mass.

The microscopic black holes needed to produce ultrahigh γ-rays may be the remnants of primordial black holes. Such black holes can be produced by

- Inflationary horizon-scale fluctuations
- Density fluctuations at phase transitions and bubble formation and collapse
- Baryon isocurvature fluctuations on small scales.

Large-mass primordial black holes ($M > 10^{15}$ gm) decaying via Hawking radiation [5] as described by the canonical ensemble in 4 space-time dimensions, $(dM/dt) \sim -M^{-2}$, would have decayed to a Planck-size mass in the present epoch. Microscopic black holes produced in this manner could be stable if quantum gravity effects terminate the decay process.

Copious microscopic black hole production can also occur if large extra dimensions exist. In this scenario black hole production can occur as the result of the collision of particles with total center of mass energy above the effective Planck scale, which can be as low as the electroweak scale $m_{ew} \sim 1\,\mathrm{TeV}$. Black holes could thus be produced in collisions of high energy cosmic rays with the Earth's atmosphere. As we show in the next Section, such black holes may live long enough to create ultrahigh γ-rays even without taking quantum gravity effects into account.

BLACK HOLES AND LARGE EXTRA DIMENSIONS

In a 4-dimensional space-time, a black hole might emerge from the collision of two particles only if its center of mass energy exceeds the Planck mass m_p (l_p will denote the Planck length). In fact, the Compton wavelength $l_M = l_p(m_p/M)$ of a point-like particle of mass $M < m_p$ would be smaller than the gravitational radius $R_H = 2\,G_N M = 2\,(l_p/m_p)\,M$ and the very (classical) concept of a black hole would lose its meaning. However, since the fundamental mass scale is shifted down to m_{ew} in the models under consideration, black holes with $M \ll m_p$ can now exist as classical objects provided

$$l_p\,(m_p/M) \ll R_H \ll L\,, \tag{1}$$

where L is the scale at which corrections to the Newtonian potential become effective. The left hand inequality ensures that the black hole behaves semiclassically, and one does not need a full-fledged theory of quantum gravity, while the right hand inequality guarantees that the black hole is small enough that its gravitational field can depart from the Newtonian behavior without contradicting present experiments.

The luminosity of a black hole in D space-time dimensions is given by

$$\mathscr{L}_{(D)}(M) = \mathscr{A}_{(D)} \int_0^\infty \sum_{s=1}^{S} n_{(D)}(\omega)\,\Gamma^{(s)}_{(D)}(\omega)\,\omega^{D-1}\,d\omega \tag{2}$$

where $\mathscr{A}_{(D)}$ is the horizon area in D space-time dimensions, $\Gamma^{(s)}_{(D)}$ the corresponding grey-body factor and S the number of species of particles that can be emitted. For the sake of simplicity, we shall approximate $\sum_s \Gamma^{(s)}_{(D)}$ as a constant. The distribution $n_{(D)}$ is the microcanonical number density [6, 7, 8]

$$n_{(D)}(\omega) = C \sum_{l=1}^{[[M/\omega]]} \exp\left[S^E_{(D)}(M - l\,\omega) - S^E_{(D)}(M)\right] \tag{3}$$

where $[[X]]$ denotes the integer part of X and $C = C(\omega)$ encodes deviations from the area law [6] (in the following we shall also assume C is a constant in the range of interesting values of M).

ADD scenario

If the space-time is higher dimensional and the extra dimensions are compact and of size L, the relation between the mass of a spherically symmetric black hole and its horizon radius is changed to [9]

$$R_H \simeq l_{(4+d)} \left(\frac{2M}{m_{(4+d)}}\right)^{\frac{1}{1+d}}, \tag{4}$$

where $G_{(4+d)} \simeq L^d\, G_N$ is the fundamental gravitational constant in $4+d$ dimensions. Eq. (4) holds true for black holes of size $R_H \ll L$, or, equivalently, of mass $M \ll M_c \equiv m_p\,(L/l_p)$. Since L is related to d and the fundamental mass scale $m_{(4+d)}$ by [1]

$$L \sim \gamma^{1+\frac{2}{d}}\, 10^{\frac{31}{d}+16}\, l_p\,, \tag{5}$$

where $\gamma \equiv m_{ew}/m_{(4+d)}$, Eq. (1) translates into

$$10^{-\frac{31+16d}{2+d}}\,\gamma m_p \sim 10^{-16}\,\gamma m_p \ll M \ll M_c\,, \tag{6}$$

where we also used the fact that $d = 1$ is ruled out by present measurement of G_N [1] and relatively high values of d (i.e., $d \sim 6$) seem to be favored (see, e.g., Refs. [10]). For $\gamma \sim 1$ (i.e., $m_{(4+d)} \sim m_{ew} \sim 1\,\mathrm{TeV}$), the left hand side above is of order m_{ew} as well.

For $R_H < L$ the Euclidean action $S^<_E \sim (M/m_{ew})^{(d+2)/(d+1)}$ and the occupation number density for the Hawking particles in the microcanonical ensemble is given by

$$n_{(4+d)}(\omega) \sim \sum_{l=1}^{[[M/\omega]]} e^{\left(\frac{M-l\,\omega}{m_{ew}}\right)^{\frac{d+2}{d+1}} - \left(\frac{M}{m_{ew}}\right)^{\frac{d+2}{d+1}}}\,. \tag{7}$$

In 4 dimensions one knows that microcanonical corrections to the luminosity become effective only for $M \sim m_p$, therefore, for black holes with $M \gg m_{ew}$ the luminosity (2)

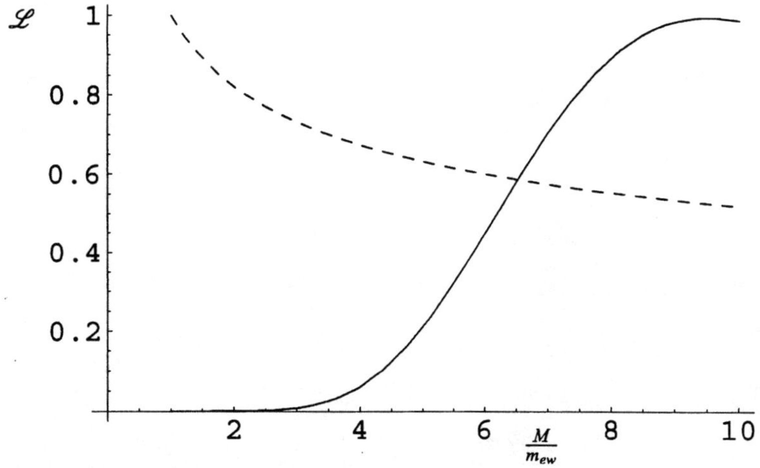

FIGURE 1. Microcanonical luminosity (solid line) for a small black hole with $d = 6$ extra dimensions compared to the corresponding canonical luminosity (dashed line). Vertical units are chosen such that the canonical luminosity $\mathscr{L}(m_{ew}) = 1$.

should reduce to the canonical result [11, 12, 13]. However, in $4+d$ dimensions, such corrections are not entirely negligible for $M \sim m_{ew}$. The decay rate corresponding to the number density (7) as calculated from Eq. (2) is exhibited for $d = 6$ in Fig. 1. For lower values of d the peak in the microcanonical luminosity shifts to smaller values of M, thus approaching the 4-dimensional canonical decay rate (also shown for comparison in Fig. 1). In all cases, the microcanonical luminosity becomes smaller for $M \sim m_{ew}$ than it would be according to the canonical luminosity, which makes the life-time of the black hole somewhat longer than in the canonical picture. In particular, for $d = 6$ one finds

$$\left.\frac{dM}{dt}\right|_{M\sim m_{ew}} \simeq -10^{-10}\,\mathscr{L}^{H}_{(10)} \sim -10^{17}\,\frac{\mathrm{TeV}}{\mathrm{s}}\,. \tag{8}$$

A black hole would therefore evaporate very quickly [2] down to $\sim 6\,m_{ew}$. Then, its life-time is dominated by the time it would take to emit the remaining $\Delta M \sim 5\,\mathrm{TeV}$, before it reaches $1\,m_{ew}$, which is approximately

$$T \sim \left(\frac{dM}{dt}\right)^{-1} \Delta M \sim 10^{-17}\,\mathrm{s}\,. \tag{9}$$

The above relatively long time also takes into account the dependence of the grey-body factor $\Gamma^{(s)}_{(4+d)}$ on d (for the details see Ref. [14]). Without a full-fledged theory of quantum gravity, nothing can be safely stated as M becomes less than m_{ew}. However, by extrapolating the microcanonical behavior, one would conclude that the evolution then proceeds roughly according to the usual exponential law of radiative decay [7].

RS scenario

In order to study this case, we shall make use of the solution given in Ref. [15]. This is one of the few known metrics on the brane which might represent such a case in the context of the RS scenario (for more candidates see Ref. [16]). Such a solution has the Reissner-Nordström form

$$-g_{tt}=\frac{1}{g_{rr}}=1-2\frac{M\,l_p}{m_p\,r}+Q^2\frac{l_p^2}{r^2}-q\frac{m_p^2\,l_p^2}{m_{(5)}^2\,r^2}\,, \tag{10}$$

and the (outer) horizon radius is given by

$$R_H=l_p\frac{M}{m_p}\left[1+\sqrt{1-Q^2\frac{m_p^2}{M^2}+\frac{q\,m_p^4}{M^2\,m_{(5)}^2}}\right]\,, \tag{11}$$

where $m_{(5)}\sim m_{ew}$ is the fundamental mass scale and q represents a (dimensionless) tidal charge. The latter can be estimated on dimensional grounds as [12, 15] $q\sim\left(\frac{m_p}{m_{ew}}\right)^\alpha\frac{M}{m_{ew}}$ and for $\alpha>-4$ the tidal term $\sim 1/r^2$ dominates over the 4-dimensional potential $\sim 1/r$ (as one would expect for tiny black holes). From Eq. (11) with $Q=0$ and $\alpha>-4$ one obtains

$$R_H\simeq l_p\left(\frac{m_p}{m_{(5)}}\right)^{1+\frac{\alpha}{2}}\sqrt{\frac{M}{m_{(5)}}}\,, \tag{12}$$

since the tidal term q dominates for both M and $m_{(5)}\ll m_p$, and one must still have Eq. (1). With one warped extra dimension [17], the length L is just bounded by requiring that Newton's law not be violated in the tested regions, since corrections to the $1/r$ behavior are of order $(L/r)^2$. This roughly constrains $l_p<L<10^{-3}\,\text{cm}$. Hence the allowed masses are, according to Eq. (1),

$$\left(\frac{m_{(5)}}{m_p}\right)^{\frac{\alpha}{3}}\ll\frac{M}{m_{(5)}}\ll\left(\frac{L}{l_p}\right)^2\left(\frac{m_{(5)}}{m_p}\right)^{2+\alpha}\,. \tag{13}$$

In particular one notices that black holes with $M\sim m_{(5)}\sim m_{ew}$ could exist only if the following two conditions are simultaneously satisfied

$$\alpha\geq 0\quad\text{and}\quad\frac{L}{l_p}\gg\left(\frac{m_p}{m_{ew}}\right)^{\frac{3+\alpha}{3}}\,. \tag{14}$$

The luminosity is now given by, for the limiting case $\alpha=0$ and taking into account the second condition in Eq. (14),

$$\mathscr{L}_{(4)}<10^{-9}\frac{M}{m_{ew}}\frac{\text{TeV}}{\text{s}}\,, \tag{15}$$

which yields an exponential decay with typical life-time $T>10^9\,\text{s}$.

Production of Black Holes by Cosmic Rays

Since the Planck mass in $4+d$ dimensions can be as small as 1 TeV, black holes can be created in cosmic ray interactions with particles in the Earth's atmosphere (or any other source of matter within the galaxy) in the processes

$$p+p \to \mathrm{BH}+X^{++}$$

$$\nu+N \to \mathrm{BH}+X\,.$$

The cross section for the production of a black hole with mass M in such processes is given to a good approximation by $\sigma \simeq \pi R_H^2$, where

$$R_H = \frac{1}{\sqrt{\pi}\, m_{(4+d)}} \left[\frac{M}{m_{(4+d)}} \left(\frac{8\,\Gamma(\frac{d+3}{2})}{d+2} \right) \right]^{\frac{1}{d+1}}, \qquad (16)$$

and we recall that $m_{(4+d)} \sim m_{ew}$. For proton and neutrino energies above $\sim 10^8$ TeV black hole production will dominate over the production of standard model particles [18]. About 100 black holes per year are created for the whole surface of the Earth by the $p+p$ process [3], while the $\nu+N$ process creates one black hole per year [18].

ACCELERATION OF A CHARGED PARTICLE BY A NEUTRAL BLACK HOLE

4-Dimensional Case

To describe the radiation a charged particle being accelerating by a neutral microscopic black hole will produce, Maxwell's equations must be solved on a curved space-time background. The perturbed electromagnetic tensor elements $f_{\mu\nu}$ are determined from the relation

$$(\sqrt{-g}f^{\mu\nu})_{,\nu} = 4\pi\sqrt{-g}\,j^{\mu}\,, \qquad (17)$$

where j^{μ} is the current associated with the falling charge. In 4 dimensions the Schwarzschild metric is spherically symmetric (in this Section the masses are in units of length, i.e. $G_N = 1$),

$$ds^2 = -\left(1+\frac{2M}{r}\right)dt^2 + \left(1+\frac{2M}{r}\right)^{-1} dr^2 + r^2\left(d\theta^2 + \sin^2\theta\, d\phi^2\right)\,, \qquad (18)$$

and, in order to solve the set of equations in Eq. (17) the perturbations $f_{\mu\nu}$ are conveniently expanded in tensor harmonics [19]. Using the field equations all of the tensor elements $f_{\mu\nu}$ can be put in terms of a single element, say f_{12}. After taking the Fourier transform ($\frac{\partial}{\partial t} \to i\omega$) the remaining element satisfies the equation

$$\frac{d^2 f_{lm}}{dr_*^2} + \left\{ \omega^2 - e^\nu \left[\frac{l(l+1)}{r^2} \right] \right\} f_{lm} = 2\sqrt{l+1/2}\, e^{i\omega T}\, e^\nu \left[e^\nu u - \frac{d}{dr}(e^\nu w) \right] \quad (19)$$

where r_* is the standard "turtle" coordinate,

$$f_{lm} = e^\nu f_{12} , \quad (20)$$

$$e^\nu = 1 - \frac{2M}{r}, \quad u = \frac{q e^{-\nu}}{r^2}, \quad w = \frac{q}{l(l+1)} \frac{dT}{dr}, \quad (21)$$

$$T = -4M\sqrt{r/(2M)} - (4/3)M\,[r/(2M)]^{3/2} - 2M \ln\left(\sqrt{r/(2M)} - 1\right)$$
$$+2M \ln\left(\sqrt{r/(2M)} + 1\right) . \quad (22)$$

The charge on the infalling particle is q and T is the time for the particle to fall from ∞ to the point r. The equation for f_{lm} cannot be solved analytically, but values for the amplitude can be obtained numerically using the Green's function $G(r,r')$. The solution of Eq. (19) is given by

$$f_{lm} = \int G(r,r')\, S(r')\, dr' \quad (23)$$

where $S(r)$ is the source term on the right hand side of Eq. (19). The energy distribution averaged over the complete solid angle is

$$\left\langle \frac{dE}{d\omega} \right\rangle = \frac{l(l+1)}{2\pi} f_{l0} f_{l0}^* . \quad (24)$$

This distribution is plotted as a solid line in Fig. 2 for unit charge q and black hole mass M ($R_H = 2M$). The energy of the radiation is

$$\Delta E \simeq \frac{C_{(4)} q^2}{M} , \quad (25)$$

where $C_{(4)}$ is the area under the curve. To produce a 100 TeV photon from a singly charged particle ($q = e$) the black hole would have to have a mass of approximately $10\, m_p$.

4+d-Dimensional Case

For the case of large extra dimensions the invariant line element is taken to be of the form

$$ds^2 = -\left[1 - \left(\frac{R_H^2}{r^2 + \sum y_i^2}\right)^{\frac{d+1}{2}}\right] dt^2 + \left[1 - \left(\frac{R_H^2}{r^2 + \sum y_i^2}\right)^{\frac{d+1}{2}}\right]^{-1} dr^2$$
$$+ r^2\left(d\theta^2 + \sin^2\theta\, d\phi^2\right) + \sum dy_i^2 , \quad (26)$$

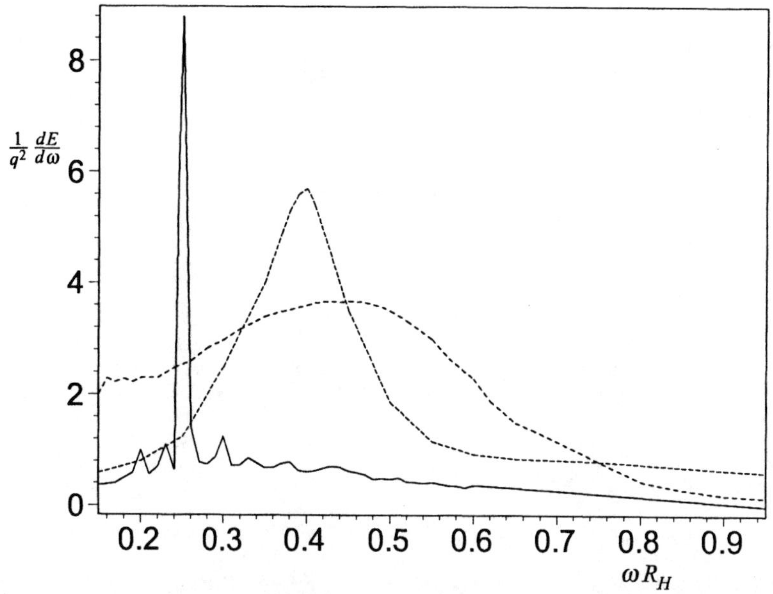

FIGURE 2. Energy distribution for a particle of unit charge falling into a 4-D black hole of mass M (solid line). Same distribution (scaled by 100) for a 5-D black hole (dashed line) and (scaled by 1000) for a 6-D black hole (dotted line). Note that secondary peaks are due to limited numerical precision.

where the y_i's are the coordinates of the d extra dimensions. We assume in this case that the electromagnetic radiation is confined to a 4-dimensional brane embedded in the $4+d$-dimensional space. Thus the y_i's are set to zero in g_{00} and g_{11}, and the harmonic tensor has the same form as in Ref. [19], but with extra rows and columns of zeros for the extra dimensions. The energy distributions for the $d = 1,2$ cases are given in Fig. 2 together with the 4-dimensional case. The corresponding energy expressions are of the same form as for the 4-dimensional case [Eq. (25)]. It is clear from the graphs that heights of the peaks decrease, that the peaks broaden and that the centers of the peaks shift to larger values of the frequency ω with increasing number of extra dimensions. The expressions used for the metric tensor elements for the $d > 0$ cases are approximations which are valid if $R_H \gg L$. The energy radiated in these cases is given by

$$\Delta E \simeq \frac{C_{(D)}q^2}{R_H} \qquad D = 5,6\,, \tag{27}$$

where the $C_{(D)}$'s are twice the areas under the corresponding curves.

The approximation used to obtain the expressions for the 5- and 6-dimensional cases is not satisfactory for $R_H \ll L$. The latter is the more realistic case for microscopic black holes, but there is no known metric which suitably describes this case. Therefore the problem of obtaining black hole parameters from the observed energies of the γ-rays remains to be solved.

ACKNOWLEDGMENTS

This work was supported in part by the U.S. Department of Energy under Grant no. DE-FG02-96ER40967.

REFERENCES

1. N. Arkani-Hamed, S. Dimopoulos and G. Dvali, Phys. Lett. **B 429**, 263 (1998); Phys. Rev. D **59**, 0806004 (1999); I. Antoniadis, N. Arkani-Hamed, S. Dimopoulos and G. Dvali, Phys. Lett. **B 436**, 257 (1998).
2. S. Dimopoulos and G. Landsberg, Phys. Rev. Lett. **87** (2001) 161602.
3. S. B. Giddings, hep-th/0110127
4. A. Borione, et al, Ap. J. **493**, 175 (1998).
5. S.W. Hawking, Nature (London) **248**, 30 (1974); Commun. Math. Phys. **43**, 199 (1975).
6. B. Harms and Y. Leblanc, Phys. Rev. D **46**, 2334 (1992); Phys. Rev. D **47**, 2438 (1993); Ann. Phys. **244**, 262 (1995); Ann. Phys. **244**, 272 (1995); Europhys. Lett. **27**, 557 (1994); Ann. Phys. **242**, 265 (1995); P.H. Cox, B. Harms and Y. Leblanc, Europhys. Lett. **26**, 321 (1994).
7. R. Casadio, B. Harms and Y. Leblanc, Phys. Rev. D **57** 1309 (1998); R. Casadio and B. Harms, Phys. Rev. D **58**, 044014 (1998); Mod. Phys. Lett. **A17**, 1089 (1999).
8. J.M. Bardeen, B. Carter, S.W. Hawking, Commun. Math. Phys. **31**, 161 (1973); G.W. Gibbons and S.W. Hawking, Phys. Rev. D **15**, 2752 (1977).
9. R.C. Myers and M.J. Perry, Ann. Phys. **172**, 304 (1986).
10. S. Cullen, M. Perelstein, Phys. Rev. Lett. **83**, 268 (1999); V. Barger, T. Han, C. Kao and R.J. Zhang, Phys. Lett, **B 461**, 34 (1999); M. Fairbairn, Phys. Lett. **B 508** (2001) 335.
11. R. Casadio and B. Harms, Phys. Lett. **B 487** (2000) 209.
12. R. Casadio and B. Harms, Phys. Rev. D **64** (2001) 024016.
13. R. Emparan, G.T. Horowitz and R.C. Myers, Phys. Rev. Lett. **85** (2000) 499.
14. R. Casadio and B. Harms, hep-th/0110255.
15. N. Dahdhich, R. Maartens, P. Papadopoulos and V. Rezania, Phys. Lett. **B 487**, 1 (2000).
16. R. Casadio, A. Fabbri and L. Mazzacurati, gr-qc/0111072; C. Germani and R. Maartens, Phys. Rev. D **64**, 124010 (2001).
17. L. Randall and R. Sundrum, Phys. Rev. Lett. **83**, 3370 (1999); Phys. Rev. Lett. **83**, 4690 (1999).
18. A. Ringwald and H. Tu, Phys. Lett. **B525**, 135 (2002).
19. F. Zerilli, Phys. Rev. D **9**, 860 (1974); M. Johnston, R. Ruffi ni, F. Zerilli, Phys. Lett. **B49**, 185 (1974).

After Acoustic Peaks: What's Next in CMB?

Asantha Cooray[1]

Theoretical Astrophysics Including Relativity Group, California Institute of Technology, Pasadena, California 91125. E-mail: asante@caltech.edu

Abstract. The advent of high signal-to-noise cosmic microwave background (CMB) anisotropy experiments has allowed detailed studies on the power spectrum of temperature fluctuations. The existence of acoustic oscillations in the anisotropy power spectrum is now established with the detection of the first two, and possibly the third, peaks. Beyond the acoustic peak structure, we consider cosmological and astrophysical information that can be extracted by pushing anisotropy observations to fine angular scales with higher resolution instruments. At small scales, a variety of contributions allow the use of CMB photons as a probe of the large scale structure: we outline possible studies related to understanding detailed physical properties such as the distribution of dark matter, baryons and pressure, and ways to measure the peculiar, transverse and rotational velocities of virialized halos such as galaxy clusters. Beyond the temperature, we consider several useful aspects of the CMB polarization and comment on an ultimate goal for future CMB experiments involving the direct detection of inflationary gravity-waves through their distinct signature in the curl-type polarization.

CMB: AT PRESENT

The cosmic microwave background (CMB) is now a well known probe of the early universe. The temperature fluctuations in the CMB, especially the so-called acoustic peaks in the angular power spectrum of CMB anisotropies, capture the physics of primordial photon-baryon fluid undergoing oscillations in the potential wells of dark matter (Hu et al. 1997). The associated physics — involving the evolution of a single photon-baryon fluid under Compton scattering and gravity — are both simple and linear, and many aspects of it have been discussed in the literature since the early 1970s (Peebles & Yu 1970; Sunyave & Zel'dovich 1970). The gravitational redshift contribution at large angular scales (e.g., Sachs & Wolfe 1968) and the photon-diffusion damping at small angular scales (e.g., Silk 1968) complete this description.

By now, there are at least five independent detections of the first and second, and possibly the third, acoustic peak in the anisotropy power spectrum (Miller et al. 1999; de Bernardis et al. 2000; Hanay et al. 2000; Halverson et al. 2001). We summarize these results in figure 1. Given the variety of experiments that are either collecting data or reducing data that were recently collected, more detections that extend to higher peaks are soon expected. The NASA's MAP mission[2] is expected to provide a significant detection of the acoustic peak structure out to a multipole of $\sim$ 1000 and, in the long

[1] Sherman Fairchild Senior Research Fellow
[2] http://map.gsfc.nasa.gov

CP624, *Cosmology and Elementary Particle Physics*, edited by B. N. Kursunoglu et al.

term, ESA's Planck surveyor[3], will extend this to a multipole of $\sim$ 2000 with better frequency coverage and polarization sensitivity.

A well known geometrical test related to the CMB temperature anisotropy power spectrum involves the location of the first acoustic peak in the multipolar space. The location depicts the projected sound horizon at the recombination and with increasing curvature, the location moves to smaller angular scales or higher l-values (Kamionkowski et al. 1994). The current data, with the first peak at a multipole of $\sim$ 200, strongly suggest a flat-universe (e.g., Jaffe et al. 2001). Additionally, alternative observations, such as the baryon fraction in galaxy clusters and details related to the large scale structure clustering, have provided strong evidence for a low matter content in the universe (about 30% to 40% of the critical density; see Turner 2001 for a summary). Note that this matter content is significantly above the baryon contribution; the latter amounts to be about 5% of the critical density based on calculations related to the big-bang nucleosynthesis and the observed abundance of light elements. The difference is the dark matter problem in astrophysics today.

Though one can effectively *invent* another form of an energy density in the universe which does not clump and does not behave as matter to preserve the geometrical flatness, there is a consistent picture from low redshift observations: the luminosity-distance diagram for type Ia supernovae out to a redshift of $\sim$ 1, and higher, indicates that the universe is accelerating (e.g., Perlmutter et al. 1999). This result implies the presence of an additional energy component in the form of the so-called cosmological constant or, more generally, in the form of a dark energy or a quintessence. This dark energy has a large negative pressure responsible for the anti-gravity behavior. A complete particle physics description of the dark energy still remains to be worked out. The physical nature of the dark energy presents the second important problem in astrophysics today.

In addition to these simple inductions on cosmology and detailed numerical estimates on parameters that define the modern cosmological models, one can also note few basic things. For example, the acoustic oscillations in the CMB anisotropy power spectrum require phases of all Fourier modes to be the same at the beginning. The inflation is expected to provide necessary adiabatic initial conditions while alternative models for structure formation, such as cosmic defects, are now strongly ruled out by the data as they predict, at most, just one broad peak in the anisotropy power spectrum due to no coherence. The shape of the first acoustic peak is also inconsistent with such alternative models and is more consistent with cold dark matter cosmologies favored by many today.

CMB: AS A PROBE OF THE LOCAL UNIVERSE

In transit to us, CMB photons also encounter the large scale structure that defines the local universe; thus, several aspects of photon properties, such as the frequency or the direction of propagation, are affected. In the reionized epoch, variations are also imprinted when photons are scattered via electrons, moving with respect to the CMB.

---

[3] http://astro.estec.esa.nl/Planck/

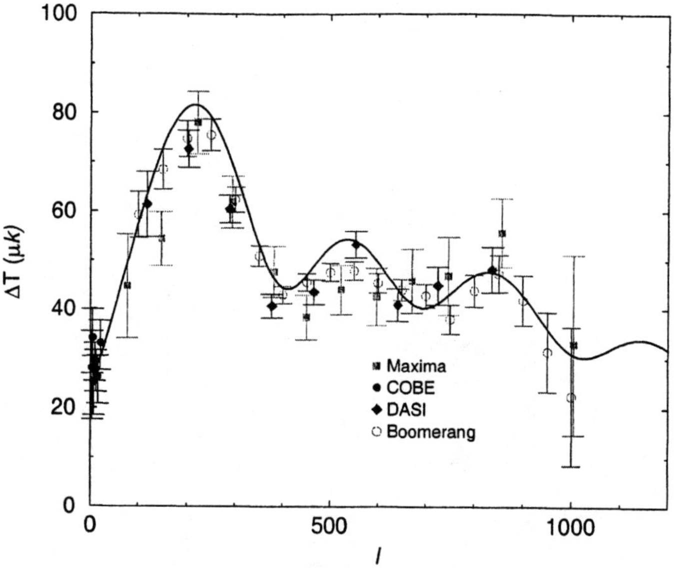

FIGURE 1. Temperature fluctuations in the cosmic microwave background as seen by COBE (filled circles; Tegmark 1996), Boomerang (open circles; Netterfield et al. 2001), MAXIMA (squares; Lee et al. 2001) and DASI (diamonds; Halverson et al. 2001). The solid line shows the theoretical expectation for a cosmology with best-fit parameters for the Boomerang data following Netterfield et al. 2001

Though these secondary contributions to the CMB temperature are also generated by the same two processes that led to primary anisotropies, gravity and Compton-scattering (see, figure 2), there is one significant difference between the simple linear description involving acoustic oscillations and the late-time modifications. In the latter case, one deals with the complex and highly non-linear large scale structure. This involves detailed astrophysics at late times, such as the evolution of gas or pressure, or the formation of structures that define the local universe. Though these secondary effects are in some cases insignificant compared to primary fluctuations, they leave certain imprints in the anisotropy structure and induce higher order correlations. These signatures can then be used to extract basic properties of the large scale structure that led to these additional changes to CMB temperature. Now, we will summarize what these signatures are, and, how they can be used to study the local universe with CMB photons.

Integrated Sachs-Wolfe Effect: The differential redshift effect from photons climbing in and out of a time-varying gravitational potential along the line of sight is called the integrated Sachs-Wolfe (ISW; Sachs & Wolfe 1967) effect. The ISW effect is important for low matter density universes $\Omega_m < 1$, where the gravitational potentials decay at low redshift, and contributes anisotropies on and above the scale of the horizon at the time of decay. The non-linear contribution to the ISW effect comes from the large scale structure momentum density field. In Cooray (2002), we presented a model for the non-linear contribution based on a simple halo based description of the dark matter distribution (see, Cooray & Sheth 2002 for a review). A useful aspect of this non-linear

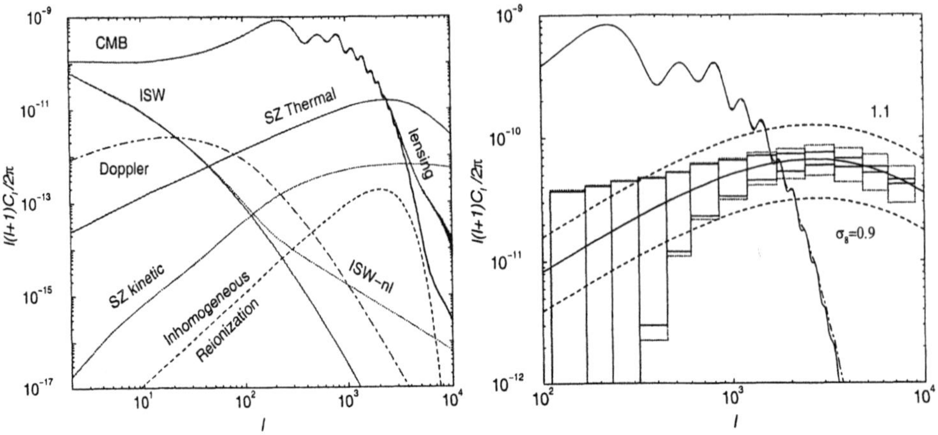

FIGURE 2. *Left*: Power spectrum for the temperature anisotropies in the fiducial ΛCDM model with $\tau = 0.1$. The curves show the local universe contributions to CMB due to gravity (ISW and lensing) and scattering (Doppler, SZ effects, patchy reionization). *Right:* The SZ thermal contribution in focus: the three curves shows the variation in the SZ effect due to a change in normalization about $\sigma_8 = 1.0$. Due to the highly non-linear behavior the SZ thermal contribution is strongly dependent on the normalization. The two sets of error bars again show the highly non-linear and Poisson behavior of the SZ contributing clusters: the small errors bars with sold likes are those under the Gaussian assumption only, while the dotted errors depict the total error with non-Gaussianities included. The non-Gaussianities generally increase the power spectrum errors by factor of few at arcminute scales. We have assumed a survey of 1 deg.^{2} and no instrumental noise contribution to the power spectrum.

effect is that it results from the transverse velocity of halos across the line of sight and leads to a dipolar temperature change aligned with the direction of motion. Though this dipolar temperature distribution towards known galaxy clusters allow this effect to be easily distinguished, a similar temperature pattern is also produced by the weak lensing effect involving the gradient of the cluster potential. Since the velocity direction on the sky need not be that of the large scale CMB gradient, which is lensed, one can effectively use a filtering scheme to separate out and extract the transverse velocity contribution from the lensing signature.

This non-linear transverse velocity contribution, however, produces fluctuations which are of order few μK when transverse velocities are of order 300 km sec^{-1}. One can attempt to reach the required subarcminute resolution required for such an observations by outfitting the focal plane of the new 100m Green Bank Telescope with a bolometer array. Assuming a sensitivity of 200 $\mu\text{K}\sqrt{\text{sec}}$ for each element, a 1000 element bolometer array can detect the effect, say towards the nearby Coma galaxy cluster, in $\sim$ 100 hours; this should be compared to the similar, or more, integration times that are currently taking to produce images of the thermal Sunyaev-Zel'dovich effect in galaxy clusters at the BIMA and OVRO arrays by Carlstrom et al. (1996) with 1-sigma noise contributions of few tens of μK. The ability to measure transverse velocities, which is generally not possible with other astrophysical methods such as spectroscopy, challenges improvements in the experimental front including novel

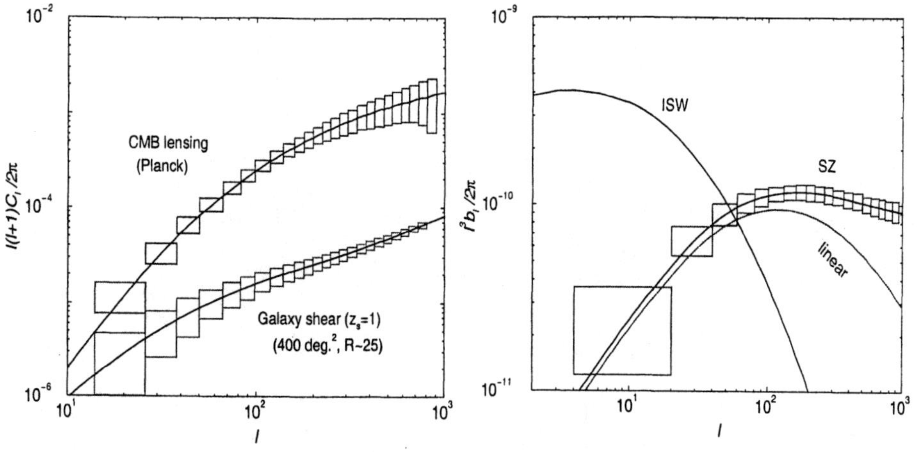

FIGURE 3. *Left*: Weak lensing as a CMB experiment. We show the errors for reconstruction of convergence, or projected mass density, to the last scattering surface with Planck temperature data. This is compared to the convergence measurements expected from observations of galaxy ellipticities and probe to redshifts of $\sim$ 1 to 2. *Right*: The construction of the dark matter-pressure correlation using the cross-correlation between lensing effect on CMB and the SZ effect. The statistic used here is the CMB^2-SZ power spectrum.

observational techniques.

Weak Gravitational Lensing: Gravitational lensing of the photons by the intervening large-scale structure both redistributes power in multipole space and enhances it due to power in the density perturbations. The most effective structures for lensing lie half way between the surface of recombination and the observer in comoving angular diameter distance. In the fiducial ΛCDM cosmology, this is at $z \sim 3.3$, but the growth of structure skews this to somewhat lower redshifts. In general, the efficiency of lensing is described by a broad bell shaped function between the source and the observer, and thus, correlates well with a large number of tracers of the large scale structure from low to high redshifts.

Since the lensing effect involves the angular gradient of CMB photons and leaves the surface brightness unaffected, its signatures are at the second order in temperature. Effectively, lensing smooths the acoustic peak structure at large angular scales and moves photons to small scales. When the CMB gradient at small angular scales are lensed by foreground structures such as galaxy clusters, new anisotropies are generated at arcminute scales. For favorable cosmologies, the mean gradient is of order $\sim$ 15 μK arcmin^{-1} and with deflection angles or order 0.5 arcmin or so from massive clusters, the lensing effect results in temperature fluctuations of order $\sim$ 5 to 10 μK. One can effectively extract this lensing contribution, and the integrated dark matter density field responsible for the lensing effect, via quadratic statistics in the temperature and the polarization. This is best achieved with the divergence of the temperature weighted temperature gradient statistic of Hu (2001) and we show errors for a construction of convergence in figure 3(a). Since lensing has a distinct non-linear mode coupling behavior, it can produce a non-Gaussian signature in CMB data. The idea behind here is that the potentials which lensed CMB also contribute to first order temperature fluctuations from

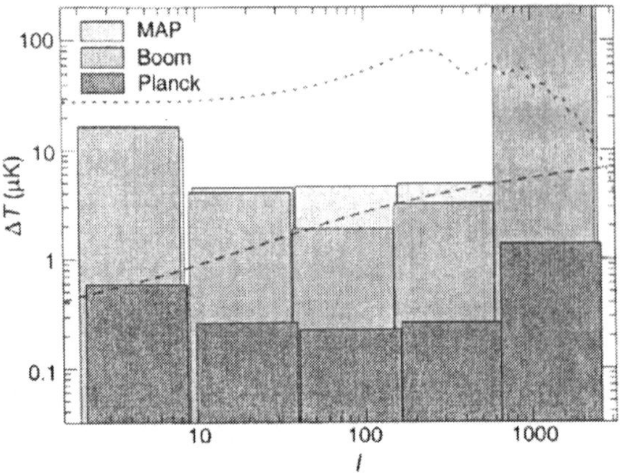

FIGURE 4. Detection thresholds for the SZ effect. Error boxes represent the 1-σ rms residual noise in multipole bands and can be interpreted as the detection threshold for MAP, Planck and Boomerang experiments. Also shown (dotted) is the level of the primary anisotropies that have been subtracted with the technique and the signal (dashed) expected in simplified models for the large scale structure SZ effect.

effects such as the ISW effect. These correlations are discussed in detail in Goldberg & Spergel (1999) and Cooray & Hu (2000).

Sunyaev-Zel'dovich Thermal Effect: At small angular scales the best known secondary contribution is the thermal Sunyaev-Zel'dovich (SZ; Sunyaev & Zel'dovich 1980) effect. The SZ effect arises from the inverse-Compton scattering of CMB photons by hot electrons along the line of sight. This effect has now been directly imaged toward massive galaxy clusters, where temperature of the scattering medium can reach as high as 10 keV producing temperature changes in the CMB of order 1 mK. Note that the effect is proportional to the temperature weighted electron density, or pressure in the large scale structure. Using a distribution of halos with gas in hydrostatic equilibrium, we can calculate the pressure clustering and project it along the line of sight to obtain the resulting angular power spectrum of temperature fluctuations (Cooray 2000; 2001a).

The general result from these halo based calculations, and confirmed in numerical simulations, is that the large scale structure pressure is heavily dominated by massive halos such as galaxy clusters. Thus, the SZ power strongly depends on the normalization of the power spectrum, such as the σ_8. The recent detection of excess power at small scales by the Cosmic Background Imager (CBI) is consistent with a high σ_8 ($\sim$ 1), but there is an additional effect involving the high non-Gaussianity. In figure 2(b), we show error bars on the SZ power spectrum from observations in a one square degree field. The smaller error bars assume Gaussian statistics, however, in reality, the non-Gaussian aspect of the SZ contribution increases errors by a factor of few at arcminute scales. This is equivalent to the statement that the SZ contribution varies significantly depending on whether one sees a massive cluster in the field of observations or not. Since massive clusters are rare, observations which are limited to small areas on the sky are biased. Any future interpretation of the SZ contribution should keep this increase in error, and any bias, in mind. To effectively obtain a fair sample of the universe, one has to survey

100 sqr. degrees or more instead of current surveys limited at most to few tens of square degrees.

In addition to these non-Gaussian aspects, the SZ thermal contribution should also be separated from the dominant primary fluctuations. Note that the SZ effect also bears a spectral signature that differs from other effects. The upscattering of photons moves them from low to high frequencies, with no effect at a frequency of $\sim$ 217 GHz. This leads to decrements at low frequencies and increments at high frequencies. An experiment such as the Planck surveyor with sensitivity beyond the peak of the spectrum can separate out the SZ contribution based on the spectral signature (Cooray et al. 2000). In figure 4, we summarize our results. This frequency separation is important since statistics related to the SZ effect can be studied separately uncontaminated by primary anisotropies and confusing foregrounds. For example, in Cooray (2001b), we introduced the CMB^2-SZ power spectrum as a probe of the lensing-SZ correlation which involves the CMB and frequency cleaned SZ maps. This statistic can be used to directly estimate how pressure correlates with dark matter; in figure 3(b), we show expected errors for the Planck mission.

Linear Doppler Effect: The bulk flow of the electrons that scatter CMB photons leads to a Doppler effect. Its effect on the power spectrum peaks around the horizon at the scattering event projected on the sky today (see figure 2). On scales smaller than the horizon at scattering, the contributions are significantly canceled as photons scatter against the crests and troughs of the perturbation. As a result, the Doppler effect is moderately sensitive to how rapidly the universe reionizes since contributions from a sharp surface of reionization do not cancel.

Kinetic Sunyaev-Zel'dovich Effect: The Ostriker-Vishniac/kinetic SZ (kSZ) effect arises from the second-order modulation of the Doppler effect by density fluctuations (Ostriker & Vishniac 1986). Due to the density weighting, kSZ effect peaks at small scales and avoids cancellations associated with the linear effect; the linear effect can also be modulated via the fraction of ionized electrons (Aghanim et al. 1996). Due to the density modulation, the kSZ effect has a distinct non-Gaussian behavior (Cooray 2001a). The bulk velocities projected along the line of sight can be extracted with higher order correlations such as the SZ^2-kSZ^2 statistic of Cooray (2001a). For a given cluster, the temperature fluctuation is proportional to the line of sight peculiar velocity and when the SZ thermal contribution is removed with frequency information, the kinetic SZ effect should dominate. With multifrequency observations of galaxy clusters at high resolution, it should, in principle, be possible to extract the bulk motions of clusters from the relative contributions of the thermal and kinetic SZ effects. In addition to the peculiar motion, any coherent rotation of the cluster will produce a dipole-like temperature distribution toward clusters, especially when the rotational axes is aligned perpendicular to the line of sight (Cooray & Chen 2001). One can again use filtering techniques to extract this dipolar signature due to the gas distribution spanning a small extent when compared to the gradient of the potential that leads to effects such as lensing. The dipolar pattern involved here allows an extremely useful probe of the cluster rotation and the angular momentum of gas distribution within galaxy clusters.

Reionization via Polarization: In addition to temperature anisotropies at smaller scales, there is also a push to observer polarization at medium to larger scales. Though polarization contribution at the recombination peaks at $l \sim 1000$, the large scale polariza-

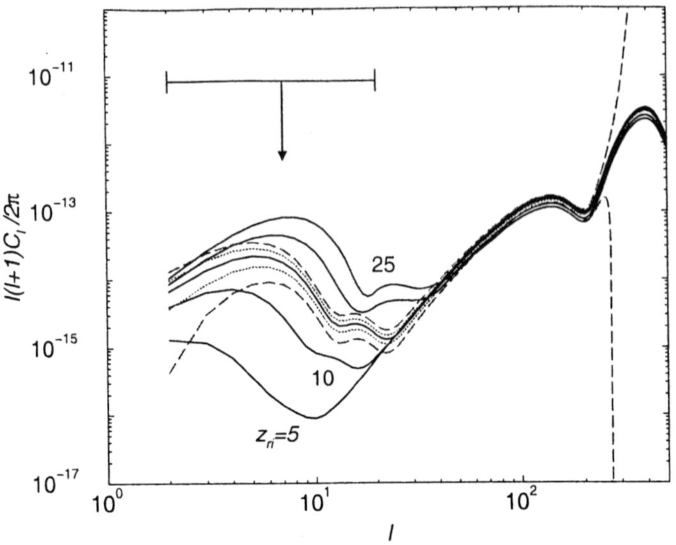

FIGURE 5. The anisotropies in the polarization. We show the power spectrum of fluctuations in the E-mode for reionization from 5 to 25 at steps of 5. The dotted lines are the expected errors for an all sky experiment, while the dashed lines are the expected errors for an experiment with 100 μK $\sqrt{\text{sec}}$ sensitivity and a beam of 2.5 degrees and 25% sky coverage. Such a dedicated experiment is possible in the near future from sites such as South Pole. For comparison, we also show a current upper limit on the polarization signal from Keating et al. (2001).

tion signal again allows a useful probe of the local universe and associated astrophysics. When the universe reionizes at low redshifts, currently believed to be some where between redshifts of 6 and 25, the temperature quadrupole rescatters at the new reionization scattering surface and produces a new contribution to the polarization. This effect leads to a new peak in the polarization anisotropy power spectra at very large angular scales and this peak scales as the square-root of the reionization redshift (Zaldarriaga 1997). In figure 5, we show the large angular scale polarization in the gradient, or the E,-mode. Since the signal is at large scales, one can easily construct a low resolution dedicated experiment for detection of the peak due to rescattering. We show expected errors for an experiment with 25% sky coverage, 100 $\mu\text{K}\sqrt{\text{sec}}$ sensitivity and a beam of 2.5 degrees.

CMB: AS A PROBE OF THE INFLATION

Though acoustic oscillations in the temperature anisotropies of CMB suggest an inflationary origin for primordial perturbations, it has been argued for a while that the smoking-gun signature for inflation would be the detection of a stochastic background of gravitational waves (e.g., Kamionkowski & Kosowsky 1999). These gravitational-waves produce a distinct signature in the CMB in the form of a contribution to the curl, or magnetic-like, component of the polarization (Kamionkowski et al. 1997; Sel-

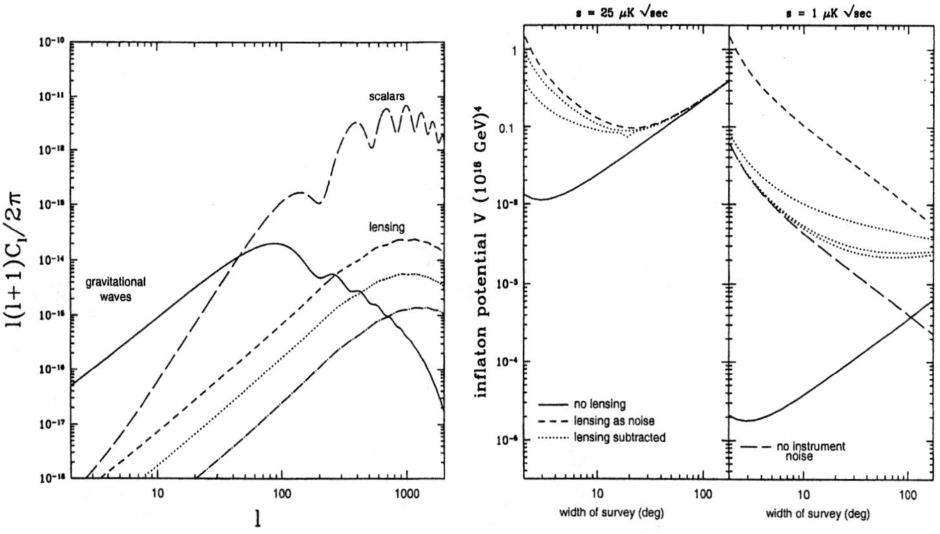

FIGURE 6. *Left:* Contributions to the CMB polarization power spectra. The long-dashed curve shows the dominant polarization signal in the gradient component due to scalar perturbations. The solid line shows the curl polarization signal from the gravitational-wave background assuming inflationary energy scale of 2.3×10^{16} GeV. The dashed curve shows the power spectrum of the curl component of the polarization due to lensing. The dotted curve is the lensing contribution to the curl component that comes from structures out to a redshift of 1. The dot-dashed line is the residual when lensing contribution is separated with a no-noise experiment and almost full sky coverage. *Right:* Minimum inflation potential observable at 1σ as a function of survey width. The left panel shows an experiment with $= 25\,\mu$K $\sqrt{\text{sec}}$ sensitivity. The solid curve shows results assuming no lensing while the dashed curve shows results including the effects of an unsubtracted lensing; we take $\theta_{\text{FWHM}} = 5'$ in these two cases. The dotted curves assume lensing is subtracted with $\theta_{\text{FWHM}} = 10'$ (upper curve) and $5'$ (lower curve). The right panel shows results assuming $1\,\mu$K $\sqrt{\text{sec}}$ sensitivity and $\theta_{\text{FWHM}} = 5', 2'$, and $1'$ (from top to bottom). The solid curve assumes $\theta_{\text{FWHM}} = 1'$ and $s = 1\,\mu$K $\sqrt{\text{sec}}$, and no lensing, while the dashed curve treats lensing as an additional noise. The long-dash curve assumes lensing subtraction with no instrumental noise.

jak & Zaldarriaga 1997). Note that there is no contribution from the dominant scalar, or density-perturbation, contribution to these curl modes. In figure 6, we show the contribution from dominant scalar modes to the polarization in the gradient component and from gravitational-waves to the curl polarization. There is, however, another source of contribution to the curl component resulting from the fractional conversion of the gradient polarization via the weak gravitational lensing effect (Zaldarriaga & Seljak 1998). The lensing-induced curl introduces a noise from which gravitational waves must be distinguished in the CMB polarization. In figure 6, we summarize these contributions following Kesden et al. (2002).

As we discussed earlier, when considering the effect of lensing on temperature anisotropies, one can use the CMB data, both temperature and polarization, to extract lensing information. This is desirable since replacing the lensing effect on CMB with what one extract from galaxy shape measurements only leads to a partial accounting of the total effect (the difference in curves in figure 3). By mapping lensing deflection from CMB data with higher order correlations, the lensing effect on polarization can be corrected to reconstruct the intrinsic CMB polarization at the surface of last scatter. Now the only curl component would be that due to gravity waves. In figure 6, we show how well the lensing signal can be extracted. With no noise all-sky map, one can remove the lensing contribution with a noise contribution that is roughly an order of magnitude lower than the original confusion. As shown, if there is no instrumental-noise limitation, the sensitivity to gravity-wave signal is maximized by covering as much sky as possible and allow the accessibility to an inflaton potential of $\sim 10^{15}$ GeV.

Acknowledgments: The author thanks Wayne Hu and Marc Kamionkowski for collaborative work and helpful advice, and acknowledges support from the Sherman Fairchild foundation and the Department of Energy.

REFERENCES

1. Aghanim, N., Desert, F. X., Puget, J.L., Gispert, R. 1996, A&A, 311, 1.
2. Carlstrom, J. E., Joy, M., & Grego, L. 1996, 461, L59.
3. Cooray, A., & Hu, W. 2000, ApJ, 534, 533.
4. Cooray, A., Hu, W., Tegmark, M. 2000, ApJ, 544, 1;
5. Cooray, A. 2002, Phys. Rev. D., in press (astro-ph/0109162).
6. Cooray, A. & Sheth, R. 2002, Physics Reports, submitted.
7. Cooray, A. 2000, Phys. Rev. D., 62, 103506.
8. Cooray, A. 2001a, Phys. Rev. D., 64, 063514.
9. Cooray, A. 2001b, Phys. Rev. D., 64, 043516.
10. Cooray, A., Chen, X. 2001, PRD submitted (astro-ph/0107544).
11. de Bernardis, P. et al., 2000, Nature, 404, 955.
12. Goldberg, D. M. & Spergel, D. N. 1999, Phys. Rev. D., 59, 103002.
13. Halverson, N. W. et al., 2001, preprint, astro-ph/0104489.
14. Hanany, S. et al., 2000, ApJ, 545, L5.
15. Hu, W. ApJ, 557, L79.
16. Hu W., Sugiyama N., Silk J. 1997, Nature 386, 37.
17. Jaffe, A. H. et al. 2001, PRL, 86, 3475.
18. Kamionkowski, M., Spergel, D. N., & Sugiyama, N. 1994, ApJ, 434, L1.
19. Kamionkowski, M. & Kosowsky, A. 1999, Ann. Rev. Nucl. Part. Sci., 49, 77.
20. Kamionkowski, M., Kosowsky, A. & Stebbins, A. 1997, PRL, 78, 2058.
21. Keating, B. et al. 2001, ApJ, 560, L1.
22. Kesden, M., Cooray, A., Kamionkowski, M. 2002, PRL submitted, astro-ph/0202434.
23. Miller, A. D. et al., 1999, ApJ, 524, L1.
24. Netterfield, C. B. et al. 2001, ApJ in press, astro-ph/0104460.
25. Ostriker, J.P., & Vishniac, E.T. 1986, Nature, 322, 804.
26. Peebles, P.J.E. and Yu, J. T. 1970, ApJ, 162. 815.
27. Perlmutter, S. et al. 1999, ApJ, 517, 565.
28. Sachs, R. K., & Wolfe, A. M., 1967, ApJ, 147, 73.
29. Seljak, U. & Zaldarriaga, M. 1997, PRL, 78, 2054.
30. Silk, J. 1968, ApJ, 151, 459.
31. Sunyaev, R.A. & Zel'dovich, Ya. B. 1980, MNRAS, 190, 413.
32. Tegmark, M. 1996, ApJ, 464, 35.
33. Turner, M. S. 2001, PASP, 113, 653.
34. Zaldarriaga, M. 1997, PRD, 55, 1822.
35. Zaldarriaga, M. & Seljak, U. 1998, PRD, 58, 023003.

Is Cosmic Acceleration Really Recent?

Philip D. Mannheim

Department of Physics, University of Connecticut, Storrs, CT 06269
mannheim@uconnvm.uconn.edu

Abstract. In the standard cosmological paradigm cosmic acceleration is to only be a very recent (viz. $z \leq 1$) phenomenon, with the universe being required to be decelerating at all higher redshifts. We suggest that this particular expectation of the standard model is to be viewed as a quite definitive test not only of the model itself but also of the fine-tuning assumption on which the expectation is based, with the expectation itself actually being readily amenable to testing once the Hubble plot can be extended out to only $z = 2$ or so. Moreover, such a modest extension of the Hubble plot will also provide for definitive testing of the non fine-tuned alternate conformal gravity theory, a theory in which the universe is to accelerate both above and below $z = 1$.

THE HUBBLE PLOT OF STANDARD COSMOLOGY

In a standard pure matter or pure radiation Friedmann cosmology the attractive nature of gravity entails the existence of an initial big bang singularity followed by a subsequent decelerating expansion. Primary evidence in general favor of such a picture is obtained from observational study of three widely separated epochs, viz. early universe nucleosynthesis, the recombination era cosmic microwave background, and the current $z \leq 1$ era (d_L, z) Hubble plot. While it had long been thought that a decelerating expansion was to occur in every epoch, data accumulated only recently [1, 2, 3, 4] now reveal the presence of an unanticipated additional repulsive component to cosmological gravity, a component most commonly attributed to the presence of a non-vanishing cosmological constant Λ, a component whose contribution to cosmic evolution is only found to start to become of consequence at around $z = 1$ or so where its presence is then central to the elucidation of the $z \leq 1$ Hubble plot data.[1] Specifically, through use of the standard Einstein-Friedmann cosmological evolution equation

$$\dot{R}^2(t) + kc^2 = \dot{R}^2(t)(\Omega_M(t) + \Omega_\Lambda(t)) \tag{1}$$

[where $\Omega_M(t) = 8\pi G\rho_M(t)/3c^2H^2(t)$ is due to ordinary $\rho_M(t) \sim 1/R^n(t)$ matter and where $\Omega_\Lambda(t) = 8\pi G\Lambda/3cH^2(t)$ is due to a cosmological constant $c\Lambda$] phenomenological data fitting has been found to yield current era values of 0.3 for $\Omega_M(t_0)$ and 0.7 for $\Omega_\Lambda(t_0)$, to thus entail that in the current era the deceleration parameter $q_0 = q(t_0) = (n/2-1)\Omega_M(t_0) - \Omega_\Lambda(t_0)$ has to take the negative value of $-1/2$, with the current era universe thus not being a decelerating one after all.

---

[1] While the cosmic microwave background data [4] are certainly compatible with the presence of a non-vanishing Λ, in and of themselves alone they can just as readily support a universe in which Λ is absent.

CP624, *Cosmology and Elementary Particle Physics*, edited by B. N. Kursunoglu et al.

While the identifying of such specific values for $\Omega_M(t_0)$, $\Omega_\Lambda(t_0)$ and $\Omega_k(t_0) = -kc^2/\dot{R}^2(t_0) = 1 - \Omega_M(t_0) - \Omega_\Lambda(t_0)$ has enabled standard cosmology to ostensibly achieve its primary purpose, namely that of determining the matter, vacuum and curvature content of the universe, and while the obtained values even provide support for the flat $\Omega_k(t) = 0$ inflationary universe paradigm [5], the particular values obtained for these parameters are nonetheless extremely perplexing. Specifically, a priori estimates for $c\Lambda \equiv \sigma T_V^4$ would suggest for T_V a value of either a typical particle physics temperature scale of order 10^{16} degrees or so or a quantum-gravitational Planck temperature scale of order 10^{33} degrees, to thereby yield an $\Omega_\Lambda(t_0)/\Omega_M(t_0)$ ratio of order 10^{60} to 10^{120}, a ratio not only overwhelmingly larger than the requisite measured value of order one, but one which would (for an $\Omega_M(t_0)$ of order one) entail that $\Omega_k(t_0)$ would have to be of order -10^{60} and thus be nowhere near flat at all.

To get round this problem the standard paradigm thus proposes that rather than use such a fundamental physics based T_V one should instead, and despite the absence of any currently known justification, fine-tune T_V down by orders and orders of magnitude so that the value $\Omega_\Lambda(t_0) = 0.7$ would then ensue. Beyond the difficulty inherent in trying to understand how this might actually be dynamically achieved, even a successful resolution of this issue would still not actually leave cosmology totally free of fine-tuning problems, since in its turn, having a current era $\Omega_\Lambda(t_0)$ of order one creates a yet further problem for the standard model, one of then having to have an expressly fine-tuned early universe. Specifically, for an $\Omega_\Lambda(t_0)$ of order one the early universe associated with Eq. (1) would need to be one in which $\Omega_M(t = t_{PL})$ would have had to have been incredibly close to one at the Planck time $t = t_{PL}$, while $\Omega_\Lambda(t = t_{PL})$ itself would have had to have been as small as $O(10^{-120})$. The early universe thus has only a one in 10^{120} chance of ever evolving into our current universe unless some explicit dynamical mechanism could be found which would naturally fix these needed initial conditions with incredible precision. In a sense this fine-tuning problem is a new variant of the venerable flatness problem. As we recall, in the pre $\Omega_\Lambda(t_0) \neq 0$ days it was very difficult to understand why an $\Omega_M(t)$ which had been redshifting for more than 10 billion years would be anywhere near one today rather than being orders and orders of magnitude smaller, with it being inflation which then provided a natural answer to this problem. Specifically, it was shown by Guth [5] that if there were to be a period of rapid de Sitter inflation (viz. rapid acceleration) prior to the onset of the current Robertson-Walker (RW) era, then such an inflationary era would precisely lead at its end to a set of initial conditions for an ensuing RW phase in which $\Omega_M(t)$ would not merely be close to one but would in fact be identically equal to one in each and every epoch and thus not susceptible to redshifting at all. However, with the advent of a non-zero cosmological constant inflation now only fixes the sum of $\Omega_M(t)$ and $\Omega_\Lambda(t)$ to be equal to one in all epochs but does not constrain their ratio. Currently then, standard cosmology stands waiting for the development of some sort of generalized version of early universe inflation which would naturally lead to initial RW era conditions which then would naturally fix the initial values of both $\Omega_M(t)$ and $\Omega_\Lambda(t)$ to the requisite precision. This then is the challenge to the standard cosmology posed by the new $z \leq 1$ Hubble plot data.

As regards actually fitting these Hubble plot data, we note that when viewed purely as a phenomenology (i.e. without regard to any of the above fine-tuning concerns) an $\Omega_M(t_0) = 0.3$, $\Omega_\Lambda(t_0) = 0.7$ standard model then performs extraordinarily well. Through use of type Ia supernovae as standard candles the authors of [1] and [2] were able to extend the Hubble plot of luminosity versus redshift out to redshifts close to one. To illustrate the quality of the fits which then ensue we follow [2] and fit 38 of their reported 42 data points together with 16 of the 18 earlier lower z points of [6], for a total of 54 data points with reported effective blue apparent magnitude m_i and uncertainty σ_i. (While we thus, following [2], leave out 6 questionable data points for the fitting, nonetheless, for completeness we still include them in the displayed Fig. (1).) For the fitting we calculate the apparent magnitude m of each supernova at redshift z via $m = 25 + M + 5log_{10}d_L$ (the luminosity distance d_L being in Mpc) where M is their assumed common absolute magnitude, and find for $\Omega_M(t_0) = 0.3$, $\Omega_\Lambda(t_0) = 0.7$ and $M = -19.37$ that $\chi^2 = \sum(m - m_i)^2/\sigma_i^2$ takes the value 57.74, with the fit itself being displayed as the lower curve in Fig. (1). [2] As the fitting shows, once one allows for the gravitational repulsion associated with a non-vanishing $\Omega_\Lambda(t_0)$ the standard model can nicely account for the supernovae data.

With the values of $\Omega_M(t_0) = 0.3$, $\Omega_\Lambda(t_0) = 0.7$ thus being established by the $z \leq 1$ Hubble plot data, we now note that since the matter density $\rho_M(t)$ redshifts while Λ of course does not, as we go to higher and higher redshift the $\Omega_M(t)/\Omega_\Lambda(t)$ ratio will get bigger and bigger, with the attractive matter density numerically being found to overcome the repulsive Λ contribution to the deceleration parameter at a redshift of only $z = 0.67$. In the standard model then the universe would be such that it would decelerate ($q(t) > 0$) continually in all epochs until the matter density contribution finally manages to redshift itself down to the cosmological constant contribution, something which is to occur at the incredibly late $z = 0.67$ when $q(t)$ would at long last finally change sign. Indeed, the particular timing of this change in sign is itself a reflection of the standard model early universe fine-tuning problem we discussed earlier, with initial conditions having to be such that this change over would occur precisely in our own epoch, neither earlier than it nor later. Now while it is very peculiar that such a turn around is to occur just in our own particular epoch, nonetheless, independent of one's views regarding the merits or otherwise of such an expectation, the prediction itself is actually readily amenable to testing, with just a modest increase in the range of z (say to $z = 2$ or so) in the Hubble plot being able to reveal its possible presence. Moreover, such a study would be a completely kinematic one, one totally independent of dynamical assumptions (such as those for instance required for the extraction of cosmological parameters from the cosmic microwave background) and would thus be completely clear cut. In this sense then study of the $z > 1$ Hubble plot can provide a completely dynamics independent test of whether or not Λ really is as small as the standard model's assumed fine-tuning would require. With the $z > 1$ Hubble plot thus being the "smoking gun" for a fine-tuned Λ, we thus exhibit in Fig. (2) the standard model expectation (the lowest curve) out to $z = 5$. In and of itself then it would be extremely informative to extend the range of the Hubble plot. However, as we now show, it would be of additional interest since it would allow for a rather unequivocal comparison between standard cosmology and the recently

[2] The very fact that the data can be fitted so well with a common M at all (and even with one of value typical of nearby supernovae) strongly suggests that type Ia supernovae are indeed good standard candles.

proposed alternate conformal cosmology, a theory which is capable of fitting the very same supernovae data without any fine-tuning at all.

THE HUBBLE PLOT OF CONFORMAL COSMOLOGY

Given both the fine-tuning problems of the standard cosmology and the absence to date of any solution to them, it is thus of value to entertain and explore possible candidate alternate cosmologies to see if any one of them might shed some light on the issue. Now while the choice of possible alternate theories is quite vast (pure metric based theories of gravity require only a general coordinate scalar action, of which there is an infinite number containing derivatives of the Riemann tensor out to arbitrarily high order), one particular such alternative is explicitly singled out. Specifically, since it possesses a symmetry which when unbroken obliges the cosmological constant to vanish identically [7], conformal gravity (viz. gravity based on the fully covariant, locally conformal invariant Weyl action

$$I_W = -\alpha_g \int d^4x(-g)^{1/2} C_{\lambda\mu\nu\kappa} C^{\lambda\mu\nu\kappa} \tag{2}$$

where $C^{\lambda\mu\nu\kappa}$ is the conformal Weyl tensor and where α_g is a purely dimensionless gravitational coupling constant) is immediately suggested and motivated. The cosmology associated with the conformal gravity theory was first presented in [8] where it was shown to both possess no flatness problem (to thereby release conformal cosmology from the need for the copious amounts of cosmological dark matter required of the standard model) and to have an effective cosmological Newton constant, G_{eff}, which actually turned out to be negative. Thus long in advance of the recent supernovae data it had been noted that conformal cosmology possessed a repulsive gravitational component.[3] Subsequently [9, 10], the cosmology was shown to also possess no horizon problem or universe age problem. And finally, it was shown [11, 12] that even after the conformal symmetry is spontaneously broken by a Λ inducing scale breaking cosmological phase transition, the theory continues to be able to keep the contribution of the induced cosmological constant to cosmic evolution under control even in the event that Λ is in fact as big as particle physics suggests, to thereby provide a completely natural solution to the cosmological constant problem without the need for any fine tuning at all. In the present paper we use the results of [11, 12] to show that conformal gravity not only controls the cosmological constant in principle, in practice it even provides for a completely acceptable accounting of the recent supernovae Hubble plot data as well.

[3] In fact, equally in advance of the supernovae data, it had also been noted [9] that in a $\Lambda = 0$ conformal cosmology the current era q_0 would then be identically equal to zero, with a $\Lambda = 0$ conformal cosmology thus possessing a repulsion not present in a $\Lambda = 0$ standard cosmology where $q_0 = 1/2$.

Analysis of the implications of conformal cosmology is greatly facilitated by considering the generic conformal matter action

$$I_M = -\hbar \int d^4x(-g)^{1/2}[S^\mu S_\mu/2 - S^2 R^\mu{}_\mu/12 + \lambda S^4 + i\bar{\psi}\gamma^\mu(x)(\partial_\mu + \Gamma_\mu(x))\psi - gS\bar{\psi}\psi] \tag{3}$$

for massless fermions and a conformally coupled order parameter scalar field. When the scalar field breaks the conformal symmetry by acquiring a non-zero expectation value S_0, the energy-momentum tensor associated with the matter action of Eq. (3) is found (for a perfect matter fluid $T^{\mu\nu}_{kin}$ of the fermions) to take the form [12]

$$T^{\mu\nu} = T^{\mu\nu}_{kin} - \hbar S_0^2(R^{\mu\nu} - g^{\mu\nu}R^\alpha{}_\alpha/2)/6 - g^{\mu\nu}\hbar\lambda S_0^4 \ , \tag{4}$$

with the complete solution to the scalar, fermionic and gravitational field equations of motion in a background RW geometry (viz. a geometry in which $C^{\lambda\mu\nu\kappa} = 0$) then reducing [12] to the remarkably simple equation $T^{\mu\nu} = 0$, i.e. reducing to

$$\hbar S_0^2(R^{\mu\nu} - g^{\mu\nu}R^\alpha{}_\alpha/2)/6 = T^{\mu\nu}_{kin} - g^{\mu\nu}\hbar\lambda S_0^4 \ , \tag{5}$$

with the vanishing of $T^{\mu\nu}$ immediately fixing the zero of energy. As we thus see, the evolution equation of conformal cosmology looks identical to that of standard gravity save only that the quantity $-\hbar S_0^2/12$ has replaced the familiar $c^3/16\pi G$, so that instead of being attractive the effective cosmological $G_{eff} = -3c^3/4\pi\hbar S_0^2$ is actually negative, and instead of being fixed as the standard low energy Newtonian G, the cosmological G_{eff} is instead fixed by the altogether different scale S_0, a scale which when large enough would yield an effective cosmological G_{eff} which would then be altogether smaller than the standard Cavendish G.[4]

Given the equation of motion of Eq. (5), the ensuing conformal cosmology evolution equation is then found (on setting $\Lambda = \hbar\lambda S_0^4$) to take a form remarkably similar to that of Eq. (1), viz.

$$\dot{R}^2(t) + kc^2 = \dot{R}^2(t)(\bar{\Omega}_M(t) + \bar{\Omega}_\Lambda(t)) \tag{6}$$

where $\bar{\Omega}_M(t) = 8\pi G_{eff}\rho_M(t)/3c^2H^2(t)$, $\bar{\Omega}_\Lambda(t) = 8\pi G_{eff}\Lambda/3cH^2(t)$. Further, unlike the situation in the standard theory where preferred values for the relevant evolution parameters (such as the magnitude and even the sign of Λ) are only determined by the data fitting itself, in conformal gravity essentially everything is already a priori known. With conformal gravity not needing dark matter to account for galactic rotation curve systematics [14], $\rho_M(t_0)$ can be determined directly from luminous matter alone, with galaxy luminosity counts giving a value for it of order $0.01 \times 3c^2H_0^2/8\pi G$ or so. Further, with $c\Lambda$ being generated by an energy density lowering particle physics vacuum breaking phase transition in an otherwise scaleless theory, $c\Lambda$ (and thus the

---

[4] In fact, with the non-relativistic terrestrial and solar system conformal gravity expectations being controlled [13] by a local G whose dynamical generation is totally decoupled [11, 12] from that of the cosmological G_{eff}, in conformal gravity cosmology is thus completely freed from the need to be controlled by the Cavendish G.

$\hbar\lambda S_0^4$ term which simulates it) must unambiguously be negative, with it thus being typically given by $-\sigma T_V^4$ where T_V is a necessarily particle physics sized scale. Then with G_{eff} also being negative, the quantity $\bar{\Omega}_\Lambda(t)$ itself must thus be positive, just as needed to give cosmic acceleration ($q(t) = (n/2-1)\bar{\Omega}_M(t) - \bar{\Omega}_\Lambda(t)$). Similarly, the sign of the spatial 3-curvature k is known from theory [11] to be negative, something which has been independently confirmed from a phenomenological study of galactic rotation curves [14]. Moreover, since G_{eff} is negative, the cosmology is singularity free and thus expands from a (negative curvature supported) finite maximum temperature T_{max}, a temperature which is necessarily greater [10, 11] (and potentially even much greater [12]) than T_V. And finally, with G_{eff} being negative, the quantity $\bar{\Omega}_M(t)$ must be negative for ordinary $\rho_M(t) > 0$ matter, with $q(t)$ thus being negative in all epochs.[5] Consequently in the conformal theory we never need to fine tune in order to make any particular epoch such as our own be an accelerating one, with repulsive cosmological gravity thus being completely natural to conformal gravity in each and every epoch.

Given only that Λ, k and G_{eff} are in fact all negative in the conformal theory, the evolution of the theory is then completely determined, with the expansion rate parameters being found [10, 11, 12] to be given by

$$R^2 = -k(\beta-1)/2\alpha - k\beta\sinh^2(\alpha^{1/2}ct)/\alpha \ , \ T_{max}^2/T^2 = 1+2\beta\sinh^2(\alpha^{1/2}ct)/(\beta-1) \ , \tag{7}$$

where $\beta = (1-16A\lambda/k^2\hbar c)^{1/2} = (1+T_V^4/T_{max}^4)/(1-T_V^4/T_{max}^4)$, $\alpha c^2 = -2\lambda S_0^2 = 8\pi G_{eff}\Lambda/3c$. In terms of the parameters T_{max} and T_V we thus obtain

$$\tanh^2(\alpha^{1/2}ct) = (1-T^2/T_{max}^2)/(T_{max}^2T^2/T_V^4+1) \ ,$$
$$H(t) = \alpha^{1/2}c(1-T^2/T_{max}^2)/\tanh(\alpha^{1/2}ct) \ ,$$
$$\bar{\Omega}_\Lambda(t) = (1-T^2/T_{max}^2)^{-1}(1+T^2T_{max}^2/T_V^4)^{-1}, \ \bar{\Omega}_M(t) = -(T^4/T_V^4)\bar{\Omega}_\Lambda(t) \tag{8}$$

at any $T(t)$ without any approximation at all. From Eq. (8) we now see that simply because T_{max} is overwhelmingly larger than the current temperature $T(t_0)$, i.e. simply because the universe is as old as it is, it automatically follows, without any fine-tuning at all, that the current era $\bar{\Omega}_\Lambda(t_0)$ has to lie somewhere between zero and one today no matter how big (or small) T_V might actually be, with conformal gravity thus having total control over the contribution of the cosmological constant to cosmic evolution. Conformal gravity thus solves the cosmological constant problem by quenching $\bar{\Omega}_\Lambda(t_0)$ rather than by quenching Λ itself (essentially by having a G_{eff} which is altogether smaller than the standard G), and with it being the quantity $\bar{\Omega}_\Lambda(t_0)$ which is the one which is actually measured in cosmology, it is only its quenching which is actually needed. With conformal gravity thus being able to naturally accommodate a large Λ we are now actually free to allow T_V to be as large as particle physics suggests. Then, for such a large $T_V/T(t_0)$ we see that the quantity $\bar{\Omega}_M(t_0)$ has to be completely negligible today[6] so that q_0 must thus, without any fine-tuning at all, necessarily lie between zero

---

[5] Included in this class of $q(t) < 0$ universes are the coasting ones in which $q(t) = 0^-$.

[6] $\bar{\Omega}_M(t_0)$ is suppressed by G_{eff} being small, and not by $\rho_M(t_0)$ itself being small, with G_{eff} being made smaller the larger rather than the smaller S_0 gets to be, to thus enable the $c\Lambda/\rho_M(t_0) = \bar{\Omega}_\Lambda(t_0)/\bar{\Omega}_M(t_0)$

and minus one today notwithstanding that T_V is huge. The essence of the conformal gravity approach then is not to change the matter and energy content of the universe at all, but rather only to change their effect on cosmic evolution, with Λ itself no longer needing to be quenched.

In order to fit the Hubble plot data we need to determine the dependence of d_L on z in the conformal theory, something we can readily do now that we have obtained the explicit form of the expansion factor $R(t)$. Thus, for temperatures well below T_{max} and for the naturally achievable [12] $T_V \ll T_{max}$ case of most practical interest to conformal gravity (viz. a case where $T^2_{max}T^2(t_0)/T^4_V$ can be of order one) we may set

$$R(t) = (-k/\alpha)^{1/2}\sinh(\alpha^{1/2}ct) \ , \tag{9}$$

so that

$$-q_0 = \tanh^2(\alpha^{1/2}ct_0) = \alpha c^2/H_0^2 \ , \ t_0 = \mathrm{arctanh}[(-q_0)^{1/2}]/(-q_0)^{1/2}H_0 \ . \tag{10}$$

For geodesics $\int_{t_1}^{t_0} cdt/R(t) = \int_0^{r_1} dr/[1-kr^2]^{1/2}$ we thus obtain

$$(-k)^{1/2}r_1 = \coth(\alpha^{1/2}ct_0)/\sinh(\alpha^{1/2}ct_1) - \coth(\alpha^{1/2}ct_1)/\sinh(\alpha^{1/2}ct_0) \ . \tag{11}$$

Then, on noting that $\sinh(\alpha^{1/2}ct_1) = (-q_0)^{1/2}/(1+q_0)^{1/2}(1+z)$ where $z = R(t_0)/R(t_1) - 1$ and where q_0 is the current value of $q(t)$, we find that we can express the general luminosity distance $d_L = r_1R(t_0)(1+z)$ entirely in terms of the current era H_0 and q_0 according to the very compact relation [16]

$$H_0d_L/c = -(1+z)^2\left\{1-[1+q_0-q_0/(1+z)^2]^{1/2}\right\}/q_0 \ . \tag{12}$$

Conformal gravity fits to the luminosity distance can thus be parametrized via the one parameter q_0, a parameter which must lie somewhere between zero and minus one, with d_L thus having to lie somewhere between $d_L(q_0 = 0) = cH_0^{-1}(z + z^2/2)$ and $d_L(q_0 = -1) = cH_0^{-1}(z+z^2)$ at temperatures well below T_{max}.

Having obtained Eq. (12) we can now turn to a data analysis. On fitting the same 54 supernovae data points as previously, our best fit is obtained for $q_0 = -0.37$, $M = -19.37$ with $\chi^2 = 58.62$. We display this fit as the upper curve in Fig. (1), and as we thus see, in the detected region the best fits of the standard and conformal models are completely indistinguishable, only in fact departing from each other at the very highest available redshifts. For comparison purposes we find that for $q_0 = 0$ a best fit value of $\chi^2 = 61.49$ is obtained with $M = -19.29$,[7] with fits for other typical values of q_0 being reported in [16].[8] Beyond the purely phenomenological fact that the conformal gravity

---

ratio to be as large as particle physics suggests while not leading to any 60 order of magnitude conflict with observation.

[7] Such high quality fitting with $q_0 = 0$ has also been noted by other authors [2, 15] though not within the context of conformal gravity.

[8] If one also takes the Δz_i errors in the reported redshifts into consideration, the conformal gravity chi squared values for the 54 points get reduced [16] to $\chi^2(q_0 = -0.33) = 54.13$ and $\chi^2(q_0 = 0) = 56.0$, while the standard model chi squared becomes $\chi^2(\Omega_M(t_0) = 0.3, \Omega_\Lambda(t_0) = 0.7) = 53.27$.

fits actually provide a good accounting of the supernovae data at all, it is important to stress that as such these fits are the first ones ever obtained in which the cosmological constant is allowed to take a large unquenched particle physics scale value, with the fits thus establishing the empirical fact that it is in fact possible to fit the supernovae data without fine-tuning.

With the standard cosmology requiring deceleration above $z = 1$ and with the conformal cosmology continuing to accelerate, extension of the Hubble plot beyond $z = 1$ will actually enable us to discriminate between the two cosmologies. We thus augment Fig. (2) by adding in the $z > 1$ conformal gravity predictions. The highest curve in Fig. (2) is the conformal gravity prediction for $q_0 = -0.37$, while the middle curve is that for $q_0 = 0$. As we see, these two typical conformal gravity curves start to depart from the standard model expectation fairly rapidly once $z > 1$, with the three curves in Fig. (2) respectively corresponding to apparent magnitudes $m = 27.17$, $m = 27.04$ and $m = 26.75$ at $z = 2$, and to $m = 30.40$, $m = 30.25$ and $m = 29.14$ at $z = 5$. A quite modest extension of the Hubble plot will thus readily enable us to discriminate between standard gravity and its conformal alternative while potentially even being definitive for both.

OUTLOOK AND CHALLENGES

While there has yet to be detailed exploration of the $z > 1$ Hubble plot using supernovae standard candles, it is of some interest to note that recently a first $z > 1$ data point was actually established [17], viz. the supernova SN 1997ff which was found to be at a redshift $z = 1.7^{+0.1}_{-0.15}$. To illustrate the data of [17] we have augmented Fig. (2) by adding in the 68% and 95% confidence region values for the measured apparent magnitude m at redshifts $z = 1.65$, $z = 1.7$ and $z = 1.75$. (In the figure the two inner horizontal bars on the vertical data points represent the extent of the 68% confidence region at each of the chosen redshifts while the two outer bars represent the 95% confidence one.) While one should not read too much into a single data point,[9] it is interesting to note that the data can accommodate both the standard and conformal theories, with it being necessary to acquire a whole set of $z > 1$ data points in order to identify any specific trend in the data that there might be, with it as yet being too early to ascertain from available supernovae data whether the $z > 1$ universe is decelerating or accelerating.

Beyond the standard candle supernovae, it has also been noted by Daly [18] that the very powerful FRII bridge radio galaxies can serve as standard yardsticks, and can thus also be used to extract cosmological parameters. As such, this technique serves to complement the supernovae analyses, and even to potentially go beyond them since radio galaxies are already being seen out to $z = 2$ or so. Interestingly, the data presented in [18] are so far found to be able to accommodate both standard and conformal gravity, with further study of this issue thus having the potential to be quite instructive.

As we noted earlier, that apart from the $z \sim 1$ region Hubble plot, cosmology can also

[9] Indeed, a shortcoming of these particular data is that SN 1997ff just happens to be gravitationally lensed by two foreground galaxies [17], with its "true" apparent magnitude thus likely to be somewhat larger (viz. dimmer) than indicated in the figure.

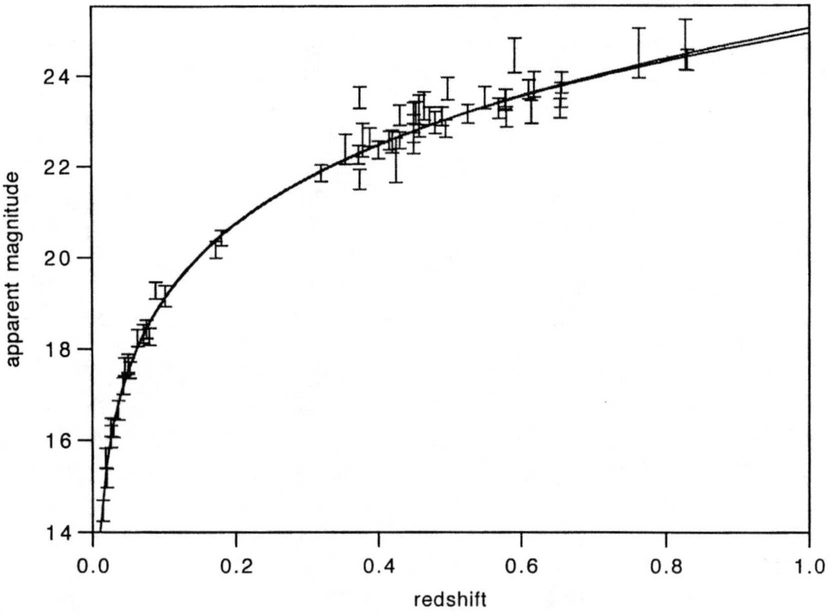

FIGURE 1. The $q_0 = -0.37$ conformal gravity fit (upper curve) and the $\Omega_M(t_0) = 0.3$, $\Omega_\Lambda(t_0) = 0.7$ standard model fit (lower curve) to the $z < 1$ supernovae Hubble plot data.

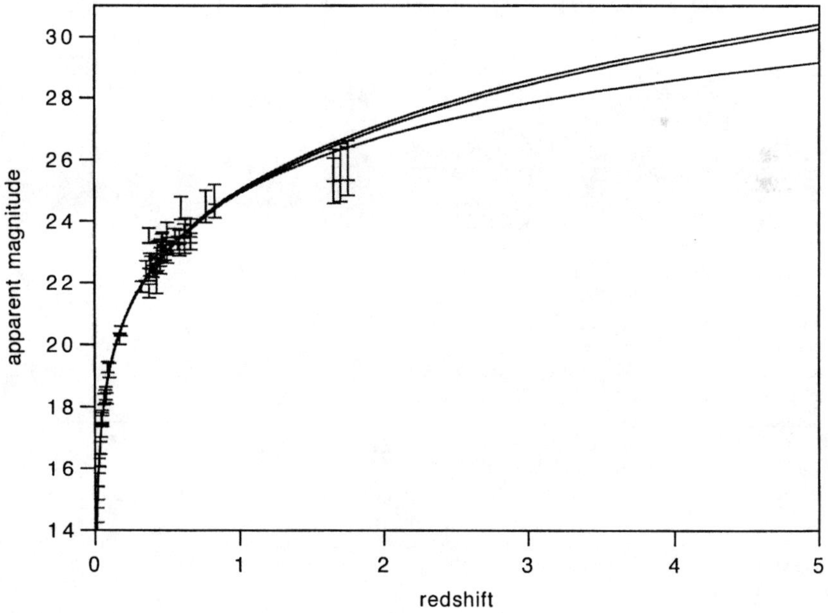

FIGURE 2. Hubble plot expectations for $q_0 = -0.37$ (highest curve) and $q_0 = 0$ (middle curve) conformal gravity and for $\Omega_M(t_0) = 0.3$, $\Omega_\Lambda(t_0) = 0.7$ standard gravity (lowest curve).

be tested at a variety of much larger redshifts, and it is thus urgent to test conformal gravity at those redshifts as well. While its predictions for the microwave background await the development of a conformal cosmology galaxy fluctuation theory, its initial predictions for nucleosynthesis have already been worked out [19]. What was found was that while the expanding and thus cooling conformal cosmology can readily generate the requisite amounts of primordial helium and lithium, because the cosmology expands altogether slower than the standard cosmology its predictions for deuterium and for ^9Be are substantially different from those of the standard model. Specifically, because of the slowness of the expansion very little primordially generated deuterium manages to survive, but because of this same slowness the cosmology is able to get passed the $A = 8$ nuclear fusion bottleneck (viz. the absence of any stable nuclei with 8 nucleons) and thus primordially produce ^9Be and elements heavier than it, in fact even producing ^9Be with its measured abundance, an abundance 8 orders of magnitude greater than that generatable in the standard theory. Now it was noted in [19] that it would be relatively easy to produce deuterium by spallation once inhomogeneities begin to develop in the universe (i.e. post-primordial but pre-galactic). With an explicit theory for such inhomogeneous deuterium production yet to be developed, conformal gravity thus remains challenged by the deuterium problem just as the standard theory remains challenged by the ^9Be problem. However, since conformal gravity so capably handles the most vexing problem facing the standard theory, viz. the cosmological constant problem, it would thus appear to merit further consideration.

The author is indebted to Drs. R. A. Daly, G. V. Dunne, K. Horne, D. Lohiya, R. Plaga and B. E. Schaefer for helpful comments. This work has been supported in part by the Department of Energy under grant No. DE-FG02-92ER40716.00.

REFERENCES

1. Riess A. G. et. al., Astronom. J. **116**, 1009 (1998).
2. Perlmutter S. et. al., Astrophys. J. **517**, 565 (1999).
3. Bahcall N. A., Ostriker J. P., Perlmutter S., and Steinhardt P. J., Science **284**, 1481 (1999).
4. de Bernardis P. et. al., Nature **404**, 955 (2000).
5. Guth A. H., Phys. Rev. D **23**, 347 (1981).
6. Hamuy M. et. al., Astronom. J. **112**, 2391 (1996).
7. Mannheim P. D., Gen. Relativ. Gravit. **22**, 289 (1990).
8. Mannheim P. D., Astrophys. J. **391**, 429 (1992).
9. Mannheim P. D., Conformal cosmology and the age of the universe, astro-ph/9601071.
10. Mannheim P. D., Phys. Rev. D **58**, 103511 (1998).
11. Mannheim P. D., Found. Phys. **30**, 709 (2000).
12. Mannheim P. D., Astrophys. J. **561**, 1 (2001).
13. Mannheim P. D., and Kazanas D., Gen. Relativ. Gravit. **26**, 337 (1994).
14. Mannheim P. D., Astrophys. J. **479**, 659 (1997).
15. Dev A., Sethi M., and Lohiya D., "Linear coasting in cosmology and SNe Ia", astro-ph/0008193 (2000).
16. Mannheim P. D., How recent is cosmic acceleration?, astro-ph/0104022 (2001).
17. Riess A. G. et. al., Astrophys. J. **560**, 49 (2001).
18. Daly R. A., these proceedings.
19. Lohiya D., Batra A., Mahajan S., and Mukherjee A., "Nucleosynthesis in a simmering universe", nucl-th/9902022 (1999); Sethi M., Batra A., and Lohiya D., Phys. Rev. D **60**, 108301 (1999).

III. ELEMENTARY PARTICLE PROCESSES

Lorentz-Violating Scattering Amplitudes

Don Colladay

New College of Florida, Sarasota, FL

Abstract. In this talk, I will discuss the effects of Lorentz-violating terms on high-energy scattering processes. The modified Feynman rules and flux factors will be given and an explicit calculation is presented for electron positron annihilation into two photons. Qualitatively new effects include siderial time variations of the cross section in the reference frame of the collider.

INTRODUCTION

The possibility of small violations of Lorentz invariance arising from more fundamental theories of nature has been of recent interest [1, 2]. An extension of the standard model that includes general terms arising from a generic spontaneous symmetry breaking mechanism has been constructed [3]. There is a deep connection between Lorentz invariance and CPT symmetry from the well-known CPT theorem as well as the more recent result that CPT violation in fact requires a violation of Lorentz invariance [4]. Therefore, tests of CPT symmetry can also be used as tests for Lorentz invariance.

So far, there has been much theoretical and experimental activity regarding the standard model extension and Lorentz symmetry issues [5], but little has been said about the high-energy behavior. This talk will address this issue in the context of high-energy scattering processes. The modified Feynman rules will be obtained and the cross section for $e^-e^+ \to 2\gamma$ will be calculated. The resulting cross section is found to contain an overall constant shift when averaged over time. Corrections are also present that vary either as once or twice the period of revolution of the earth. A more complete analysis relating to this talk can be found in [6].

LORENTZ-VIOLATING QED

Here we restrict attention to the QED sector of the full standard model extension for illustrative purposes. The QED extension is obtained by restricting the full standard model extension to the electron and photon sectors. Imposing gauge invariance and restricting to power-counting renormalizable terms in the standard model extension yields a lagrangian of

$$\mathcal{L} = \tfrac{1}{2} i \overline{\psi} \Gamma^\nu \overleftrightarrow{D_\nu} \psi - \overline{\psi} M \psi + \mathcal{L}_{\text{photon}} \quad , \tag{1}$$

where Γ and M denote

$$\Gamma^\nu = \gamma^\nu + c^{\mu\nu}\gamma_\mu + d^{\mu\nu}\gamma_5\gamma_\mu \quad , \tag{2}$$

CP624, *Cosmology and Elementary Particle Physics*, edited by B. N. Kursunoglu et al.

$$M = m + a_\mu \gamma^\mu + b_\mu \gamma_5 \gamma^\mu + \tfrac{1}{2} H^{\mu\nu} \sigma_{\mu\nu} \quad . \tag{3}$$

The parameters a, b, c, d, and H are related to fixed background expectation values of tensor fields. In this work we neglect possible Lorentz-violating contributions to the photon sector since these terms are already known to be stringently bounded using cosmological birefringence tests [7].

An immediate difficulty arises in the analysis of the above lagrangian since there are extra time derivative couplings arising from the Γ^0 term. This means that the field ψ does not satisfy a conventional Schrödinger time evolution with a hermitian hamiltonian. To circumvent this problem, a spinorial field redefinition is used to eliminate the extra time derivatives [8]. The relevant redefinition is

$$\psi = A\chi \quad , \tag{4}$$

where the matrix A satisfies

$$\overline{A}\Gamma^0 A = \gamma^0 \quad ; \quad (\overline{A} = \gamma^0 A^\dagger \gamma^0)) \quad , \tag{5}$$

therefore conjugating away the time derivatives. Such a transformation is always possible provided the observer is in a concordant frame, a frame in which the violation parameters are small [9]. The redefined field χ therefore obeys the standard Schrödinger time evolution equation, $i\partial_0 \chi = H\chi$, with a modified hamiltonian given by

$$H = \gamma^0 \overline{A}(-i\Gamma_j D^j + M)A \quad . \tag{6}$$

FEYNMAN RULES WITH LORENTZ VIOLATION

The general structure of the computation of cross sections in the presence of Lorentz violation parallels the conventional approach. Development of the perturbative solution to a general scattering problem leads to a set of modified Feynman rules. The general method of the application of the rules is similar to the conventional case, but with several important modifications.

First, translational invariance of the theory implies that p^μ is conserved at all vertices in the diagrams as in the usual case. However, care must be taken to use the modified dispersion relation satisfied by $p^0(\vec{p})$. This modification has an effect on the kinematics of the scattering process and can alter conventional particle trajectories. For example, we will see that for electron-positron annihilation in the center of momentum frame, the incoming group velocities are not in fact equal and opposite while the outgoing photons go off back to back (when lorentz-violating photon terms are neglected as is discussed below).

Second, the spinor solutions to the modified Dirac equation must be included on each external leg of the diagrams to diagonalize the asymptotic states. This fact can also be deduced through application of the standard LSZ reduction procedure for the fermions.

Third, the fermion propagator used on all internal lines in any diagram takes the form

$$S_F(p) = \frac{1}{\gamma^0 E - \overline{A}\vec{\Gamma}A \cdot \vec{p} - \overline{A}MA} \quad . \tag{7}$$

These rules can be used to generate the relevant S-matrix elements for any given cross section in the QED extension yielding the transition probability per unit volume per unit time. This probability is dependent upon the normalization of the incident beams and must be divided by a flux factor F to account for the properties of the initial state and yield a physical cross section.

Consideration of two colliding beams (not necessarily collinear) motivates the following definition

$$F = N_1 N_2 |\vec{v}_1 - \vec{v}_2| \quad , \tag{8}$$

in terms of the beam densities, N_1 and N_2, and the magnitude of the beam velocity difference.

In the absence of Lorentz violation, this factor may be expressed in the Lorentz-covariant form

$$F = 4[(p_1 \cdot p_2)^2 - m^4]^{\frac{1}{2}} \quad , \tag{9}$$

valid in any reference frame. In the Lorentz-violating case, a complication is encountered since the form of the field redefinition $\psi = A\chi$ depends on the reference frame used to describe the scattering process. This means that the asymptotic physical states will appear to different observers with fundamentally different properties. For example, the effective mass of an electron with a c coupling term (shown in the next section) is $\tilde{m} \equiv m(1 - c_{00})$ which depends on observer reference frame through the frame-dependent parameter c_{00}.

To circumvent difficulties associated with the field redefinition we have found it to be most convenient to perform the entire calculation of any given cross section in a single observer reference frame. This implies that the calculation of the flux factor should be performed in the same frame that the S-matrix element is computed in. The velocities of the beams are calculated using the expression for the group velocity of a wave packet

$$\vec{v}_g \equiv \vec{\nabla}_p E(\vec{p}) \quad . \tag{10}$$

Note that the modified dispersion relation may in general cause E to depend on the direction of $\vec{p}$ yielding a velocity that is not in fact parallel to the momentum.

RELATIVISTIC ELECTRON-POSITRON PHYSICS

To gain more insight into the specifics of the general procedure outlined in the previous sections, attention will now be restricted to ultrarelativistic electron and positron physics. The relevant energy scale here is taken to be that of a high-energy collider, still much lower than the scale where causality or stability of the low-energy effective theory break down [9].

The full QED lagrangian in Eq.(1) can be simplified in the relativistic limit as many of the terms are in fact negligible. For instance, the derivative couplings c and d will dominate over the nonderivative a, b, and H couplings at high energies and momenta. Moreover, unpolarized beams are present, so the effects of the d terms will average out in the sum over right- and left-handed particles. In short, only c will contribute to ultrarelativistic, unpolarized electron and positron scattering experiments.

The lagrangian of Eq.(1) can therefore be taken as

$$\mathrm{L} = \tfrac{1}{2} i(\eta_{\mu\nu} + c_{\mu\nu})\overline{\psi}\gamma^\mu \overleftrightarrow{D^\nu} \psi - m\overline{\psi}\psi \quad , \tag{11}$$

in the above limit. The field redefinition in Eq.(4) used for removal of the time derivatives takes the specific form (to lowest order in c)

$$\psi \equiv A\chi = (1 - \tfrac{1}{2} c_{\mu 0}\gamma^0\gamma^\mu)\chi \quad . \tag{12}$$

The lagrangian expressed in terms of the redefined field χ is

$$\mathrm{L} = \tfrac{1}{2} i(\eta_{\mu\nu} + \mathrm{C}_{\mu\nu})\overline{\chi}\gamma^\mu \overleftrightarrow{D^\nu} \chi - \tilde{m}\overline{\chi}\chi \tag{13}$$

with the definitions

$$\begin{aligned} \tilde{m} &\equiv m(1 - c_{00}) \quad , \\ \mathrm{C}_{\mu\nu} &\equiv c_{\mu\nu} - c_{\mu 0}\eta_{0\nu} + c_{\nu 0}\eta_{0\mu} - c_{00}\eta_{\mu\nu} \quad , \end{aligned} \tag{14}$$

or, in matrix form,

$$\mathrm{C} = \begin{pmatrix} 0 & c_{01} + c_{10} & c_{20} + c_{02} & c_{30} + c_{03} \\ 0 & c_{11} + c_{00} & c_{12} & c_{13} \\ 0 & c_{21} & c_{22} + c_{00} & c_{23} \\ 0 & c_{31} & c_{32} & c_{33} + c_{00} \end{pmatrix} \quad . \tag{15}$$

Note that all elements in the first column of the matrix are zero showing explicitly the removal of the time derivative couplings. The effective electron mass $\tilde{m}$ also depends on observer reference frame through the parameter c_{00}.

The next step is to quantize the field χ and define the single-particle asymptotic states for the theory. The approach here follows a general construction previously presented in the literature [9]. First, the relativistic quantum mechanics is constructed by solving for the dispersion relation and the spinor components. Following this, quantization conditions are imposed on the solutions, a procedure yielding a positive definite hamiltonian.

The modified Dirac equation for χ can be solved exactly using plane-wave solutions for particles and antiparticles as

$$\chi(x) = e^{-ip_\mu x^\mu} u(\vec{p}) \quad ; \quad \chi(x) = e^{ip_\mu x^\mu} v(\vec{p}) \quad . \tag{16}$$

Focusing on the particle momentum-space spinor $u(\vec{p})$, it is found to satisfy

$$[(\eta_{\mu\nu} - \mathrm{C}_{\mu\nu})\gamma^\mu p^\nu - \tilde{m}]u(\vec{p}) = 0 \quad . \tag{17}$$

A nontrivial solution implies that the dispersion relation is

$$(p^\mu + \mathrm{C}^\mu{}_\nu p^\nu)(p_\mu + \mathrm{C}_{\mu\alpha} p^\alpha) - \tilde{m}^2 = 0 \quad , \tag{18}$$

which yields (to lowest order in C)

$$E(\vec{p}) \approx \sqrt{\vec{p}^2 + \tilde{m}^2} - \frac{p^j \mathrm{C}_{jk} p^k}{\sqrt{\vec{p}^2 + \tilde{m}^2}} - \mathrm{C}^0{}_j p^j \quad . \tag{19}$$

The same development applies to the antiparticle spinors $v(\vec{p})$ and yields the same dispersion relation as is expected from the CPT symmetry of the c terms. The energy is therefore degenerate for particles and antiparticles. It can also be seen from Eq.(19) that the energy depends explicitly on the direction of $\vec{p}$ allowing the group velocity to have a different direction than the momentum.

The free-field theory is constructed by expanding $\chi(x)$ in terms of fourier components and promoting the amplitudes to operators as in the conventional case:

$$\chi(x) = \int \frac{d^3\vec{p}}{(2\pi)^3 N(\vec{p})} \sum_{\alpha=1}^{2} \left[b_{(\alpha)}(\vec{p}) e^{-ip\cdot x} u^{(\alpha)}(\vec{p}) + d^{\dagger}_{(\alpha)}(\vec{p}) e^{ip\cdot x} v^{(\alpha)}(\vec{p}) \right], \tag{20}$$

where the spinors are normalized to

$$\begin{aligned} u^{(\alpha)\dagger}(\vec{p}) u^{(\alpha')}(\vec{p}) = \delta^{\alpha\alpha'} N(\vec{p}) \quad &, \quad v^{(\alpha)\dagger}(\vec{p}) v^{(\alpha')}(\vec{p}) = \delta^{\alpha\alpha'} N(\vec{p}), \\ u^{(\alpha)\dagger}(\vec{p}) v^{(\alpha')}(-\vec{p}) = 0 \quad &, \quad v^{(\alpha)\dagger}(-\vec{p}) u^{(\alpha')}(\vec{p}) = 0. \end{aligned} \tag{21}$$

Quantization is implemented by imposing

$$\begin{aligned} \{b_{(\alpha)}(\vec{p}), b^{\dagger}_{(\alpha')}(\vec{p}\,')\} &= (2\pi)^3 N(\vec{p}) \delta_{\alpha\alpha'} \delta^3(\vec{p} - \vec{p}\,'), \\ \{d_{(\alpha)}(\vec{p}), d^{\dagger}_{(\alpha')}(\vec{p}\,')\} &= (2\pi)^3 N(\vec{p}) \delta_{\alpha\alpha'} \delta^3(\vec{p} - \vec{p}\,'), \end{aligned} \tag{22}$$

on the field mode operators. Translational invariance implies that there exists a conserved energy-momentum tensor explicitly given by

$$\Theta^{\mu\nu} = \frac{i}{2} \tilde{\eta}_{\alpha}{}^{\mu} \overline{\chi} \gamma^{\alpha} \overset{\leftrightarrow}{\partial^{\nu}} \chi. \tag{23}$$

The corresponding conserved four-momentum is diagonal in the mode operators:

$$\begin{aligned} P^{\mu} &= \int d^3\vec{x} : \Theta^{0\mu} : \\ &= \int \frac{d^3\vec{p}}{(2\pi)^3 N(\vec{p})} p^{\mu} \sum_{\alpha=1}^{2} \left[b^{\dagger}_{(\alpha)}(\vec{p}) b_{(\alpha)}(\vec{p}) + d^{\dagger}_{(\alpha)}(\vec{p}) d_{(\alpha)}(\vec{p}) \right]. \end{aligned} \tag{24}$$

Note that this would not have been the case if the field redefinition of Eq.(4) had not been implemented.

The single particle states can be defined as in the conventional manner using the raising mode operators acting on the vacuum state. The resulting normalization of these states is $\langle p', \alpha' | p, \alpha \rangle = (2\pi)^3 N(\vec{p}) \delta_{\alpha\alpha'} \delta^3(\vec{p}\,' - \vec{p})$. It follows that the number density of an incident plane wave is $N(\vec{p})$ particles per unit volume.

EXPLICIT EXAMPLE OF A CROSS SECTION

In this section the theory developed for relativistic electron-positron physics is applied to the explicit process of pair annihilation into two photons. The explicit cross section is obtained and various properties due to the Lorentz-violating effects are discussed.

The relevant modifications to the Feynman rules for this case include modified fermion propagator

$$S_F(p) = \frac{i}{p_\mu(\gamma^\mu + C_\nu{}^\mu\gamma^\nu - \tilde{m})} \quad , \tag{25}$$

and the modified vertex factor of

$$-ie(2\pi)^4\delta^4(\sum p)[\gamma^\mu + C_\nu{}^\mu\gamma^\nu] \quad , \tag{26}$$

arising from the modified lagrangian.

The tree-level diagrams for the process $e^-e^+ \to 2\gamma$ are similar the conventional case, with the modified propagator and vertex factors included. The resulting S-matrix element is

$$S_{fi} \;=\; -ie^2(2\pi)^4\delta^4(k_1+k_2-p_1-p_2)\bar{v}(p_2)\left[\tilde{\not{\varepsilon}}_2\frac{1}{\tilde{\not{p}}_1-\tilde{\not{k}}_1-\tilde{m}}\tilde{\not{\varepsilon}}_1+(1\leftrightarrow 2)\right]u(p_1). \tag{27}$$

In this equation, p_1, p_2 are the electron and positron momenta, and k_1, k_2 are the photon momenta. The spinors u and v solve the modified Dirac equation after the reinterpretation, while ε_1, ε_2 are the two photon polarization vectors. The notation $\tilde{\not{p}} \equiv (\eta_{\mu\nu}+C_{\mu\nu})\gamma^\mu p^\nu$ is used to simplify the expression. In calculating with the above expression, it is important to recall that the electron and positron energies satisfy modified dispersion relations and the spinors are exact solutions to the modified Dirac equation satisfied by χ.

To define the physical cross section, the factor F in Eq.(8) is calculated in the same frame that the above S-matrix element is evaluated in. We choose the center of momentum frame for the evaluation of both of these quantities. Note that the group velocities of the beams are not necessarily equal and opposite in this frame due to the modified dispersion relations as is illustrated in figure 1. The scattering angle is defined using the incoming electron momentum and one outgoing photon momentum as $\cos\theta = \hat{p}\cdot\hat{k}$. The magnitude of the velocity difference between the incident beams in the center of momentum frame in the relativistic limit is $|\vec{v}_1 - \vec{v}_2| \approx 2(1-\hat{p}_j C^{jk}\hat{p}_k)$.

To obtain the cross section, the conventional steps are now followed. The electron and positron spins are averaged and the final state photon polarizations are summed over. For simplicity, the momentum of the electron beam is chosen to point in the 3-direction. The incident flux factor F is divided out and the final state photon phase space is included. As a final step, the azimuthal angle is integrated over for simplification.

The resulting cross section is

$$\frac{d\sigma}{d\cos\theta} = \int_0^{2\pi} d\phi \frac{d\sigma}{d\Omega} = \left(\frac{d\sigma}{d\cos\theta}\right)_{QED}[1+\Delta] \quad , \tag{28}$$

where the first factor represents the conventional QED cross section and the Lorentz-violating correction Δ is given by

$$\Delta = c_{00} + c_{33} - 2\frac{\cos^2\theta}{1+\cos^2\theta}(c_{11}+c_{22}-2c_{33}) \quad . \tag{29}$$

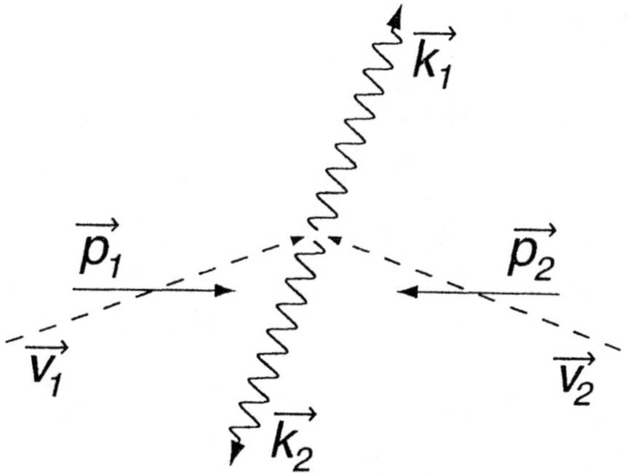

FIGURE 1. Kinematics for the process $e^- e^+ \to 2\gamma$ in the center of momentum frame. Note that the velocities of the incident beams are not in general equal and opposite due to the difference in directions of the group velocity $\vec{v}$ and the momentum $\vec{p}$.

The correction consists of a piece that is a scaling of the cross section and another part with an altered angular dependence. Note that the cross section now explicitly depends on time since the c_{ij} components change as the Earth rotates. This occurs since the 3-direction points along the electron beam momentum that rotate with the earth.

To understand the time dependence, it is useful to transform to a nonrotating basis $(\hat{X}, \hat{Y}, \hat{Z})$ where the $\hat{Z}$-direction is along the axis of the earth and the other two directions are at rest with respect to the background stars [10]. The components of c with respect to this basis are constant so that the time-dependence can be explicitly extracted from the cross section. For example, the term c_{33} expressed in terms of the nonrotating basis is

$$\begin{aligned} c_{33} = \quad & c_{ZZ} + \tfrac{1}{2}(c_{XX} + c_{YY} - 2c_{ZZ}) \sin^2\chi \\ & + \tfrac{1}{2}(c_{YZ} + c_{ZY}) \sin 2\chi \sin \Omega t + \tfrac{1}{2}(c_{XZ} + c_{ZX}) \sin 2\chi \cos \Omega t \\ & + \tfrac{1}{2}(c_{XY} + c_{YX}) \sin^2\chi \sin 2\Omega t + \tfrac{1}{2}(c_{XX} - c_{YY}) \sin^2\chi \cos 2\Omega t. \end{aligned} \tag{30}$$

In the above expression, χ is the angle between $\hat{p}$ and $\hat{Z}$, and Ω is the sidereal rotation frequency of the earth. Note that there are three components to the time dependence: a time-independent factor, a piece that varies with period $T = 2\pi/\Omega$ and another piece with a period of $T = \pi/\Omega$. The conventionally measured cross section gives the time integrated constant piece as the time dependent terms average to zero.

SUMMARY

In this talk, a framework has been discussed for calculation of cross sections and decay rates within the context of the Lorentz-violating standard model extension. The resulting

Feynman rules and kinematical factors have been presented. It was shown that it is most convenient to perform an analysis in a single frame due to complications arising from the frame-dependent field redefinition. The cross section for $e^- e^+ \to 2\gamma$ was calculated explicitly using the framework presented. It was found that the modifications to the conventional cross section contain an overall scaling as well as a sidereal time dependence due to the rotation of the earth. This time dependence is a qualitatively new feature generic to cross sections and decay rates in the Lorentz-violating QED extension.

ACKNOWLEDGMENTS

This work was supported in part by a University of South Florida Division of Sponsored Research grant.

REFERENCES

1. V.A. Kostelecký and S. Samuel, Phys. Rev. D **39**, 683 (1989); *ibid.* **40**, 1886 (1989); Phys. Rev. Lett. **63**, 224 (1989); *ibid.* **66**, 1811 (1991); V.A. Kostelecký and R. Potting, Nucl. Phys. B **359**, 545 (1991); Phys. Lett. B **381**, 89 (1996); Phys. Rev. D **63**, 046007 (2001); V.A. Kostelecký, M. Perry, and R. Potting, Phys. Rev. Lett. **84**, 4541 (2000).
2. S. Carroll, *et al.*, Phys. Rev. Lett. **87**, 141601 (2001).
3. D. Colladay and V.A. Kostelecký, Phys. Rev. D **55**, 6760 (1997); Phys. Rev. D **58**, 116002 (1998).
4. O.W. Greenberg, hep-ph/0201258, (2002).
5. For a summary of recent theoretical models and experimental tests see, for example, *CPT and Lorentz Symmetry*, V.A. Kostelecký, ed., World Scientific, Singapore, 1999; *CPT and Lorentz Symmetry II*, V.A. Kostelecký, ed., World Scientific, Singapore, 2002.
6. D. Colladay and V.A. Kostelecký, Phys. Lett. B **511**, 209 (2001).
7. S.M. Carroll, G.B. Field, and R. Jackiw, Phys. Rev. D **41**, 1231 (1990); V.A. Kostelecký and M. Mewes, preprint IUHET 438, July 2001; M. Mewes, these proceedings.
8. R. Bluhm *et al.*, Phys. Rev. Lett. **79**, 1432 (1997); Phys. Rev. D **57**, 3932 (1998).
9. V.A. Kostelecký and R. Lehnert, Phys. Rev. D **63**, 065008 (2001).
10. V.A. Kostelecký, and C.D. Lane, Phys. Rev. D **60**, 116010 (1999); J. Math. Phys. **40**, 6245 (1999).

Extra Compact Dimensions and Fermion Mass Hierarchy

P. Q. Hung

Department of Physics, University of Virginia, 382 McCormick Road, P. O. Box 400714, Charlottesville, Virginia 22904-4714

Abstract. In this talk, I will present a new pespective on the origin of a phenomenologically very successful model of quark mass hierarchy: the pure phase mass matrix. Two extra compact spatial dimensions are introduced. In addition, a scenario is presented, in which standard fermions "live" inside a thick brane whose thickness is of the size of the compact dimension. Fermions are "localized" inside the thick brane by domain walls. These "locations" along the sixth dimension determine the nature of the pure phase mass matrix: In one instance, it is hermitian and at another instance, it is non hermitian. The latter case requires the breaking of both the symmetry between left-handed and right-handed fermions, as well as the family symmetry.

QUARK MASS HIERARCHY

The problem of the origin of fermion masses and, in particular, its hierarchy is certainly one of the most mysterious puzzles in physics. In confronting this problem, one usually turns one's attention first to the quark sector since this is where experimental "informations" are most "complete". Why does one have $m_u \ll m_c \ll m_t$ and $m_d \ll m_s \ll m_b$? What is the origin of the mixing angles and the CP phase in the CKM matrix? It goes without saying that the answers to these questions certainly lie outside the framework of the Standard Model (SM).

There exists two approaches to this problem. The first one is the construction of models based on one's own prejudice on what the physics beyond the SM might be. The other approach is purely phenomenological and is based on some simple ansätz. It is fair to say that each approach suffers from various shortcomings. For instance, our knowledge of the physics beyond the SM is practically nil at this point: We know there is something out there but we do not know what it is. As a result, it is hard to choose a particular approach among a wide variety of models, each of which is endowed with strong motivations. In addition, in order to be sufficiently "realistic", the construction of these models is generally rather complicated. Nevertheless, some of the lessons learned there can be quite useful. On the other side of the coin, some of the ansäzes used in the phenomenological approach are quite attractive and successful. A drawback of such an approach is the lack of a theoretical foundation. Nevertheless, it has its own merit for it is, in some sense, a form of bottom-up approach. One of such ansäzes is particularly successful from a phenomenological point of view: the so-called pure phase mass matrix [1].

A pure phase mass matrix has the form $\mathcal{M} = g_Y(v/\sqrt{2})\{\exp(i\theta_{ij})\}$, where $i, j =$

CP624, *Cosmology and Elementary Particle Physics*, edited by B. N. Kursunoglu et al.

1,2,3. The attractiveness of this ansätz is the assumption of a *single universal* Yukawa strength g_Y, of a single mass scale v, and of matrix elements with *unit moduli* $\exp(i\theta_{ij})$. The only problem is to find a model for it. Attempts have been made along this line for the *special case* [2] of a symmetric matrix with the results being that the Higgs structure turns out to be extremely complicated. This tends to obscure the physics behind a pure phase mass matrix.

It is my humble opinion that new ways of thinking about this problem are sorely needed. Not only one has to understand the origin of the mass hierarchy, but also the origin of the mixing angles and the CP phase in the CKM matrix. In this talk, I will report on one of such attempt made in collaboration with Marcos Seco [3]. We followed a bottom-up philosophy with, as a starting basis, the adoption of the pure phase mass matrix ansätz. We shall try to explain its origin by building a four-dimensional effective theory out of six dimensions, with two extra compact spatial dimensions.

Why six dimensions? It turns out that by adding two extra compact dimensions, some interesting features concerning chiral fermions emerge such as the appearance of phases in the fermion wave functions along the sixth dimension. It is these special features that will help in the construction of the pure phase mass matrix (and beyond).

CHIRAL FERMIONS IN A BRANE WORLD

In this section, I will describe our bottom-up approach to large compact extra dimensions. We add extra spatial dimensions one by one, compactify them, and see what kind of fermions one obtains in a four-dimensional effective theory. In doing this, one should keep in mind one important requirement: the existence of massless chiral fermions in four dimensions which would describe SM quarks and leptons. This requirement depends on the way the extra spatial dimensions are compactified. It turns out that the proper way to compactify is through the so-called S_1/Z_2 orbifold which also comes under the name of line compactification. To set the stage for an eventual construction of a pure phase mass matrix, I will first start with one extra spatial dimension denoted by y.

Fermions in Five Dimensions

One useful note: The notations used for fermions in five dimensions in this section will be changed later on, i.e. they are to be restricted solely to this section.

The Lorentz group in five dimensions is $SO(4,1)$. Denoting the fifth (spatial) dimension by y, compactification on an orbifold S_1/Z_2 means that the length of the line $L = \pi R$, where R is the radius of the circle and where one has the identification $y \to -y$ (modulo $2\pi R$). One now has what one calls a "thick" brane- a membrane with thickness L along the fifth direction.

The next step is to start with free fermions in five dimensions which are simply 4-component Dirac spinors. An excellent exposition of this exercise can be found in Ref. [4] and reviewed in Ref. [3]. In this talk, I will simply summarize some of the important points.

One of the key points is the Z_2 symmetry. Roughly speaking, under this symmetry, a fermion field ψ transforms like $\psi \to \pm\gamma_5\psi$ while a scalar field transforms as $\phi \to \pm\phi$ (where the dependence on y is omitted for the time being). The free Lagragian looks like

$$\begin{aligned} \mathscr{L} &= \bar{\psi}\left(i\,\partial\!\!\!/ + i\gamma_y\frac{\partial}{\partial y}\right)\psi, \\ &= \bar{\psi}\left(i\,\partial\!\!\!/ - \gamma_5\frac{\partial}{\partial y}\right)\psi. \end{aligned} \tag{1}$$

The "line" compactification restricts the fifth coordinate y to be in the interval $[0,L]$. It can easily be seen that $\mathscr{L}$ is invariant under the Z_2 symmetry: $\psi(x,y) \to \Psi(x,y) = \pm\gamma_5\psi(x,L-y)$. The plus sign refers to the field with *even* Z_2 parity while the minus sign refers to *odd* parity. Upon imposing a modified periodic condition: $\psi(x,y) = \Psi(x,L+y) = \psi(x,2L+y)$, one obtains $\psi(x,-y) = \Psi(x,L-y) = \pm\gamma_5\psi(x,y)$ and $\psi(x,L+y) = \Psi(x,y) = \pm\gamma_5\psi(x,L-y)$. It is straightforward to see that ψ is unchanged under the Z_2 transformation at $y = 0, L$ (the so-called orbifold "fixed points").

The important point in the above exercise is the emergence of massless chiral fermions in four dimensions. The 4-component Dirac fermion ψ can be decomposed into a right-handed fermion, ψ_+, and a left-handed fermion, ψ_-, namely, $\psi = \psi_+ + \psi_-$, where $\gamma_5\psi_\pm = \pm\psi_\pm$. From the above discussion, one can see the following features: (a) for an *even* Z_2 parity field, $\psi \to +\gamma_5\psi$, one has $\psi_+(x,y=0,L) \neq 0$ and $\psi_-(x,y=0,L) = 0$, (b) for an *odd* Z_2 parity field, $\psi \to -\gamma_5\psi$, one has $\psi_+(x,y=0,L) = 0$ and $\psi_-(x,y=0,L) \neq 0$. One can write $\psi_\pm(x,y)$ as $\psi_\pm(x,y) = \psi_\pm(x)\xi_\pm(y)$. Since a zero mode has no y-dependance, it is then clear that the even and odd parity cases imply respectively (a) $\xi_+^0(y) = constant$, $\xi_-^0(y) = 0$; (b) $\xi_-^0(y) = constant$, $\xi_+^0(y) = 0$. In other words, for an even field (case (a)), only a right-handed zero mode survives while for an odd field (case (b)), only a left-handed field survives. The massive states- the so-called Kaluza-Klein modes- have masses $M_n = n\pi/L$ with $\xi_+^n(y) \propto \cos n\pi y/L$ or $\sin n\pi y/L$ depending on whether the fermion is even or odd, with the reverse situation for $\xi_-^n(y)$.

Fermion localization inside a thick brane

As discussed above, one notices that the zero mode chiral fermion is *uniformly* spread over y (being a constant along that direction). For reasons such as the avoidance of fast proton decay, etc.., as expounded in Ref. [5] for example, the massless chiral fermions, which would represent the SM quarks and leptons, should have the possibility to be "localized" inside the thick brane. To see why it is desirable to have that feature, let us couple ψ with a scalar field Φ which has an *odd* Z_2 parity, namely $\Phi(x,y) \to -\Phi(x,L-y)$. Why an odd scalar field? This is because of the following point: When combined with the periodic boundary condition similar to that for ψ, one would obtain $\Phi(x,-y) = -\Phi(x,y)$ and $\Phi(x,L-y) = -\Phi(x,L+y)$ which, following a reasoning similar to the one above, would imply that there is no massless (zero) mode for this scalar field in four dimensions. It merely serves as a background field whose purpose is to provide a mechanism for the localization of the massless chiral fermion inside the thick brane as we shall see below.

The Yukawa coupling between ψ and Φ is written as $f\bar{\psi}\Phi\psi$. Φ has a kink solution $\langle\Phi(x,y)\rangle = h(y) = V\tanh(\mu y)$, where V and $\mu = (\lambda/2)^{1/2}V$ come from $\frac{\lambda}{4}(\Phi^2 - V^2)^2$. From the Dirac equation for ψ, one can obtain an equation for $\xi(y)$. Let us concentrate on an odd parity field so that only a left-handed zero mode survives in four dimensions. It can be seen that the solution for $\xi_-^0(y)$ in the presence of the kink is

$$\xi_-^0(y) = C\,exp(-\int_0^y f h(y')dy'). \tag{2}$$

For $f > 0$, one can easily see that $\xi_-^0(y)$ is localized at $y = 0$. In fact, $\xi_-^0(y)$ is trapped inside a domain wall (little brane) of thickness $((\lambda/2)^{1/2}V)^{-1}$. Inside that domain wall, one can make a simple harmonic oscillator (SHO) approximation, namely $fh(y) \approx 2\tilde{\mu}^2 y$ leading to a properly normalized wave function $\xi_-^0(y) = (\sqrt{\tilde{\mu}}/(\pi/2)^{1/4})\exp(-\tilde{\mu}^2 y^2)$ which clearly shows the localization of the fermion around $y = 0$.

At this moment, it is not clear what the deep mechanism for localizing fermions at different places inside the brane might be. However, one can assume that this mechanism exists. Various possibilities have been pointed out in Refs. ([4], [3]). Here, I will simply assume that it exists and that one can arrange various kinks such that the locations of different fermions, namely y_i, are determined by $fh(y_i) - m_i = 0$. Fermion i would have a wave function $\xi_{-,i}^0(y) \propto \exp(-\tilde{\mu}^2(y - y_i)^2)$. What are the implications of this separation of fermions inside the brane?

In all of this, we would like to build an effective Lagrangian in four dimensions. This means that we will integrate out the extra fifth dimension. In so doing, we will inevitably encounter wave function overlaps of two fermions along y. For example, an effective interaction containing fermions i and j would be proportional to $O_{ij}(x)\int dy\xi_{-,i}^0(y)\xi_{-,j}^0(y)$. The overlap $\int dy\xi_{-,i}^0(y)\xi_{-,j}^0(y)$ would represent an effective interaction strength, which could, in principle, be very small if, for some reasons, there is little overlap between the two fermions. This scenario (wave function overlap) has numerous implications, one of which is the *fermion mass hierarchy* issue presented at this conference.

Another very important assumption is that other SM fields such as gauge and Higgs fields have zero modes which are uniformly spread along y. (From the point of view of the Z_2 parity, these are *even* fields, and hence the existence of the zero modes.)

Before embarking on such an endeavor, a few remarks are in order here concerning the possibility that the mass hierarchy could simply be accomplished by a hierarchy of wave function overlaps along the fifth dimension [5]. It is a very attractive idea but, unfortunately, a "realistic" model with a small number of assumptions has yet to be invented. The question of the origin of the mixing angles and the CP phase of the CKM matrix is problematic in this approach.

PURE PHASE MASS MATRICES FROM 6 DIMENSIONS

Our objective is to understand the origin of the mass matrix $\mathcal{M} = g_Y(v/\sqrt{2})\{\exp(i\theta_{ij})\}$. We have seen the behavior of the wave function of a domain wall fermion in five dimensions: $\xi_{-,i}^0(y) \propto \exp(-\tilde{\mu}^2(y - y_i)^2)$. If we couple the SM left-handed (LH) and right-handed (RH) fermions to the Higgs field, the effective Yukawa coupling in four dimensions will be proportional to the overlap of the LH and RH fermions. This would give g_Y. If one sticks to just five dimensions, the phases factors $\exp(i\theta_{ij})\}$ would have to come from some complicated Higgs structure, as Ref. [2] have shown for the special case of a symmetric matrix. One gains nothing by going to a higher dimension in this case. Would another compact dimension help? More importantly, what is the four-dimensional effective theory obtained from compactifying two extra spatial dimensions? This last question is interesting in its own right, and I will describe it below. However let me give the punchline for this talk. What we obtained was two things: (a) The effective Yukawa strength described above comes from the wave function overlap along the fifth dimension; (b) The phase factors come from the wave function overlap along the sixth dimension.

Fermions in six dimensions

In this section, I will summarize some of the salient features of fermions in six dimensions. A detailed discussion can be found in Ref. [3]. The Lorentz group is $SO(5,1)$. As in [3], the two extra spatial coordinates are labeled as y and z.

In this section and hereon, I will use ψ to denote a spinor in *six dimensions*. As a warning, one should not be confused with a similar notation used in five dimensions.

There are two 4-component irreducible spinors: ψ_+ and ψ_-, which are complex conjugates of one another. Grouping these spinors into a reducible Dirac spinor $\psi = (\psi_+, \psi_-)$, the free Lagrangian can be written as

$$\mathcal{L}_\psi = i\bar{\psi}\Gamma^N \partial_N \psi, \tag{3}$$

where $N = 0, 1, 2, 3, y, z$. The metric used here is simply (-+++++). The most important point here is the part of (3) which contains the derivatives with respect to y and z, namely

$$\mathcal{L}_\psi = -i\bar{\psi_+}\gamma^\mu\partial_\mu\psi_+ - i\bar{\psi_-}\gamma^\mu\partial_\mu\psi_- + \bar{\psi_+}\gamma^5\partial_y\psi_+ + \bar{\psi_-}\gamma^5\partial_y\psi_- - i\bar{\psi_+}\partial_z\psi_+ + i\bar{\psi_-}\partial_z\psi_- . \tag{4}$$

Notice the factor i appearing in Eq. (4) for the derivatives along z. In a nutshell, this is due to the fact that spinors in $SO(5)$ are real while those in $SO(6)$ are complex. This will play an important role in the emergence of the phases. But let us first discuss the symmetry and boundary conditions here. The sixth direction is assumed to be compactified on an S_1/Z_2 orbifold.

The Z_2 symmetry $\psi_\pm(x^\alpha, z) \to \Psi_\pm(x^\alpha, z) = \psi_\mp(x^\alpha, L_6 - z)$, when combined with the modified boundary conditions $\psi_\pm(x^\alpha, z) = \Psi_\pm(x^\alpha, L_6 + z) = \psi_\pm(x^\alpha, 2L_6 + z)$, yields $\psi_\pm(x^\alpha, -z) = \Psi_\pm(x^\alpha, L_6 - z) = \psi_\mp(x^\alpha, z)$ and $\psi_\pm(x^\alpha, L_6 + z) = \Psi_\pm(x^\alpha, z) = \psi_\mp(x^\alpha, L_6 - z)$. One recognizes that $z = 0, L_6$ are the orbifold fixed points.

To see what zero modes survive when we compactify the sixth dimension, it is convenient to rewrite $\psi_\pm$ as $\psi_\pm = \frac{1}{\sqrt{2}}(\chi \pm i\eta)$. It follows that the boundary conditions can now be expressed as $\chi(x^\alpha, -z) = \chi(x^\alpha, z)$, $\eta(x^\alpha, -z) = -\eta(x^\alpha, z)$, $\chi(x^\alpha, L_6 + z) = \chi(x^\alpha, L_6 - z)$, $\eta(x^\alpha, L_6 + z) = -\eta(x^\alpha, L_6 - z)$. From this, one can immediately see that η vanishes at the fixed points. It follows that only χ has a zero mode χ^0 which uniformly spreads along z. Notice that, from a notation point of view, χ^0 is identical to ψ_{5dim} used in the previous section. As in that section, one can write χ^0 in terms of right-handed and left-handed fermions as $\chi^0 = \chi_R + \chi_L$.

What happens when we couple $\psi_\pm$ to a background scalar field?

Just as in the five-dimensional case, one can write down such an interaction, with the requirements of Z_2 invariance and reality of the Yukawa interactions. In Ref. [3], two scalar fields were invoked, both being odd fields. They are denoted by Φ and Φ', with Φ having *no* zero mode in five dimensions while Φ' has no zero mode in four dimensions. The Yukawa interactions which satisfy these requirements are

$$\mathscr{L}_{Yuk} = \mathscr{L}_\Phi + \mathscr{L}_{\Phi'}, \tag{5}$$

where

$$\mathscr{L}_\Phi = f\,(\bar{\psi}_+\psi_+ - \bar{\psi}_-\psi_-)\Phi, \tag{6}$$

$$\mathscr{L}_{\Phi'} = f'\,(\bar{\psi}_+\psi_- + \bar{\psi}_-\psi_+)\Phi'. \tag{7}$$

Next, one rewrites the above interactions in terms of χ and η. One has

$$\mathscr{L}_\Phi = i f\,(\bar{\chi}\Phi\eta - \bar{\eta}\Phi\chi), \tag{8}$$

and

$$\mathscr{L}_{\Phi'} = f'\,(\bar{\chi}\Phi'\chi - \bar{\eta}\Phi'\eta). \tag{9}$$

As usual, the minimum energy solutions (kinks) for Φ and Φ' are $\langle\Phi\rangle = h(z)$ and $\langle\Phi'\rangle = h'(y)$. We then notice that $-i\,\bar{\psi}_+\partial_z\psi_+ + i\,\bar{\psi}_-\partial_z\psi_- = \bar{\chi}\partial_z\eta - \bar{\eta}\partial_z\chi$. Furthermore, $\bar{\psi}_+\gamma^5\partial_y\psi_+ + \bar{\psi}_-\gamma^5\partial_y\psi_- = \bar{\chi}\gamma^5\partial_y\chi + \bar{\eta}\gamma^5\partial_y\eta$. Next we write the zero mode $\chi_{L.R}$ as

$$\chi_{L,R}(x,y,z) = \chi_{L,R}(x)\xi^0_{L,R}(y)\xi^0_\chi(z). \tag{10}$$

It is then straigthforward to find the equations for $\xi^0_{L,R}(y)$ and $\xi^0_\chi(z)$, namely

$$(\partial_z + i f h(z))\xi^0_\chi(z) = 0;\ (\partial_y + f'h'(z))\xi^0_{L,R}(y) = 0. \tag{11}$$

Eq. (11) clearly shows the difference in behavior between $\xi^0_\chi(z)$ and $\xi^0_{L,R}(y)$. The former will be an oscillating function of z while the latter is a real function of y. We have already encountered the solution for $\xi^0_\pm(y)$ (Eq. (2)). The solution for $\xi^0_\chi(z)$ is

$$\xi_{\chi,0}(z) = \frac{1}{\sqrt{L_6}}e^{is(z)}, \tag{12}$$

where

$$s(z) = f\int_0^z dz' h(z'). \tag{13}$$

A massless fermion in four dimensions has an *oscillating* phase along the sixth dimension! This is the phase factor we have been searching for. However, this is not the end of the story. To obtain an effective lagrangian in four dimensions, we still have to integrate out the two extra dimensions. We have seen earlier the case with five dimensions.

the new feature here is the oscillating phase along the sixth dimension. What would an overlap of the wave functions along z imply? Well, it implies a number of interesting things, one of which was the subject of investigation reported in this talk, namely a pure phase mass matrix.

Pure Phase Mass Matrix

In four dimensions, the SM Yukawa term responsible for giving masses to fermions is typically of the form $g_Y \bar{\psi}_L H \psi_R + h.c.$, where H is the Higgs field. In the parlance of extra compact dimensions, these fields would be zero modes which survive in four dimensions. In our case, we would have to write down a Yukawa interaction in six dimensions which, leaving aside any symmetry other than the Z_2 symmetries, should be invariant under Z_2 and which is real as we have done in Eqs. (6, 7). In contrast with the background scalars discussed above, the SM Higss field is required to be an *even* field in order to have a zero mode in four dimensions. With this in mind, one can now write such an interaction in six dimensions. The notation used in this section will be slightly different from the one used in Ref. [3].

From hereon, I will reserve the subscripts $\pm$ to denote the two complex spinors in six dimensions, and the superscripts (L,R) to denote higher-dimensional fermions which will reduce to left and right-handed fermions in four dimensions.

For the SM, we want to have both left-handed zero modes ($SU(2)_L$ doublets) and right-handed zero modes ($SU(2)_L$ singlets). As we have seen earlier, along y, an *even* parity fermion transforms as $\psi^{(R)} \to \gamma_5 \psi^{(R)}$, $\bar{\psi}^{(R)} \to -\bar{\psi}^{(R)}\gamma_5$, while an *odd* parity fermion transforms as $\psi^{(L)} \to -\gamma_5 \psi^{(L)}$, $\bar{\psi}^{(L)} \to \bar{\psi}^{(L)}\gamma_5$. If we denote the SM Higgs field by $H = H(x,y,z)$ and requiring H to be an *even* field, the only Yukawa coupling between $\psi^{(L,R)}$ and H which respects the two Z_2 symmetries is found to be

$$\mathscr{L}_Y = f_Y(\bar{\psi}_+^{(L)}\psi_-^{(R)} + \bar{\psi}_-^{(L)}\psi_+^{(R)})H + h.c. \tag{14}$$

In terms of χ and η, Eq. (14) becomes

$$\mathscr{L}_Y = f_Y(\bar{\chi}^{(L)}\chi^{(R)} - \bar{\eta}^{(L)}\eta^{(R)})H + h.c. \tag{15}$$

In five dimensions one would be left with the first term of (15) written in terms of the χ and H zero modes (with the appropriate wave function overlap).

The effective Yukawa Lagrangian in four dimensions can be now written as (ϕ is the zero mode of H):

$$\mathscr{L}_{Y,eff} = g_Y f_6(\bar{\chi}_L^{(L)}\chi_R^{(R)})\phi + h.c., \tag{16}$$

where

$$g_Y = \int dy' \xi_L^{0,(L)}(y')\xi_R^{0,(R)}(y'), \tag{17}$$

and

$$f_6 = \int dz' \xi_\chi^{0,(L)}(z')\xi_\chi^{0,(R)}(z'). \tag{18}$$

Under what conditions will f_6 be a pure phase of the form $\exp(i\theta)$? Although each $\xi_\chi^0(z)$ has a pure phase form (Eq. (12)), the integral of two such factors does not necessarily give rise to a pure phase. These conditions have been studied in details in Ref. [3]. Here, I will simply summarize a few points concerning this issue.

With the kink solution for Φ, ξ_χ^0 can be written explicitely as (for fermion i)

$$\xi_\chi^{i,0}(z) = \frac{1}{\sqrt{L_6}} e^{i(f v \ln(\cosh(\mu z))/\mu - m_i z)}, \tag{19}$$

where I have taken into account a general location for the kink, namely $f h(z) - m_i = 0$. The oscillatory behavior of ξ_χ is determined by the location of the domain wall! When (19) is plugged into (18) to calculate the overlap between fermion i and fermion j, a necessary step in the construction of the matrix elements of the mass matrix, it was found [3] that $f_6^{ij} \approx \exp(i\theta_{ij})$ if the widths of the domain walls along z are approximately comparable to the thickness L_6 of the brane, namely $1/\mu_6 \sim L_6$. This is very interesting because the thickness of the brane along the fifth dimension y is presumed to be much larger than the width of any domain wall. (Of course, if one relaxes the pure phase requirement, this statement will no longer hold.)

From hereon, I will refer loosely to the "locations" of fermions along z as actually the locations of the associated domain walls.

From the above discussions, one might have guessed by now that the phase angles will depend on the locations of the domain walls. What these locations are is beyond the scope of our present investigation. However, one can ask the following question: Given these locations, what kind of mass matrix would one obtain? (Of course, one can envision other phenomena, but the mass matrix is our present focus.) The results obtained in Ref. [3] are summarized below. We assumed a family symmetry $S_{3L} \otimes S_{3R}$ in [3] which is subsequently broken and is represented by the different locations of the domain walls associated with different families. The mass matrix is (i, j are family indices)

$$\mathcal{M} = g_Y \frac{v}{\sqrt{2}} \begin{pmatrix} a_{11} & a_{12} & a_{13} \\ a_{21} & a_{22} & a_{23} \\ a_{31} & a_{32} & a_{33} \end{pmatrix}, \tag{20}$$

where

$$a_{ii} \approx 1, \tag{21}$$

and

$$a_{ij} \approx \exp\left(i \frac{(\Delta m_{iR,jL})^2}{4\Delta\mu_{ij}^2} \right). \tag{22}$$

$\mathcal{M}$ is a Pure Phase Mass Matrix! Here $\langle\phi\rangle = v/\sqrt{2}$. The various quantities present in (22) are

$$\Delta m_{iR,jL} = m_{iR} - m_{jL}, \tag{23}$$

and

$$\Delta\mu_{ij}^2 \equiv (1/2)(a_i \mu_i^2 - a_j \mu_j^2), \tag{24}$$

where $f v_i/\mu_i \equiv a_i$ are parameters related to various kinks.

As discussed in Ref. [3], there are a number interesting properties associated with the above mass matrix.

(I) When $m_{iR} = m_{jL}$, i.e. the "locations" of all left-handed and right-handed fermions, regardless of of the family index, are the same along z, the phase angles *vanish*. All matrix elements become *unity*: the well-known Democratic Mass Matrix, which unfortunately does not quite work.

A mechanism is needed to separate these "locations"! One needs to break the Lef-Right symmetry and/or the Family symmetry.

(II) When $m_{iR} = m_{iL}$, i.e. for each family, left and right-handed fermions are "located" at the same place, it is straightforward to see that $\mathcal{M}$ becomes a *hermitian matrix*.

(III) When $m_{iR} = m_R$ and $m_{iL} = m_L$, $\mathcal{M}$ is also *hermitian*.

(IV) When all "locations" are different (Left, Right, Family), $\mathcal{M}$ is *non-hermitian*.

From a phenomenological point of view, case (IV) is the correct one. However, one can recognize that, e.g. (IV) is an extension of (III) in that whatever mechanism, which might be responsible for the family symmetry breaking, is also responsible for splitting the "locations" of different families and hence transforming a hermitian matrix into a non-hermitian one. (An example was given in [3].) This has an interesting implication. A hermitian mass matrix has the following property: $\arg(\det\mathcal{M}) = 0$. Does this have anything to do with a possible solution to the strong CP problem? After all one has $\arg(\det\mathcal{M}) \neq 0$ for the realistic case (IV). How can one keep the parameter $\bar{\theta}$ small (less than 10^{-9})? Does family symmetry breaking have anything to do with this? This interesting problem is under investigation.

The above discussion is a generic one which can be applied to both the Up and Down quark sectors. A very important question still remains: What separate the mass scales of the two sectors? If there is only one SM Higgs field with VEV $v/\sqrt{2}$, the differences in mass scales would come from the difference in the effective Yukawa couplings $g_{Y,U}$ and $g_{Y,D}$. This would imply that U_R and D_R are widely separated along y, assuming that U_L and D_L are located at the same place to insure universality of the weak interactions. The other attractive possibility is to have two Higgs doublets, with two different VEV's, v_1 and v_2. In this case, U_R and D_R can be at the same locations. This latter case is very reminescent of a typical Left-Right model. It is amusing that a Pure Phase mass Matrix constructed from six dimensions has implications far beyond its original aim.

Finally, I would like to apologize for a very incomplete list of references. A more extensive list can be found in [3].

ACKNOWLEDGMENTS

I would like to thank Professor Behram Kursunoglu and the Organizers of the Coral Gables conference for a pleasant and exciting atmosphere at the meeting. This work is supported in parts by the US Department of Energy under grant No. DE-A505-89ER40518.

REFERENCES

1. G. C. Branco, J. I. Silva-Marcos and M. N. Rebelo, *Phys. Lett. B*, **237**, 446 (1990).
2. P. M. Fishbane and P. Q. Hung, *Phys. Rev. D*, **57**, 2748 (1998).
3. P Q. Hung and M. Seco, *hep-ph/0111013* (2001).
4. H. Georgi, A. K. Grant and G. Hailu, *Phys. Rev. D*, **63**, 064027 (2001).
5. N. Arkani-Hamed and M. Schmaltz, *Phys. Rev. D*, **61**, 033005 (2000).

Searching for Exceptional Physics

P. Ramond

Institute for Fundamental Theory, Physics Department,
University of Florida, Gainesville, FL 32611, USA

Abstract. We investigate an intriguing mathematical structure in connection with eleven-dimensional M-theory: Euler triplets associated with the coset $F_4/SO(9)$ may provide a generalization of eleven-dimensional supergravity.

INTRODUCTION

Unique mathematical structures appear ubiquitously in the description of Nature. M-theory and Superstring theories [1] which offer the best hope for resolving quantum mechanics and general relativity rely on miracles often traced to such algebraic "coincidences".

Consistency of superstring [2] theories in $9+1$ dimensions relies on the triality of the light-cone little group $SO(8)$, which links its tensor and spinor representations via a Z_3 symmetry. The exceptional group F_4 is the smallest which realizes this triality explicitly. It was surprising to find another consistent theory in one more space dimension since the $SO(9)$ little group has very different spinor and tensor representations. A possible hint for fermion-boson confusion is the anomalous embedding of $SO(9)$ into an orthogonal group in which the vector representation of the bigger group is identified with the spinor of the smaller group

$$SO(16) \supset SO(9)\,, \qquad \mathbf{16} \doteq \mathbf{16}\,.$$

The exceptional group F_4 is the smallest group in which $SO(8)$ triality is explicitly realized, but the use of exceptional groups to describe space-time symmetries has not been so evident because exceptional algebras relate tensor and spinor representations of their orthogonal subgroups, while Spin-Statistics requires them to be treated differently. Yet there are some mathematical curiosities worth noting. For one, there is the anomalous Dynkin embedding of F_4 inside $SO(26)$ and $SO(9)$ inside F_4

$$SO(26) \supset F_4 \supset SO(9)\,,$$

or its non-compact variety

$$SO(25,1) \supset F_{4(-20)} \supset SO(9)\,.$$

This chain might point to a (M)-heterotic construction from the bosonic string to M-theory.

CP624, *Cosmology and Elementary Particle Physics*, edited by B. N. Kursunoglu et al.

A formulation of finite-dimensional Hilbert spaces in terms of the algebra of observables, proposed by P. Jordan [3], has not yet proven fruitful in Physics, in spite of many attempts. In all but one case, it is akin to rewriting the familiar Dirac ket description in terms of density matrices, but it also unearthed a unique structure on which Quantum Mechanics can be implemented, even though it cannot be described by kets in Hilbert space. Our interest lies in the fact that its automorphism group is F_4 and its natural description lies in the sixteen-dimensional (Cayley) projective space $F_4/SO(9)$.

Finally, we note that the exceptional group F_4 appears in the light-cone description of the eleven-dimensional supergravity supermultiplet.

SUPERGRAVITY IN ELEVEN DIMENSIONS

Eleven dimensional $N = 1$ Supergravity [4] is an awesome field theory that includes gravity, butas it is not renormalizable it does not stand on its own as a physical theory. However the eleven-dimensional theory is the limit of M-theory which, like characters on the walls of Plato's cave, has revealed itself through its compactified version onto lower-dimensional manifolds.

$N = 1$ supergravity in eleven dimension contains three massless fields, the familiar symmetric second-rank tensor, $h_{\mu\nu}$ which represents gravity, a three-form field $A_{\mu\nu\rho}$, and the Rarita-Schwinger spinor $\Psi_{\mu\,\alpha}$. From its Lagrangian, one can derive the expression for the super Poincaré algebra, which in the unitary transverse gauge assumes the particularly simple form in terms of the nine (16×16) γ_i matrices which form the Clifford algebra

$$\{\gamma^i, \gamma^j\} = 2\delta^{ij} , \qquad i, j = 1, \ldots, 9 .$$

Supersymmetry is generated by the sixteen real supercharges

$$\mathcal{Q}^a_\pm = \mathcal{Q}^{a\,*}_\pm ,$$

which satisfy

$$\{\mathcal{Q}^a_+, \mathcal{Q}^b_+\} = \sqrt{2}p^+\delta^{ab} , \qquad \{\mathcal{Q}^a_-, \mathcal{Q}^b_-\} = \frac{\vec{p}\cdot\vec{p}}{\sqrt{2}p^+}\delta^{ab} , \qquad \{\mathcal{Q}^a_+, \mathcal{Q}^b_-\} = -(\gamma_i)^{ab}p^i ,$$

and transform as Lorentz spinors

$$[M^{ij}, \mathcal{Q}^a_\pm] = \frac{i}{2}(\gamma^{ij}\mathcal{Q}_\pm)^a , \qquad [M^{+-}, \mathcal{Q}^a_\pm] = \pm\frac{i}{2}\mathcal{Q}^a_\pm ,$$

$$[M^{\pm i}, \mathcal{Q}^a_\mp] = 0 , \qquad [M^{\pm i}, \mathcal{Q}^a_\mp] = \pm\frac{i}{\sqrt{2}}(\gamma^i\mathcal{Q}_\pm)^a .$$

A very simple representation of the 11-dimensional super-Poincaré generators can be constructed, in terms of sixteen anticommuting real χ's and their derivatives, which transform as the spinor of $SO(9)$, as

$$\mathcal{Q}^a_+ = \partial_{\chi^a} + \frac{1}{\sqrt{2}}p^+\chi^a , \qquad \mathcal{Q}^a_- = -\frac{p^i}{p^+}\left(\gamma^i\mathcal{Q}_+\right)^a ,$$

$$
\begin{aligned}
M^{ij} &= x^i p^j - x^j p^i - \frac{i}{2}\chi\,\gamma^{ij}\partial_\chi\,,\\
M^{+-} &= -x^- p^+ - \frac{i}{2}\chi\,\partial_\chi\,,\\
M^{+i} &= -x^i p^+\,,\\
M^{-i} &= x^- p^i - \frac{1}{2}\{x^i, P^-\} + \frac{ip^j}{2p^+}\chi\gamma^i\gamma^j\partial_\chi\,.
\end{aligned}
$$

The light-cone little group transformations are generated by

$$
S^{ij} = -\frac{i}{2}\chi\gamma^{ij}\partial_\chi\,,
$$

which satisfy the $SO(9)$ Lie algebra. In order to examine the spectrum, we rewrite the supercharges in terms of eight complex Grassmann variables

$$
\theta^\alpha \equiv \frac{1}{\sqrt{2}}\left(\chi^\alpha + i\chi^{\alpha+8}\right)\,, \qquad \overline{\theta}^\alpha \equiv \frac{1}{\sqrt{2}}\left(\chi^\alpha - i\chi^{\alpha+8}\right)\,,
$$

where $\alpha = 1, 2, \ldots, 8$. The eight complex θ transform as the $(\mathbf{4}, \mathbf{2})$, and $\overline{\theta}$ as the $(\overline{\mathbf{4}}, \mathbf{2})$ of the $SU(4)\times SU(2)$ subgroup of $SO(9)$. The eight complex supercharges

$$
\begin{aligned}
\mathbf{Q}^\alpha_+ &\equiv \frac{1}{\sqrt{2}}\left(\mathcal{Q}^\alpha_+ + i\mathcal{Q}^{\alpha+8}_+\right) = \frac{\partial}{\partial\overline{\theta}^\alpha} + \frac{1}{\sqrt{2}}p^+\theta^\alpha\,,\\
\mathbf{Q}^{\alpha\dagger}_+ &\equiv \frac{1}{\sqrt{2}}\left(\mathcal{Q}^\alpha_+ - i\mathcal{Q}^{\alpha+8}_+\right) = \frac{\partial}{\partial\theta^\alpha} + \frac{1}{\sqrt{2}}p^+\overline{\theta}^\alpha\,,
\end{aligned}
$$

satisfy

$$
\{\mathbf{Q}^\alpha_+, \mathbf{Q}^{\beta\dagger}_+\} = \sqrt{2}p^+\delta^{\alpha\beta}\,.
$$

They act irreducibly on chiral superfields which are annihilated by the covariant derivatives

$$
\left(\frac{\partial}{\partial\overline{\theta}^\alpha} - \frac{1}{\sqrt{2}}p^+\theta^\alpha\right)\Phi(y^-, \theta) = 0\,,
$$

where

$$
y^- = x^- - \frac{i\theta\overline{\theta}}{\sqrt{2}}\,.
$$

Expansion of the superfield in powers of the eight complex θ's yields 256 components, with the following $SU(4)\times SU(2)$ properties

$$
1 \sim (\mathbf{1}, \mathbf{1})\,,
$$

$$\begin{array}{lcl} \theta & \sim & (\mathbf{4},\mathbf{2})\ , \\ \theta\theta & \sim & (\mathbf{6},\mathbf{3}) \oplus (\mathbf{10},\mathbf{1})\ , \\ \theta\theta\theta & \sim & (\overline{\mathbf{20}},\mathbf{2}) \oplus (\bar{\mathbf{4}},\mathbf{4})\ , \\ \theta\theta\theta\theta & \sim & (\mathbf{15},\mathbf{3}) \oplus (\mathbf{1},\mathbf{5}) \oplus (\mathbf{20}',\mathbf{1})\ , \end{array}$$

and the higher powers yield the conjugate representations by duality. These make up the three representations of $N = 1$ supergravity

$$\begin{array}{lcl} \mathbf{44} & = & (\mathbf{1},\mathbf{5}) \oplus (\mathbf{6},\mathbf{3}) \oplus (\mathbf{20}',\mathbf{1}) \oplus (\mathbf{1},\mathbf{1})\ , \\ \mathbf{84} & = & (\mathbf{15},\mathbf{3}) \oplus (\overline{\mathbf{10}},\mathbf{1}) \oplus (\mathbf{10},\mathbf{1}) \oplus (\mathbf{6},\mathbf{3}) \oplus (\mathbf{1},\mathbf{1})\ , \\ \mathbf{128} & = & (\mathbf{20},\mathbf{2}) \oplus (\overline{\mathbf{20}},\mathbf{2}) \oplus (\mathbf{4},\mathbf{4}) \oplus (\bar{\mathbf{4}},\mathbf{4}) \oplus (\mathbf{4},\mathbf{2}) \oplus (\bar{\mathbf{4}},\mathbf{2})\ . \end{array}$$

The highest weights of the supergravity representations are

$$\begin{array}{lclcl} \mathbf{44} & : & \theta^1\theta^4\theta^5\theta^8 = (0,2,0;0) & \sim & (\mathbf{20}'\ \mathbf{1}) \\ \mathbf{84} & : & \theta^1\theta^8 = (2,0,0;0) & \sim & (\mathbf{10},\mathbf{1}) \\ \mathbf{128} & : & \theta^1\theta^4\theta^8 = (1,1,0;1) & \sim & (\mathbf{20},\mathbf{2})\ , \end{array}$$

together with their $SU(4) \times SU(2)$ properties. All other states are generated by acting on these highest weight states with the lowering operators. The highest weight chiral superfield that describes $N = 1$ supergravity in eleven dimensions is simply

$$\Phi = \theta^1\theta^8\, h(y^-,\vec{x}) + \theta^1\theta^4\theta^8\, \psi(y^-,\vec{x}) + \theta^1\theta^4\theta^5\theta^8 A(y^-,\vec{x})\ ,$$

which summarizes the spectrum of the super-Poincaré algebra in eleven dimensions of either a free field theory or a free superparticle.

Since the little group generators act on a 256-dimensional space, we can express them in terms of sixteen (256×256) matrices, Γ^a, which satisfy the Dirac algebra

$$\{\Gamma^a\,,\Gamma^b\} = 2\delta^{ab}\ .$$

This leads to an elegant representation of the $SO(9)$ generators

$$S^{ij} = -\frac{i}{4}(\gamma^{ij})^{ab}\,\Gamma^a\Gamma^b \equiv -\frac{i}{2}f^{ijab}\Gamma^a\Gamma^b\ .$$

The coefficients

$$f^{ijab} \equiv \frac{1}{2}(\gamma^{ij})^{ab}\ ,$$

naturally appear in the commutator between the generators of $SO(9)$ and any spinor operator T^a, as

$$[T^{ij}, T^a] = \frac{i}{2}\left(\gamma^{ij} T\right)^a = i f^{ijab} T^b .$$

But there is more to it, the $(\gamma^{ij})^{ab}$ can also be viewed as structure constants of a Lie algebra. Manifestly antisymmetric under $a \leftrightarrow b$, they can appear in the commutator of two spinors into the $SO(9)$ generators

$$[T^a, T^b] = \frac{i}{2}(\gamma^{ij})^{ab} T^{ij} = f^{abij} T^{ij} ,$$

and one easily checks that they satisfy the Jacobi identities. Remarkably, the 52 operators T^{ij} and T^a generate the exceptional Lie algebra F_4, showing explicitly how an exceptional Lie algebra appears in the light-cone formulation of supergravity in eleven dimensions.

Character Formula

The $SO(9)$ representations which label the degrees of freedom of supergravity in eleven dimensions, $h_{(ij)} \sim (2000)$, $A_{[ijk]} \sim (0010)$, $\Psi_{i\alpha} \sim (1001)$, display remarkable group-theoretic kinships, summarized in the following table

irrep	(1001)	(2000)	(0010)
D	128	44	84
I_2	256	88	168
I_4	640	232	408
I_6	1792	712	1080
I_8	5248	2440	3000

where D is the dimension of the representation, and I_n are the Dynkin indices of the representations, related to the four Casimir operators of $SO(9)$. We note that the dimension and Dynkin indices of the fermion is the sum over those of the bosons, except for I_8, indicating that these three representations have much in common.

Surprisingly, the supergravity fields are the first of an infinite number of triplets [5] of $SO(9)$ representations which display the same group-theoretical relations: equality of dimension and all Dynkin indices except I_8 between one representation and the sum of the other two. Quantum theories of these Euler triplets may have very interesting divergence properties, as these numbers typically occur in higher loop calculations, and such equalities usually increase the degree of divergence, and the failure of the equality for I_8 is probably related to the lack of renormalizability of the theory [6].

This mathematical fact has been traced to a character formula [7] related to the three equivalent embeddings of $SO(9)$ into F_4

$$V_\lambda \otimes S^+ - V_\lambda \otimes S^- = \sum_c \mathrm{sgn}(c)\, U_{c\bullet\lambda} .$$

On the left-hand side, V_λ is a representation of F_4 written in terms of its $SO(9)$ subgroup, $S^\pm$ are the two spinor representations of $SO(16)$ written in terms of its anomalously embedded subgroup $SO(9)$, $\otimes$ denotes the normal Kronecker product of representations, and the $-$ denotes the naive substraction of representations. On the right-hand side, the sum is over c, the elements of the Weyl group which map the Weyl chamber of F_4 into that of $SO(9)$. In this case there are three elements, the ratio of the orders of the Weyl groups (it is also the Euler number of the coset manifold), and $U_{c\bullet\lambda}$ denotes the $SO(9)$ representation with highest weight $c\bullet\lambda$, where

$$c\bullet\lambda = c(\lambda+\rho_{F_4})-\rho_{SO(9)} ,$$

and the ρ's are the sum of the fundamental weights for each group, and sgn(c) is the index of c. Thus to each F_4 representation corresponds a triplet, called Euler triplet.

The Euler triplet corresponding to the F_4 representation $[a_1\,a_2\,a_3\,a_4]$ is made up of the following three $SO(9)$ representations listed in order of increasing dimensions:

$$(2+a_2+a_3+a_4, a_1, a_2, a_3) \ , \ (a_2, a_1, 1+a_2+a_3, a_4) \ , \ (1+a_2+a_3, a_1, a_2, 1+a_3+a_4)$$

The spinor representations appear with odd entries in the fourth place. Euler triplets with the largest spinor and two bosons must have both a_3 and a_4 even or zero.

Kostant [8] interprets this character formula as the index formula of a Dirac-like operator formed over the coset $F_4/SO(9)$. This coset is the sixteen-dimensional Cayley projective plane, over which we introduce the Clifford algebra

$$\{\Gamma^a, \Gamma^b\} = 2\delta^{ab} \ , \ \ a,b=1,2,\ldots,16 ,$$

generated by (256×256) matrices. The Kostant equation is defined as

$$\mathcal{K}\!\!\!/\ \Psi = \sum_{a=1}^{16} \Gamma^a T^a \Psi = 0 ,$$

where T_a are F_4 generators not in $SO(9)$, with commutation relations

$$[T^a, T^b] = i f^{abij} T^{ij} .$$

Although it is taken over a compact manifold, it has non-trivial solutions. Kostant's operator commutes with the sum of generators,

$$[\mathcal{K}\!\!\!/ , L^{ij}] = 0 ,$$

where

$$L^{ij} \equiv T^{ij} + S^{ij} ,$$

allowing its solutions to be labelled by $SO(9)$ quantum numbers.

The same construction of Kostant's operator applies to all equal rank embeddings, and its trivial solutions display supersymmetry [7, 9, 10, 11]. In particular we note the cases

$E_6/SO(10)\times SO(2)$, with Euler number 27, $E_7/SO(12)\times SO(3)$ with Euler number 63, and $E_8/SO(16)$, where the Euler triplets contain 135 representations [5]. These cosets with dimensions 32 ,64, and 128 could be viewed as complex, quaternionic and octonionic Cayley plane [12]

Oscillator Representation of F_4

In order to display explicit solutions of Kostant's equation, it is convenient to use an oscillator representation of F_4, originally constructed by Fulton [13].It requires three sets of oscillators transforming as **26**, labelled by $A_0^{[\kappa]}$, $A_i^{[\kappa]}$, $i=1,\cdots,9$, $B_a^{[\kappa]}$, $a=1,\cdots,16$, and their hermitian conjugates, and where $\kappa=1,2,3$. Under $SO(9)$, the $A_i^{[\kappa]}$ transform as **9**, $B_a^{[\kappa]}$ transform as **16**, and $A_0^{[\kappa]}$ is a scalar. They satisfy the commutation relations of ordinary harmonic oscillators

$$[A_i^{[\kappa]}, A_j^{[\kappa']\dagger}] = \delta_{ij}\,\delta^{[\kappa][\kappa']} , \qquad [A_0^{[\kappa]}, A_0^{[\kappa']\dagger}] = \delta^{[\kappa\,\kappa']} .$$

Note that the $SO(9)$ spinor operators satisfy Bose-like commutation relations

$$[B_a^{[\kappa]}, B_b^{[\kappa']\dagger}] = \delta_{ab}\,\delta^{[\kappa][\kappa']} .$$

The generators T_{ij} and T_a

$$T_{ij} = -i\sum_{\kappa=1}^{4}\left\{\left(A_i^{[\kappa]\dagger}A_j^{[\kappa]} - A_j^{[\kappa]\dagger]}A_i^{[\kappa]}\right) + \frac{1}{2}B^{[\kappa]\dagger}\gamma_{ij}B^{[\kappa]}\right\} ,$$

$$T_a = -\frac{i}{2}\sum_{\kappa=1}^{4}\left\{(\gamma_i)^{ab}\left(A_i^{[\kappa]\dagger}B_b^{[\kappa]} - B_b^{[\kappa]\dagger}A_i^{[\kappa]}\right) - \sqrt{3}\left(B_a^{[\kappa]\dagger}A_0^{[\kappa]} - A_0^{[\kappa]\dagger}B_a^{[\kappa]}\right)\right\} ,$$

satisfy the F_4 algebra,

$$\begin{aligned}
[T_{ij}, T_{kl}] &= -i(\delta_{jk}T_{il} + \delta_{il}T_{jk} - \delta_{ik}T_{jl} - \delta_{jl}T_{ik}) , \\
[T_{ij}, T_a] &= \frac{i}{2}(\gamma_{ij})_{ab}T_b , \\
[T_a, T_b] &= \frac{i}{2}(\gamma_{ij})_{ab}T_{ij} ,
\end{aligned}$$

so that the structure constants are given by

$$f_{ijab} = f_{abij} = \frac{1}{2}(\gamma_{ij})_{ab} .$$

The last commutator requires the Fierz-derived identity

$$\frac{1}{4}\theta\gamma^{ij}\theta\,\chi\gamma^{ij}\chi = 3\theta\chi\,\chi\theta + \theta\gamma^i\chi\,\chi\gamma^i\theta ,$$

from which we deduce

$$3\,\delta^{ac}\delta^{db} + (\gamma^i)^{ac}(\gamma^i)^{db} - (a \leftrightarrow b) \;=\; \frac{1}{4}(\gamma^{ij})^{ab}(\gamma^{ij})^{cd}\,.$$

To satisfy these commutation relations, we have required both A_0 and B_a to obey Bose commutation relations (Curiously, if both are anticommuting, the F_4 algebra is still satisfied). One can just as easily use a coordinate representation of the oscillators by introducing real coordinates u_i which transform as transverse space vectors, u_0 as scalars, and ζ_a as space spinors which satisfy Bose commutation rules

$$\begin{aligned} A_i &= \frac{1}{\sqrt{2}}(u_i + \partial_{u_i})\,, & A_i^\dagger &= \frac{1}{\sqrt{2}}(u_i - \partial_{u_i})\,, \\ B_a &= \frac{1}{\sqrt{2}}(\zeta_a + \partial_{\zeta_a})\,, & B_a^\dagger &= \frac{1}{\sqrt{2}}(\zeta_a - \partial_{\zeta_a})\,, \\ A_0 &= \frac{1}{\sqrt{2}}(u_0 + \partial_{u_0})\,, & A_0^\dagger &= \frac{1}{\sqrt{2}}(u_0 - \partial_{u_0})\,. \end{aligned}$$

Using square brackets $[\cdots]$ to represent the Dynkin label of F_4, and round brackets $(\cdots)$ to represent those of $SO(9)$, we list some of the combinations which will be used for investigating the solutions of Kostant's equation

$$\begin{aligned} u_1 + iu_2 &\sim [0\;0\;0\;1] \sim (\;1\;0\;0\;0)\,, \\ u_3 + iu_4 &\sim [1\;0\;0\,{-1}] \sim ({-1}\;1\;0\;0)\,, \\ \zeta_1 + i\zeta_9 &\sim [0\;0\;1\,{-1}] \sim (\;0\;0\;0\;1)\,, \\ \zeta_8 + i\zeta_{16} &\sim [0\;1\,{-1}\;0] \sim (\;0\;0\;1\,{-1})\,, \\ \zeta_3 - i\zeta_{11} &\sim [1\,{-1}\;1\;0] \sim (\;0\;1\,{-1}\;1)\,, \\ \zeta_6 - i\zeta_{14} &\sim [1\;0\,{-1}\;1] \sim (\;0\;1\;0\,{-1})\,. \end{aligned}$$

Hence $u_1 + iu_2$ and $\zeta_1 + i\zeta_9$ are the highest weights of the $SO(9)$ representations **9**, and **16**, respectively.

Solutions of Kostant's Equation

As we have seen, for every representation of F_4, $[a_1, a_2, a_3, a_4]$, there is one $SO(9)$ Euler triplet solution of Kostant's equation. The trivial solution with $a_1 = a_2 = a_3 = a_4 = 0$, yields the $N = 1$ supergravity multiplet in eleven dimensions, $(2000) \oplus (0010) \oplus (1001)$ with the highest weight components $\theta^1\theta^4\theta^5\theta^8$, $\theta^1\theta^8$, and $\theta^1\theta^4\theta^8$, described by the chiral superfield

$$\Phi_{0000} \;=\; \theta^1\theta^8\,h_{0000}(y^-,\vec{x}) \;+\; \theta^1\theta^4\theta^8\,\psi_{0000}(y^-,\vec{x}) \;+\; \theta^1\theta^4\theta^5\theta^8\,A_{0000}(y^-,\vec{x})\,.$$

In general, the highest weight solutions appear in the form of $f(u_i,\zeta_a)\Theta(\theta)$, where both $f(u_i,\zeta_a)$ and $\Theta(\theta)$ are the highest weights $SO(9)$ states with respect to the earlier defined T_{ij} and S_{ij}. The solutions have the quantum numbers of their sum $L_{ij}=S_{ij}+T_{ij}$, which commutes with Kostant's operator. $\Theta(\theta)$ is one of the three polynomials above, $\theta^1\theta^4\theta^5\theta^8$, $\theta^1\theta^8$, or $\theta^1\theta^4\theta^8$. Since the θ parts describe a superparticle in eleven dimensions, it is tempting to interpret states in the other Euler triplets as superparticles dressed with fields described by these new variables, vector coordinates $u_i^{[\kappa]}$ and twistor coordinates $\zeta_a^{[\kappa]}$.

The supergravity Euler triplet is supersymmetric with an equal number of fermions and bosons. None of the other Euler triplets display space-time supersymmetry, although a subclass does contain equal number of fermions ($SO(9)$ spinors) and bosons ($SO(9)$ tensors): those for which the Dynkin indices a_3 and a_4 are even, with no constraints on a_1 and a_2. Curiously, these are precisely the Euler triplets where the twistors ζ_a appear quadratically, exactly what spin-statistics requires since odd powers would generate fermions ($SO(9)$ spinors) with Bose properties. In this framework, spin-statistics is linked with equality of bosons and fermions.

Although there are four different families of these special Euler triplets, here we mention only the case when only $a_1 \neq 0$.

In this case, the three supergravity states are dressed the same way by something with the quantum number of a 2-form, (0100), made up of two vectors and two spinors. It is described by the superfield

$$\begin{aligned}\Phi_{(1000)} &= \left(\phi_{(1000)}\theta^1\theta^4\theta^5\theta^8 + A_{(1000)}\theta^1\theta^8 + \psi_{(1000)}\theta^1\theta^4\theta^8\right)\times \\ &\times\left([u_1+iu_2\,,u_3+iu_4]+[\zeta_1+i\zeta_9\,,\zeta_6-i\zeta_{14}]+[\zeta_8+i\zeta_{16}\,,\zeta_3-i\zeta_{11}]\right)\,,\end{aligned}$$

which requires only two oscillator copies as

$$[a,b] \equiv a^{[1]}b^{[2]}-a^{[2]}b^{[1]}\,,$$

is the determinant of 2 copies of a and b states. This doubling may indicate the presence of E_6, the complex extension of F_4.

We have shown elsewhere [14] that Poincaré symmetries can be implemented on the Euler triplets only if they describe massless states. As these states represent higher spin particles, one can expect difficulties in implementing their interactions [15, 16], although there is an infinite number of triplets [17].

This family of triplets may be generated by the coupling of a two-form potential [18] to the superparticle in eleven-dimensions. This coupling breaks supersymmetry but it may be that it generates an infinite-dimensional representation, each component being an Euler triplet. The 2-forms would act as ladder operators between them. This geometrical possibility will be investigated in a future work [19].

ACKNOWLEDGMENTS

I wish to thank Professor B. Kursunoglu for his tireless efforts to organize such a pleasant conference. This research is supported in part by the US Department of Energy under grant DE-FG02-97ER41029.

REFERENCES

1. For a beautiful review, see J. H. Schwarz *Superstring Theory: An Overview* Apr 2000. In Mitra, A.N. (ed.): Quantum field theory 605-610.
2. P. Ramond, *Phys. Rev.* **D3**, 2415 (1971); A. Neveu and J. H. Schwarz, *Nucl.Phys.* **B31**, 86(1971)
3. P. Jordan *Nachr. Ges. Wiss. Gothingen*, 209(1933).
4. E. Cremmer, B. Julia, J. Scherk *Phys. Lett* **B76**, 409(1978)
5. T. Pengpan and P. Ramond,*Phys. Rep.* **315**. 137(1999)
6. T. Curtright, *Phys. Rev. Lett.* **48**, 1704(1982)
7. B. Gross, B. Kostant, P. Ramond, and S. Sternberg, *Proc. Natl. Acad. Scien.*, 8441 (1998)
8. B. Kostant, "*A Cubic Dirac Operator and the Emergence of Euler Number Multiplets of Representations for Equal Rank Subgroups*", *Duke J. of Mathematics* **100**, 447(1999).
9. Lars Brink, P. Ramond, *Dirac Equations, Light-Cone Supersymmetry, and Superconformal Algebras* In Shifman, M.A. (ed.): The many faces of the superworld 398-416;HEP-TH 9908208.
10. Lars Brink, *Euler Multiplets, Light-Cone Supersymmetry and Superconformal Algebras*, in Proceedings of the International Conference on Quantization, Gauge Theory, and Strings: Conference Dedicated to the Memory of Professor Efim Fradkin, Moscow 2000.
11. P. Ramond, *Boson Fermion Confusion: The String Path to Supersymmetry, Nucl.Phys.Proc.Suppl.* **101**, 45(2001); Hep-Th 0102012.
12. M. Atiyah, talk at the Swedish Royal Society, Stockholm, September 2001.
13. T. Fulton, *J. Phys. A:Math. Gen.* **18**, 2863(1985)
14. P. Ramond, Algebraic Dreams. UFIFT-HET-01-27 (Dec 2001), Hep-th/0112261.
15. E. Witten and S. Weinberg, *Phys. Lett.* **B96**, 59(1980); Loyal Durand III, *Phys. Rev.* **128**, 434(1962); K. Case and S. Gasiorowicz, *Phys. Rev.* **125**, 1055(1962)
16. C. Aragone and S. Deser, *Nuovo Cim.* **57B**, 33(1980)
17. M. A. Vasiliev, hep-th/0104246, and references therein.
18. M. Kalb, P. Ramond, *Phys. Rev.* **D9**, 2273(1974)
19. L. Brink, P. Ramond, and X. Xiong, in preparation.

Acceleration and Use for Physics of Polarized Proton Beams at RHIC

Yousef I. Makdisi

Collider-Accelerator Department, Brookhaven National Laboratory, Upton NY 11973

Abstract: The Relativistic Heavy Ion Collider completed its second year for physics with gold beam collisions at center of mass energy of 200 GeV per nucleon. An important component of the physics program is the acceleration and collisions of polarized proton beams at center of mass energies up to 500 GeV, luminosity of 2.10^{32} cm^{-2}. sec^{-1} and beam polarization of 70%. The primary focus of the spin physics program is the study of the spin structure of the nucleon. Particular emphasis is on the polarized gluon structure function through direct photon production and inclusive jet production as well as the study of flavor separated quark and antiquark structure functions from $W^{+/-}$ production. In this presentation I will discuss our progress in acceleration and preservation of polarized proton beams at high energies and the expected physics reach using the installed STAR and PHENIX detectors.

Introduction

I would like to thank the symposium organizers for the invitation to make this presentation on behalf of my colleagues who are in the trenches toiling with the rigors of commissioning the polarized proton beams in RHIC while I enjoy the physics and hospitality of this wonderful location.

The study of spin has a long history that started at low energies using primarily polarized targets as well as rescattering techniques from carbon analyzers. It was always thought that spin effects are a low energy manifestation that will go away as we proceed to high energies. The Chair of this session Alan Krisch along with his collaborator the late Larry Ratner pioneered into the high energy domain with their success at accelerating a polarized proton beam to 12 GeV/c first at the ANL ZGS and later at the AGS. One of the significant findings was the large analyzing power in pp elastic scattering that increased with transverse momentum, see H. Neal in these proceedings. Also that the pp elastic scattering at 90° in the center of mass system was approximately 4 times more likely with spins aligned in the same direction compared to spins aligned in the opposite direction. Elsewhere, at Fermilab energies, large polarization was observed in inclusive lambda production and later in other hyperons.

The late eighties and early nineties witnessed the interesting results from Deeply Inelastic Scattering (DIS) experiments using both electron and muon beams to probe

CP624, *Cosmology and Elementary Particle Physics*, edited by B. N. Kursunoglu et al.

the polarized structure of the nucleon. The finding that the quarks carried a small portion of the nucleon spin confounded conventional thinking and set in motion a revival in spin physics interest and led to the proposal to collide polarized protons at RHIC[1]. In this presentation I discuss our current knowledge of the nucleon spin structure and the role of the RHIC spin physics program. I describe the process and techniques employed in the acceleration, preservation of the polarization of the beams at high energies as well as measuring of the beam polarization. Finally I discuss the progress that we attained as of this writing.

The Nucleon Spin Puzzle

A brief summation of the history of our understanding of the nucleon spin is described in the following: The naive quark model naturally summed over the quark spins or helicities $\Sigma = 1 = \Delta u+\Delta d+\Delta s$. Where $\Delta q = \int g_1^q dx$ refers to the quark helicity. In 1973 the Ellis-Jaffe sum rule assumed the sea component $\Delta s = 0$ and subsequently predicted the values of $\int g_1^p dx$ & $\int g_1^n dx$. In 1974 Sehgal suggested that $\Delta s = 0 \rightarrow \Sigma = 0.58 \neq 1$. However, over the span from 1989-99 the EMC/ SLAC/ HERMES experiments provided what was termed a "spin crisis" as they found $\Sigma = 0.3$ and $\Delta s = -0.1$. The Bjorken sum rule, derived about 30 years ago in the asymptotic limit of $Q^2 \rightarrow \infty$ used quark current algebra and predicted $\int g_1^p dx - \int g_1^n dx = 1/6[g_A/g_V]$ the axial and vector couplings in weak interactions.

Indeed the series of DIS experiments at SLAC (E80, E143, E155) with polarized electron beams, CERN (NMC and SMC) using polarized muon beams and later at HERA (HERMES) provided us a significant amount of insight as to the spin structure of the nucleon. These experiments used a combination of polarized proton, deuteron, and helium targets to measure the asymmetries $A_1(x,Q^2)$ and extract the corresponding g_1^p and g_1^n functions which when integrated over the Bjorken x range provide the valence quark contribution to the nucleon spin.

The Ellis-Jaffe sum rule did not survive as the sea component was non zero at –10%. The Bjorken sum rule, on the other hand, was validated to 10%. Nevertheless the striking finding that the quarks contributed a mere 30% was a little unsettling.

A sample of the quality of the data is shown in figure 1. It is worth noting that while the lower reach of the data is approximately 10^{-3} in x_{Bj}, there is no consensus as to the behavior below that which is important to get the integral.

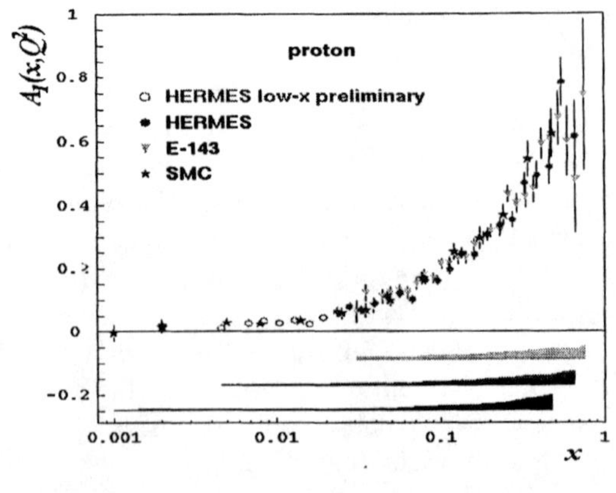

Figure1. Representative data from the DIS experiments

Adding the gluon content or the orbital angular momentum modifies the relationship

$$1/2 = (1/2)\,\Delta\Sigma + \Delta G + L_q + L_G$$

Since the gluon does not couple directly to the leptons, the DIS experiments have to revert to an indirect technique and look for a hadron in the final state in semi-inclusive processes. This, along with assumptions utilizing QCD evolution of the data versus Q^2, provided an early glimpse to the role of the gluon albeit with large errors Figure 2. The good news is that ΔG is large.

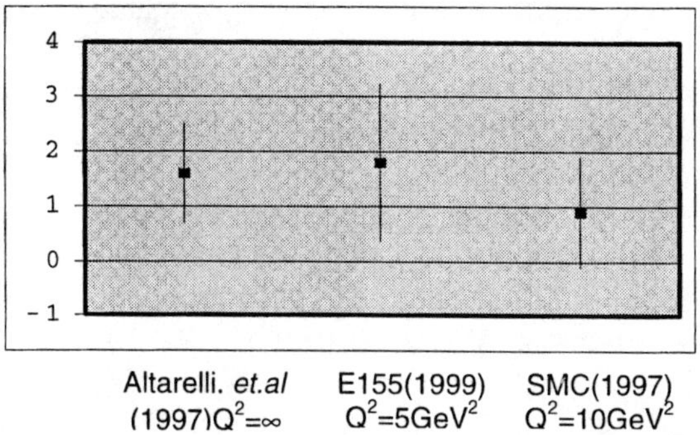

Figure 2. A chart indicating the results (theory and experimental) of the efforts to determine the gluon structure function.

The HERMES experiment at HERA is in the process of carrying out a measurement of the gluon polarization using open charm production.

The RHIC spin program[2] will complement these efforts by providing a more direct link to the gluon component using reactions such as direct photon production, inclusive jet and di-jet production as well as charm production. The RHIC energy reach provides additional access to quark and antiquark polarization via the $W^{+/-}$ production at the extreme rapidity distributions.

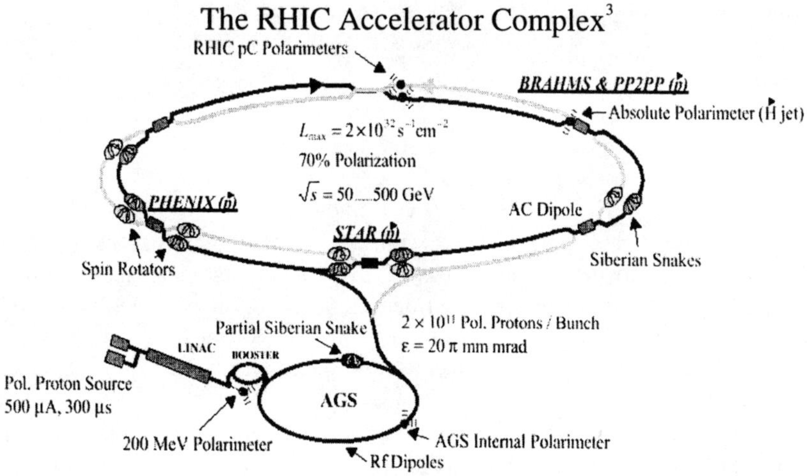

Figure 3. The RHIC polarized protons accelerator complex

The source an Optically Pumped Polarized Ion Source (OPPIS), originally from KEK, was upgraded and improved through a BNL, KEK, TRIUMF collaboration. It delivers 500 uA over a 300 usec pulse duration. The polarization is measured after the 200 MeV LINAC using a p-Carbon inclusive polarimeter that attains a 1% statistical accuracy in a few minutes. The source has consistently delivered 70 percent beam polarization into the Accumulator/ Booster with an intensity of 2.10^{11} per bunch. The beam polarization in the booster is preserved using simple harmonic dipole corrections. The beam is then injected into the AGS and the polarization measured first at $G\gamma = 7.5$ using a pp elastic scattering polarimeter. Generally we observe a similar polarization value at this stage comparable to what is seen at 200 MeV.

The AGS beam is then accelerated and injected into RHIC where each of the two rings is filled with 55 bunches in a 60-bunch pattern of about 1.10^{11} polarized protons per bunch. The fill pattern also sets each bunch polarization direction to assure that the experiments observe collisions with all possible combinations of beam polarization orientations. The beam polarization is measured at injection energy of 24 GeV. Beams in both rings are accelerated simultaneously to top energy of 100 GeV and stored. The beam polarization is then measured at the beginning of the store and subsequently at two-hour intervals. Beam stores in excess of 10 hours were routine.

Acceleration of polarized proton beams

The stable spin direction is determined by the magnetic structure of the accelerator. In a circular accelerator, this is in the vertical direction to assure minimal impact on the polarization vector from the guiding dipole field. Several conditions conspire to cause beam depolarization. When a resonance condition holds beam depolarization can occur. These resonances are generally expressed in terms of the spin tune $\nu_{sp.}$ *Imperfection resonances* arise from sampling errors in the magnetic fields or due to closed orbit errors. These occur when

$$\nu_{sp} = G\gamma = \text{integer}$$

G = 1.7928 being the anomalous magnetic moment of the proton and $\gamma = E/m$. Such resonances occur at approximately 0.45 GeV intervals during the acceleration process. *Intrinsic resonances* arise from sampling the focusing quadrupole fields due to finite beam emittance. They occur when

$$\nu_{sp} = G\gamma +/- \nu_y = \text{integer}$$

ν_y being the vertical betatron tune of the machine. For the AGS the vertical tune is approximately 8.75 and the beam crosses seven such resonances before reaching RHIC injection energy. Traditionally, imperfection resonances were overcome by flattening the field using harmonic dipole corrections. Intrinsic resonances were crossed quickly by a sudden change in the tune using strong fast quadrupoles. More recently we replaced the correction dipoles with a local single spin rotator that amplifies these resonances to force a spin flip. This rotation δ alters the spin tune conditions such that $\cos(\pi\nu_{sp}) = \cos(\delta/2) . \cos(\pi G\gamma)$. If $\delta = 0$ we revert to the above conditions. If $\delta = \pi$ the we have a 180° rotation and $\nu_{sp} = ½$. With δ = any other value then ν_{sp} cannot be an integer. In the AGS[4] this rotator is a partial snake (δ=5%) solenoidal magnet. The action of this solenoid is depicted in figure 4.

Similarly, an alternate scheme was employed to deal with intrinsic resonances. A strong rf dipole[5] forces large betatron beam oscillations near the location of the offending intrinsic resonance. This creates an artificial spin resonance that is excited coherently for the whole beam.

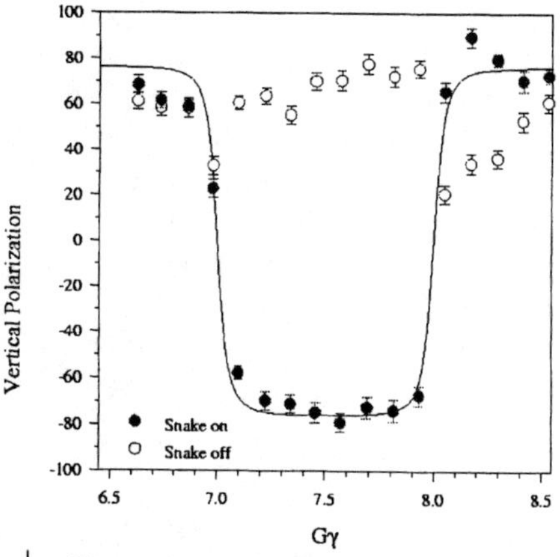

Figure 4. Spin flipping at $G\gamma = 7$ & 8 w/snake

This resonance then dominates and the spin of the beam adiabtically follows the rf dipole, Figure 5. This technique was used to spin flip four resonances during the acceleration cycle. Record beam polarization nearly 40% was attained in the AGS, Figure 6.

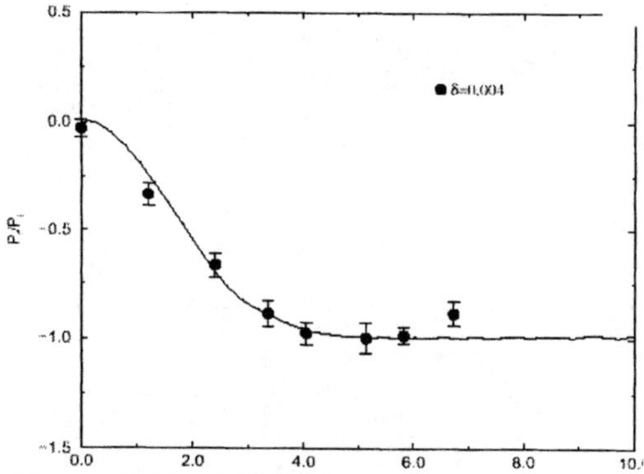

Figure 5. The AC dipole preserves the beam polarization near the resonance $G\gamma = 12+\nu$ Vs the coherent amplitude size.

However, the goal of achieving the 70% polarization was hampered by excessive beam emittance and coupling resonances that are induced by the solenoidal partial snake. A new concept is being studied which employs a significantly stronger super conducting helical partial snake (30% of a full 180° rotation) to overcome both the imperfection as well as the intrinsic resonances of strengths encountered in the AGS. No coupling resonances occur with a helical snake.

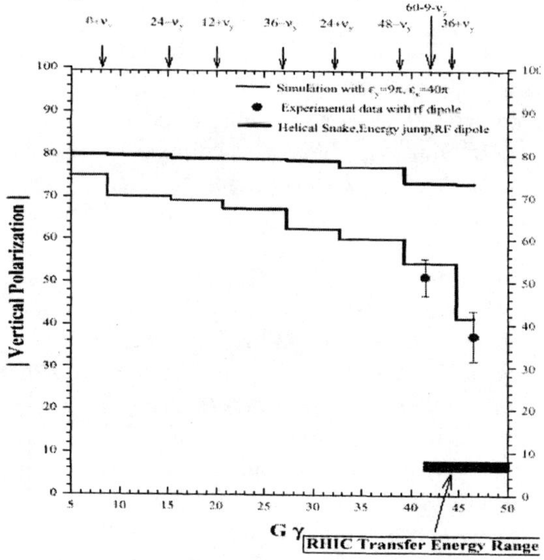

Figure 6. AGS beam polarization measurements and tracking simulations Vs intrinsic resonance locations.

At higher energies, full snakes (180° spin flip) are employed. At RHIC, two snakes at opposite sides in each ring have been installed. Super conducting helical snakes at 25 tesla-meters, were chosen for RHIC in order to effect this local spin flip. The helical design minimizes the transverse beam excursion and results in a smaller aperture. The snake action flips the vertical spin vector while maintaining the other components at zero. The snake is also transparent to the beam dynamics. Spin rotators will be installed during the shutdown in summer 2002 around the PHENIX and STAR detectors to allow for manipulation of the proton spin direction transverse or longitudinal with respect to the beam travel depending on the physics program.

Measurement of the Proton Beam Polarization

Measuring the beam polarization is an integral component of any polarized proton effort. In the AGS the polarimeter measures the recoil pp elastic scattering at t = 0.1-0.3 GeV^2/c^2. The analyzing power at 24 GeV is ≈ 3% with a 10% error and falls as the inverse of the beam momentum. Thus it is less useful at higher energies. For RHIC we chose the analyzing power in p-Carbon elastic scattering in the Coulomb Nuclear Interference (CNI) region Figure 7. The calculated analyzing power[6] is about 4%. The calculation does not take into account a single spin flip hadronic component which is not easily determined. The process is expected to depend minimally on beam energy.

A prototype AGS p-C polarimeter[7] (E950) measured the clearly identified carbon at 90 degrees in the lab system. The CNI asymmetry was measured near the RHIC injection energy and calibrated with an external polarimeter in an extracted beam line, Figure 7. A similar polarimeter was installed in RHIC. The RHIC Blue ring data taken during the FY 2000 commissioning run is overlaid.

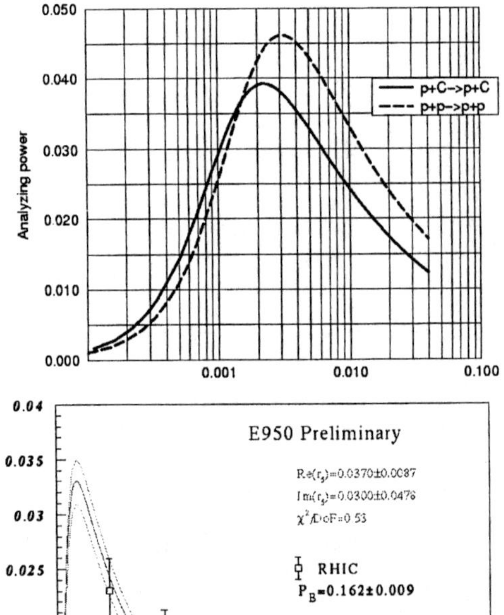

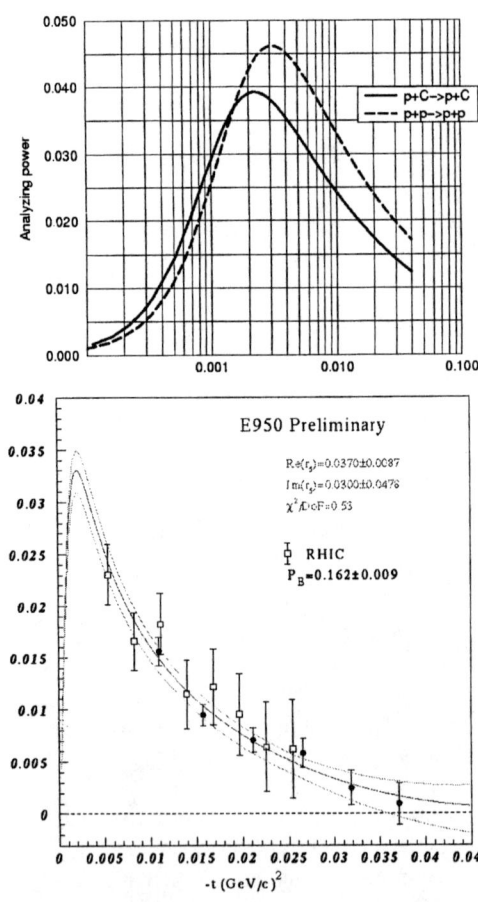

Figure 7. The calculated analyzing power Vs t in pp and p-C CNI. Measurements at the AGS and RHIC.

Status

The RHIC schedule this year called for a three-week commissioning period of the polarized proton beams followed by a five-week run for physics. This process was underway during the conference and has been completed as of this writing. Thus I can provide some details of the achievements. The OPPIS source ran quite reliably and delivered 70% beam polarization and intensity of 2.10^{11} protons per bunch to the AGS. The gymnastics of preserving the beam polarization in the AGS did not fare as well, but resulted in a 25-30% polarization at extraction. The lower AGS polarization is due the to the slower ramping Westinghouse power supply instead of the Siemens Motor Generator set which was undergoing repairs.

Polarized bunches were injected into RHIC near $G\gamma = 46.5$ approximately 24 GeV, the sweet spot for transfer with minimal loss in beam polarization. Polarization measurement at RHIC injection confirmed this. The bunch polarization pattern in each ring was set to assure collisions in sets of all possible spin combinations. The beams were then accelerated to 100.4 GeV, stored, and the polarization measured at 2-hour intervals. The two snakes in each ring, functioned as designed. We were able to measure comparable asymmetries at store compared to injection energy in many of the ramps. Despite the fact that we do not know the analyzing power at 100 GeV, indications are that we were able to preserve the polarization at store. The beam polarization was affected by two parameters, the betatron tune had to be kept away from the ¼ integer during acceleration, and beam orbit corrections to ½ mm rms were in order. The CNI polarimeters worked quite reliably. A 2% statistical measurement took about 2 minutes.

The RHIC spin flipping AC (rf) dipoles saw limited action. As described earlier, the dipoles excite an artificial resonance that flips the spins of all bunches in a ring. This initial attempt was successful in that the spins were flipped. However, we retained only 50% of the beam polarization after the flip.

Attempts to decelerate the beam from 100 GeV back to injection energy were unsuccessful. This is needed to calibrate the analyzing power of the CNI polarimeters. The plan was to measure the asymmetry at injection, accelerate and measure at store, decelerate back to injection and measure again. The difference in the two asymmetry measurements at injection could be ascribed equally to the acceleration and deceleration processes. This technique can provide an acceptable calibration of the analyzing power at any energy. The beam loss during deceleration was ascribed to the hysteresis of the magnets.

The run had two goals. One was to take pp data to normalize the heavy ion data taken with gold beams at the same energy per nuceon, and the other to collect spin physics data using both transverse and longitudinal polarization. Due to the relatively low 20% beam polarization, it was decided to stay with the transverse beams only. Spin physics measured the asymmetry in inclusive π^0 and γ production at large x_f with an eye towards local polarimetery (STAR and PHENIX). Asymmetry in inclusive π^0 production at relatively high p_t at large x_t, and asymmetry in inclusive charged hadron production (PHENIX), the asymmetry in pp elastic scattering at low t (the pp2pp experiment). All in all this proved to be a successful run.

The major RHIC spin physics program will utilize the two large heavy ion detectors PHENIX and STAR, Figure 8, which were constructed to handle large multiplicities from heavy ion collisions. Additional apparatus was added to enhance the spin physics capabilities. In PHENIX this included, a second muon arm, a special electromagnetic trigger, and the addition of more beam-beam counters. STAR added a barrel as well as one end cap electromagnetic calorimeters in addition to beam-beam trigger counters. Furthermore local polarimeters at each detector are contemplated for future runs. The two detectors are complementary in their physics capabilities. While STAR comprises a large volume TPC with jet physics capability, PHENIX has high rate capability, high-resolution calorimeters, and muon detection. The smaller detectors BRAHMS and PHOBOS as well as the newly installed pp2pp elastic scattering experiment will also utilize the spin physics aspects.

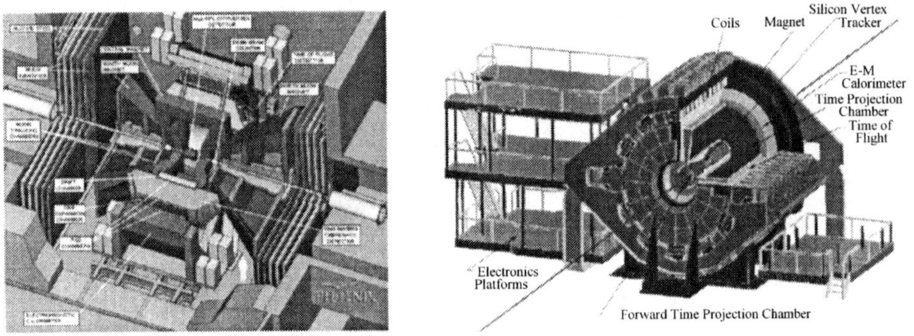

Figure 8. The PHENIX and STAR detectors at RHIC

The primary quest is the direct measurement of the gluon polarization and contribution to the spin content of the nucleon. The primary tools are the clean direct photon production, inclusive jet production, and J/ψ and charm production[2]. The reach of RHIC, see Figure 9, in this domain compares favorably with the planned efforts and will add significantly to our knowledge of the gluon polarization. This will be carried out at center of mass energies of 200 GeV as well as 500 GeV.

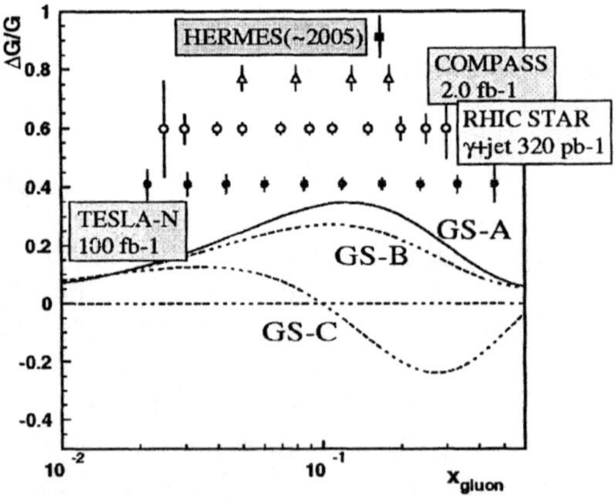

Figure 9. Projected Gluon polarization measurements

The second goal is the direct measurement of the quark and antiquark polarization using $W^{+/-}$ production[2] and decay into electrons and muons at 500 GeV cms. This is particularly true as one probes the extremes of the rapidity distributions where the single spin asymmetry in W production directly probes the quark polarization. An example of the statistical accuracy in a 10-week of RHIC running at design luminosity is shown in Figure 10.

Finally, with the flexibility that RHIC provides with the spin rotators at each experiment, the plan also calls for studying transversity distributions, which is a new arena. Another exciting possibility is to use parity violation in hard jet production to probe potential new contact interactions beyond the Standard Model. Of course the pp2pp experiment is also poised to utilize spin and measure the single and double spin asymmetries in pp elastic scattering in the Coulomb and medium t regions.

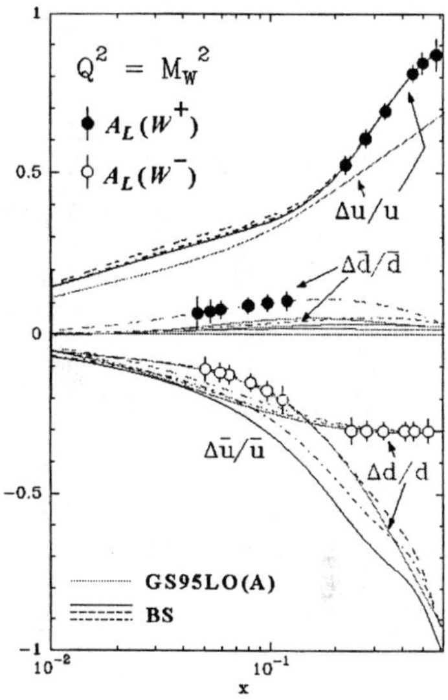

Figure 10: The range and statistical accuracy that can be measured by PHENIX of the quark and antiquark polarization.

Conclusions

The recent success achieved in accelerating and colliding polarized proton beams at RHIC provides a unique opportunity and a new laboratory to enrich our understanding of spin structure functions of the nucleon and probe spin effects in related hadronic QCD processes. We are at the early stages of this exciting era. More work is in store to install the spin rotators, attain higher beam polarization, and install a polarized jet target to calibrate the beam polarization to better than 5% accuracy at all energies.

Acknowledgements

I am indebted to the members of the RHIC SPIN Collaboration, experimentalists, theorists and accelerator physicists whose work is highlighted. Of course collectively, we are indebted to RIKEN, Japan for their financial support for the Polarized Proton Collider hardware at RHIC as well as the RIKEN BNL Research Center. This work is performed under the Department of Energy Contract DE-AC02-98CH10886.

References

1. Proceedings of the Polarized Collider Workshop, University Park PA, 1990, J. Collins, S. Heppelman, and R.W. Robinett, editors.
2. For a comprehensive review of physics precursors and motivation. Prospects For Spin Physics at RHIC, G. Bunce, N. Saito, J. Soffer, W. Vogelsang, *Annual Reviews of Nuclear and Particle Science,* 2000, **50**, pp 525-575.
3. I. Alexeev et al. Polarized Proton Collider at RHIC, Design Manual 1998, unpublished.
4. H. Huang et al. Phys. Rev. Lett. **73**, 2982 (1994)
5. M. Bai et al. Phys. Rev. Lett. 80, 4673 (1998)
6. For a discussion on the latest theoretical developments see B.Z. Kopeliovich and T.L. Trueman, Phys. Rev. **D64**, 34004 (2001)
7. J. Tojo et al. Measurement of Analyzing Power for Proton-Carbon Elastic Scattering in CNI Region with a 22 GeV/c Polarized Proton Beam, Submitted to Phys. Rev. Lett.

Spin Effects in High Energy Scattering in a Simple Constituent Model

Homer A. Neal

Department of Physics, University of Michigan, Ann Arbor, Michigan 48109-1120 USA

Talk Presented at the Coral Gables Conference on Cosmology and Elementary Particle Physics, December 14, 2001

Abstract: In this paper the author presents an overview of the evolution of proton-proton spin physics over the past quarter century and describes recent work on interpreting high energy polarization effects in a constituent scattering model.

INTRODUCTORY REMARKS

It is with considerable trepidation that I take the floor to speak to you today about advances in spin physics because I recall having been asked to address this same subject at the Coral Gables Conference some 24 years ago [1]. Clearly, the audience will have every reason to expect me to address the question "what has happened in the intervening years". Indeed, in keeping with the request of the conference organizers, my presentation today will be more of a retrospective one than a detailed technical review. I will provide references to sources where the original data may be found and I will give a brief overview of some of my ongoing work in studying polarization phenomena in a quark-quark scattering model. My focus will be on reflections related to the evolution of spin physics through the window of proton-proton scattering.

SPIN: ITS HISTORY

In his reflections, Dirac -- who, parenthetically, quizzed me vigorously about some new data I presented at my talk here in 1977 -- indicated that one of the earliest suggestions that electrons had spin came from Arthur Compton in 1921 [2]. Compton noted that if electrons were not symmetrical many properties of magnetism could be explained. He also noted that tracks in a Wilson Cloud Chamber had unexpected kinks. He reasoned that if the electron had a magnetic moment it would induce a magnetic moment in the surrounding medium and lead to the observed helical motion of the electron.

In the following years there were other indications that new quantum numbers were needed to explain observations. The number of spectral lines observed in atomic spectroscopy seemed to be double what one would expect from standard Bohr-orbit calculations. German physicists labeled the problem the "duplexity problem":

CP624, *Cosmology and Elementary Particle Physics*, edited by B. N. Kursunoglu et al.

"Whenever one took an atom and brought in a further electron, the number of states that one got was double what one expected" [2].

In this environment of puzzlement, it was Kronig in early 1925 who first suggested more formally that the electron had an intrinsic spin. The idea was not embraced by Pauli, and Kronig did not pursue the concept further [3]. Additional insights on the Pauli-Kronig exchanges can be found in the Pauli Festschrift volume edited by Fierz and Weisskopf [4].

Later in 1925, however, two Dutch students, Goudsmit and Uhlenbeck, produced a short paper which suggested electrons had spin and gave it to their advisor, Ehrenfest. They then traveled to Haarlem and showed the paper to Lorentz. Lorentz argued that the concept was not sound and could not possibly be correct because of energy considerations. Goudsmit and Uhlenbeck went back to Leiden and asked Ehrenfest to withdraw the paper. But it was too late. Ehrenfest had submitted the paper to Naturwissenshaften [5]. And, as you know, the rest is history.

Surprisingly, a couple of years later, the discovery of the spin of the proton had a similar tentative emergence. David Dennison had submitted a paper on June 3, 1927 on the specific heat of the hydrogen molecule. Two weeks after that submission (June 16, 1927, to be exact), he sent an addendum to the paper in which he pointed out for the first time that "the mixing ratio of 1/3:1 means that the spin of the proton is 1/2 ...." [3]. Dennison's personal reflections about this work can be found in his memoirs in reference [6].

In the spirit of full disclosure, I must warn the audience that there will be many references in my talk that seem to be directed toward the University of Michigan and its role in the evolution of spin physics. That is no accident, since I am drawing from materials used when I delivered the inaugural talk on the occasion of my being named the Samuel A. Goudsmit Distinguished University Professor at the University of Michigan a couple of years ago. I used that occasion to celebrate the 75 years of excellence of Michigan in spin physics and I want to share examples of those achievements with the audience here today.

Samuel Goudsmit and George Uhlenbeck, the discoverers of electron spin, spent the principal part of their professional careers on the faculty of the University of Michigan. David Dennison, who, as noted above, was the first to publish the finding that the proton had spin of 1/2, spent almost all of his career at the University of Michigan and was, I might add, one of my most revered professors when I did my graduate work in Ann Arbor. Richard Crane, who is one of our distinguished Michigan emeritus faculty, made the first measurement of g-2 for the electron - a measurment that many at the time had labeled as impossible. Oliver Overseth, a University of Michigan emeritus faculty member, made the first alpha measurment for the lambda hyperon. Michael Longo, who is here at the Conference today, and who was my thesis advisor at Michigan, has made a series of important polarization studies, including his recent studies of the cascade hyperon. Finally, another Michigan faculty member who is well known to the Conference attendees is Alan Krisch, who over the decades has led the advances in both polarized targets and polarized beam technologies that have resulted in so many of the new precision data that are now available to the physics community. So, as you can see, celebrating the contributions of Michigan researchers to spin physics seems appropriate. However, by no means do I intend to slight today the enormous contributions made by others.

THE EXPERIMENTAL CHALLENGES

At each level of aggregation, from molecules to atoms to nuclei, there are interesting questions about spin to be explored -- and are being explored. In elementary particle physics we are focused on a more fundamental set of issues. We want to know what happens when two elementary particles collide and what part of the scattering process is due to the spins of the colliding particles. We demand that any rigorous theory of particle interactions also properly explain the role of spin in the interactions. Spin determines the statistics that the interacting particles must obey and we are ever alert to any indication that our understanding of the spin-statistics relationship is not correct. We are also vigilant in insuring that every opportunity is exploited to use spin as a tool in determining the extent to which of our revered conservation laws are obeyed, and these are most directly tested at the level of the elementary particles.

The set of studies I will review here involves the examination of spin effects in a few well defined scattering processes. Namely, what is the role of spin in elastic proton-proton scattering and in inclusive hyperon production.

MEASUREMENT TECHNIQUES

I will use a set of my earlier experiments to provide examples of the techniques that were available for making spin studies prior to the recent advances in polarized beam and polarized target technologies. Then I will comment on the increased sensitivity made possible by these advances. The ultimate challenge faced by experimenters is, succinctly, how to fix the spins of the initial state particles and measure the spins of the final state particles -- and to know the identities and origins of each particle studied in the reaction. Not every spin experiment has to know all spins in the reaction, but the more spin information you demand, the more insight you gain about the role of spin in the interaction, and the more difficult the experiment becomes.

When I gave my previous talk at the Coral Gables Conference, the premiere tools for fixing the spin of target protons were the lanthanum magnesium nitrate (LMN) and ethylene glycol polarized targets. The operating principle for the LMN target was to polarize magnetic centers in the target and to stimulate via microwave radiation those centers to couple to the "free" protons in the target and flip the proton's spin. A LMN center does this through a joint spin flip, but then it relaxes back to its original spin state in the presence of a large magnetic field and is ready to couple to another "wrong" state proton and flip it. Meanwhile the flipped protons remain in their polarized state for a long period, because of their weak magnetic moment and long relaxation times. The ethylene glycol targets operated on a slightly different principle of so-called thermal mixing.

In our Argonne ZGS intermediate energy polarization study [7] we employed an ethylene glycol target with a maximum target polarization of approximately 40%. This experiment was the first to use a polarized proton target in an external proton beam at high energies. We used a beam of ~ 5 x 10^9 protons per pulse, limited to avoid target damage. Moreover, we had to contend with background events because free protons comprised only ~10% of the target. Today one has targets with polarizations well in excess of 90% and that are much more radiation resistant. This is a remarkable advance over the time since my last presentation. I must give credit for this to the team of collaborators led by Alan Krisch [8].

In 1977 the concept of high energy polarized proton beams was in its infancy. It was also the determined effort of Alan Krisch and colleagues that made it possible to achieve beams of highly polarized protons at many accelerators (the Argonne ZGS, the Brookhaven AGS, RHIC, and the Indiana Cyclotron). Since there are other talks at this Conference on polarized beams, and since it is a subject well described in the literature [8], I will not go into detail here about the design of these beams. I will mention, however, that one of the great nemeses of polarization beam acceleration is that the beam can become rapidly depolarized by machine imperfections as a given particle passes numerous times through the same imperfection. That problem has been cleverly addressed by two Russian physicists (Derbenev and Kondratenko [9]), who use the machine tune to cause the proton spin to rotate by 180 degrees on each turn around the machine, which causes the effects of the imperfection to be cancelled. The technology to accomplish this is commonly referred to as the "Siberian Snake" [10]. By the way, relative to the Michigan involvement in spin physics, Dr. Derbenev has been on the Michigan staff for more than a decade.

Advances in polarized beams and targets over the decades have been phenomenal -- and this is what has made possible several of the physics results I will review below.

In an experiment at the Brookhaven Cosmotron Mike Longo and I made pioneering measurements of the polarization in p-p elastic scattering at energies of approximately 1 GeV using only a carbon analyzer and a polarimeter that tracked its alignment relative to the incoming beam by using optical spark chambers. An example of the results is shown in Figure 1 [11]. The significance of these measurements is that, for the first time, precision measurements were made beyond 90 degrees cms for |t| values greater than 1 $(GeV/c)^2$. What we observed was that the polarization assumed a value of zero well before it needed to in satisfying symmetry conditions at 90 degrees cms. We believe this was the first observation of the zero in p-p polarizations at -t =~ 1 $(GeV/c)^2$, a feature which we now know persists all the way into the hundreds of GeV/c incident beam momentum. As noted later, I believe this to be due to a transition from the scattering of one quark off one quark to two quarks off two. Unfortunately, we did not know that quarks existed at the time of this measurement. This advance was made simply by using a carbon analyzer, with no polarized targets or beams.

While preparing for additional spin studies early in my years as a faculty member at Indiana University, I noted that the existing cross section data in p-p elastic scattering was still of poor statistical quality. It seemed to make no sense to try to understand polarization effects in an energy region where even the cross section was poorly known, so I temporarily refocused my efforts toward attaining a precise measurement for the p-p cross sections in the intermediate energy regime. At that time the technology of magnetostrictive spark chambers was just being perfected, and we decided to use these chambers for the cross section measurements. The results are shown in Figure 2. One special outcome of these measurements was the discovery of the unusual behavior of the fixed angle cross section, as portrayed in Figure 3. At the time these findings were just regarded as another bizarre rendering of the stubborn and complex p-p system. As discussed later, however, my interpretation now of these breaks is that they too are indicative of the quark structure of the proton.

As previously mentioned, in another experiment I led we used for the first time an ethylene polarized target in an external proton beam to study the polarization in p-p elastic scattering in the 6-12 GeV/c range. One of our findings from that experiment was a fractal "bump" structure in the polarization [7]. I will refer to this odd feature in a later section as one that gives further credence to our picture of the constituent structure of the proton.

CAN WE SEE QUARK POLARIZATION?

As I departed for a sabbatical at the Bohr Institute in 1973, in addition to my normal luggage I was burdened by the data described above. The cross section data had shown discontinuities that I did not understand. The polarization data displayed structure which seemed strange and directly correlated with the cross section breaks. As I took up these issues with Holger Nielsen, whom I met early in my stay at the Institute, we rather quickly became interested in the possibility that the effects were a manifestation of the quark structure of the proton. Indeed, we had become aware of the work of Landshoff who was having success explaining cross section magnitudes by using an independent quark model [12].

The principal postulate we advanced in a Physics Letter article [13] was that within t ranges where the fixed angle cross section was smooth and continuous in slope, the scattering process was dominated by a fixed number of quark-quark scatters, and as one crossed one smooth region to another the slope change that occurred was due to a change in the number of q-q scatters involved (see Figure 3). We also assumed that, to the extent permitted by the quarks combining to make a spin-1/2 proton, the quarks behaved independently within the proton, they contributed their own polarizations to make up the overall polarization of the proton, and the internal processes that occurred after the quark scattering resulted in the appropriate redistribution of quark momenta but did not affect the overall proton polarization.

This further suggested that in the region we ascribed to a single quark scattering from a single quark ($-t < 1$ $(\mathrm{GeV/c})^2$), the proton polarization observed was actually the polarization in single q-q scattering. Armed with the knowledge of what the single q-q scattering polarization was, we could then predict what the polarization would be in regions where the dominant scattering mechanism was for 2, 3 or more scatters. Indeed, in our 1974 paper we focussed on using the region #1 data to predict what would happen to the polarization in region #2 ($1 < |t| < 3.5$ $(\mathrm{GeV/c})^2$). In that region one would expect the polarization to be roughly 2 P(t/4), where P(t) is the polarization in region #1. We were impressed with the success of the model in this test. The data beyond region #2 was so poor at that time that we were content to note that we expected the polarization to be even larger in that region, but we attempted no detailed fits.

As the decades passed and the technological developments for improved polarized targets and beams proceeded, the Krisch group began to accumulate high statistics data in what we call region #3 ($-t > 3.5$ $(GeV/c)^2$), the region where we might expect that three quarks would scatter from three quarks [8]. These measurements were the subject of much skepticism in the community in that no mainstream model could explain how spin effects of > 30% could be present at high energies and large $|t|$. Indeed, most models predicted that in that regime, the polarization should be zero. Our quark-quark scattering model was, to our knowledge, alone in predicting that there should not only be significant spin effects in that region, they should be large, and be simply a scaled version of the polarization in the previous regions. A fit from our 1998 Physics Letter publication [14] (Figure 4) shows how the model is able to account for the significant structure in the existing p-p polarization data at the energy that, at present, has the largest span of t values in existence. The number of multiple scatters was allowed to vary when the fit shown was generated.

The success of this simple model encourages us to believe that it is the individual quarks inside the proton that are being seen, that there are only three types of scatters (since there seems to be no significant shape changes in the cross section that would signal a fourth region), and that even the spin information in each q-q scattering is preserved and passed on to the scattered proton.

We are often asked if the so-called "spin crisis" [15], which suggests that quarks do not carry all of the proton spin, would invalidate our model. The answer is it does not. The only features of the scattering process that we have invoked is that there are three "clusters" inside the proton that are responsible for the scattering. Each cluster could, for example, be a valence quark plus gluons and the general concept of the model would remain unchanged. The site of the spin is not important.

ARE SPIN CORRELATION EFFECTS MEASUREABLE AT THE QUARK LEVEL?

The next challenge we presented the model was to explain the extraordinary behavior observed in the p-p cross section when the two initial protons have their spins parallel vs. anti-parallel. What one notices in the precise A_{nn} data from the Krisch group (Figure 5) is that the two cross sections are comparable up until $|t|$ values are reached that represent the entry into our region #3. In a paper by Neal and Nielsen published last year in Physics Letters [16], we have argued that, if the basic tenets of the quark interchange model are assumed to be valid, the only way to have three quarks scatter from three quarks and not pay the penalty for spin flips of the protons, is to have the initial state protons have their spins parallel. In such a picture the anti-parallel cross section in region #3 would be expected to just be an extrapolation of the region #2 cross section. However, with the additional channels available to the parallel cross section, it would be expected to be larger than the anti-parallel values, starting near $-t = 3.5(GeV/c)^2$. So our model even goes some distance in explaining this dramatic effect.

QUARK POLARIZATIONS AND SPIN EFFECTS IN INCLUSIVE LAMBDA PRODUCTION

I now would like to say a few words about some current work I am pursuing that seeks to test the applicability of our model to Λ inclusive polarizations.

At the time of my last appearance at this Conference it had just been reported that huge polarizations had been observed in Λ hyperons which were inclusively produced when 400 GeV/c protons struck a Be target [17]. This was very puzzling since both the very high energy and the inclusive nature of the production should suggest that spin effects would be small, if they existed at all. To date, this observation, and the many similiar observations for other hyperons, remains a mystery.

I have attempted in recent months to determine just how different the inclusive lambda polarization is from what one would expect based on p-p data. More precisely, since our model purports to be able to extract the q-q polarization parameter, one might simply ask the question "at a given Λ p_t , what would one expect the Λ polarization to be if that polarization were due solely to the u,d quarks coming from the incident proton?" Heretofore, this has not been a question that could be simply broached, because mainstream models have no way of asserting directly what the q-q polarization parameter should be. What almost all models have done so far in dealing with the Λ polarization matter is to assume that the polarization comes mainly from the s-quark, which is produced from the vacuum as part of a s-sbar pair, as illustrated in Figure 6.

Figure 7 illustrates the remarkable features of the lambda inclusive data, including the prominent features of the large polarization values and a plateau for different values of the Feynman x [18]. Figure 8 shows the measured lambda polarization along with values predicted for the lambda polarization using the 300 GeV/c p-p polarization from Synder *et al* in conjunction with our model [19]. Some simple assumptions have been made about how to scale the momentum of the proton data before overlaying those data on the lambda plot. In essence we require that the p_t of the u,d scatters be sufficient to provide the p_t needed by the u,d,s and sbar quarks in producing the lambda. The fact that the measured and predicted results agree so well further encourages us in believing the general validity of our model. By allowing some polarization contribution by the s-quark and allowing it the ability to flip spins with the u,d quarks at low pt, but then decreasing that coupling as pt grows -- and noting the requirement for the u,d quarks to ultimately be in a singlet state in the final lambda -- one has the basic ingredients for achieving a flattening in the lambda polarization, also in keeping with observation. This work is ongoing and additional steps are required to rigorously underpin or alter the assumptions advanced above.

ANTICIPATED SPIN STUDIES AT LHC ENERGIES

Throughout the last few decades, whenever the high energy physics community has faced the prospects of moving to the next higher energy or $|t|$ range there have been many voices suggesting that in the new range spin effects will have died away. Such assertions have always been wrong. As we face the prospects of running at the

LHC in a few years, I fully expect that we will again be in an interesting arena where spin effects will be measurable and important in many analyses.

The LHC will produce an enormous number of t-tbar pairs, for example. The relative helicities of these particles will set considerable constraints on the decay distributions of these particles, and will provide sensitive tests of the Standard Model [20].

Also, Λ_b should be produced in prodigious quantities at the LHC. By using the measured decay properties of $\Lambda_b \rightarrow \psi\Lambda$ and $\psi \rightarrow \mu\mu$, one should be able to determine the polarization of the Λ_b to < 2%. This may be sufficient to gain some additional insights into how hyperons are inclusively produced, particularly since most analyses suggest that, given the massive nature of the b-quark, polarization effects associated with the Λ_b should be significantly larger than even those of the Λ [20].

CONCLUDING REMARKS

In concluding, I hope I have given you some flavor of the evolution of spin physics, the role it has played in shaping our thoughts about the interactions of particles, and our expectation that it will continue to provide insights as data are acquired at the LHC.

I want to stress that the window I have offered here on spin physics is very narrow. This subfield has so many other components these days, ranging from probes in deep inelastic scattering, e+ e- studies, and polarized structure functions to anticipated new discoveries at RHIC and UNK. Indeed, it is the explosion of the richness of spin studies that may be one of the most notable developments in this area of science over the past quarter-century.

I want to again salute the members of our community who have advanced the technology in this complex area of our field. I believe that such efforts will eventually allow us to make the measurements necessary to really learn the role of the mysterious spin quantum variable in Nature.

I wish to acknowledge Jens Zorn for his assistance with several of the historical issues referenced in the article. Finally, I wish to acknowledge the conference organizers for sustaining this very special set of scientific meetings over the decades.

REFERENCES:

1. H. A. Neal, Advances in the Study of Spin Effects in Nucleon-Nucleon Scattering at Small and Intermediate Momentum Transfers in New Frontiers in High Energy Physics, Edited by Behram Kursunoglu, Arnold Perlmutter, and Linda F. Scott (Plenum Publishing Corporation, 1978).
2. Dirac, P. 1975. In Proc. Summer Study High Energy Physics. *Polarized Beams*, ed. J. Roberts, *Argonne National Lab. Rep. HEP-75-02.*
3. 3)The Story of Spin, Sin-Itiro Tomonaga, University of Chicago Press, 1997, pp. 258.
4. 4) "Theoretical Physics in the 20th Century, A Memorial Volume to Wolfgang Pauli", Edited by M. Fierz and V.F. Weisskopf. New York, Interscience (1960) 328p
5. Goudsmit, S. , Uhlenbeck, G. 1925. Naturwissenschaften 13:953.
6. D.M. Dennison, "Recollections of Physics and of Physicists During the 1920s", Am. J. Phys. 42, 1051-1056 (1974).
7. Abshire, G.W. et al, Physical Review D 9 (1974) 555.

8. R.C. Fernow, A.D. Krisch, High Energy Physics with Polarized Beams, Ann. Rev. Nucl. Part. Sci. 1981, 31:107-44.
9. Y. S. Derbenev and A. M. Kondratenko, Part. Accel. 8 115 (1978).
10. A. D. Krisch, Physics of Atomic Nuclei, 62, Issue 3, 1999 pp. 479-484.
11. H. A. Neal and M. J. Longo, Phys. Rev. 161, 1374 (1967).
12. P.V. Landshoff, Phys. Rev. D 10 (1974) 1024.
13. H.A. Neal, H.B. Nielsen, Physics Letters B 51 (1974) 79.
14. H.A. Neal, J.B. Kuah, H.B. Nielsen, Physics Letters B 439 (1998) 407. (Note: correct coefficient of t in Fig. 2 in this reference is -1.522; correct units for cross section for Figs. 5, 6 are $\mu b/(GeV/c)^2$.
15. John Ellis, Proceedinsg of the 12th International Conference on High Energy Spin Physics, pg. 7, 1997, World Scientific.
16. Homer A. Neal and Holger B. Nielsen, Physics Letter B 508 (2001) 251-258.
17. G. Bunce et al, PRL36, 1113 (1976).
18. J. Lach, Hyperon Polarization: An Experimental Overview, Fermilab-Conf-92/378, December, 1992.
19. B. Lundberg et al., Phys. Rev. D40, 3557 (1989); C. Wilkinson et al., Phys. Rev. Let. 46, 803 (1981); J.H. Synder et al, Phys. Rev. Lett. 41, 781 (1978).
20. "ATLAS Detector and Physics Performance Technical Design Report", Vol. II, CERN/LHCC/99-15, ATLAS Collaboration, 25 May, 1999.

Precision measurement of the muon anomalous magnetic moment

E.P. Sichtermann[11], H.N. Brown[2], G. Bunce[2], R.M. Carey[1], P. Cushman[9], G.T. Danby[2], P.T. Debevec[7], M. Deile[11], H. Deng[11], W. Deninger[7], S.K. Dhawan[11], V.P. Druzhinin[3], L. Duong[9], E. Efstathiadis[1], F.J.M. Farley[11], G.V. Fedotovich[3], S. Giron[9], F. Gray[7], D. Grigoriev[3], M. Grosse-Perdekamp[11], A. Grossmann[6], M.F. Hare[1], D.W. Hertzog[7], V.W. Hughes[11], M. Iwasaki[10], K. Jungmann[6], D. Kawall[11], M. Kawamura[10], B.I. Khazin[3], J. Kindem[9], F. Krienen[1], I. Kronkvist[9], R. Larsen[2], Y.Y. Lee[2], I. Logashenko[1,3], R. McNabb[9], W. Meng[2], J. Mi[2], J.P. Miller[1], W.M. Morse[2], D. Nikas[2], C.J.G. Onderwater[7], Y. Orlov[4], C.S. Özben[2], J.M. Paley[1], C. Polly[7], J. Pretz[11], R. Prigl[2], G. zu Putlitz[6], S.I. Redin[11], O. Rind[1], B.L. Roberts[1], N. Ryskulov[3], S. Sedykh[7], Y.K. Semertzidis[2], Yu.M. Shatunov[3], E. Solodov[3], M. Sossong[7], A. Steinmetz[11], L.R. Sulak[1], C. Timmermans[9], A. Trofimov[1], D. Urner[7], P. von Walter[6], D. Warburton[2], D. Winn[5], A. Yamamoto[8], D. Zimmerman[9]

[1] *Department of Physics, Boston University, Boston, MA 02215, USA*
[2] *Brookhaven National Laboratory, Upton, NY 11973, USA*
[3] *Budker Institute of Nuclear Physics, Novosibirsk, Russia*
[4] *Newman Laboratory, Cornell University, Ithaca, NY 14853, USA*
[5] *Fairfield University, Fairfield, CT 06430, USA*
[6] *Physikalisches Institut der Universität Heidelberg, 69120 Heidelberg, Germany*
[7] *Department of Physics, University of Illinois at Urbana-Champaign, IL 61801, USA*
[8] *KEK, High Energy Accelerator Research Organization, Tsukuba, Ibaraki 305-0801, Japan*
[9] *Department of Physics, University of Minnesota, Minneapolis, MN 55455, USA*
[10] *Tokyo Institute of Technology, Tokyo, Japan*
[11] *Department of Physics, Yale University, New Haven, CT 06520, USA*

Abstract. The Muon $g-2$ collaboration has measured the anomalous magnetic g value of the positive muon with an uncertainty of 1.3 parts per million. The result $a_\mu^+(\text{expt}) = 11\,659\,202(14)(6) \times 10^{-10}$, based on data collected in 1999 at Brookhaven National Laboratory, is in good agreement with the preceding data on a_μ^+ and a_μ^-, and improves the combined uncertainty by a factor of about three. The analysis of data collected in 2000 and 2001 is well underway and, when combined with data from a requested, final run in the fall of 2002 and winter of 2003, is expected to further improve the experimental uncertainty by a factor of about three to 0.4 ppm. The measurement tests standard theory and has the potential to discover new physics.

CP624, *Cosmology and Elementary Particle Physics*, edited by B. N. Kursunoglu et al.

INTRODUCTION

The magnetic moment $\vec{\mu}$ of a particle with charge e, mass m, and spin $\vec{s}$ is given by

$$\vec{\mu} = \frac{e}{2mc} g \vec{s} \tag{1}$$

in which g is the gyromagnetic ratio. Dirac theory predicts $g = 2$ for point particles with spin $\frac{1}{2}$.

Historically, precision measurements of the anomalous magnetic g values

$$a = \frac{g-2}{2} \tag{2}$$

of leptons and baryons have played an important role in the development of particle theory. For example, the g value of the proton is found to differ sizably from 2, which provides evidence for a rich internal proton (spin) structure. The lepton g values deviate from 2 only by about one part in a thousand, consistent with the current evidence that leptons are point particles. The anomalous magnetic g value of the electron, a_e, is among the most accurately measured quantities in physics, and is known with about four parts per billion (ppb) uncertainty. Its value is described in terms of Standard Model field interactions, with a leading order contribution $\alpha/(2\pi)$ from the Schwinger term. QED processes involving virtual electrons, positrons, and photons contribute almost all of the measured value of a_e, whereas processes involving particles heavier than the electron contribute only at the ppb level, similar in size to the present experimental uncertainty.

The anomalous magnetic g value of the muon, a_μ, is more sensitive to processes involving massive particles than a_e of the electron, typically by a factor $(m_\mu/m_e)^2 \sim 4 \cdot 10^4$. A series of three experiments at CERN [1, 2, 3] measured a_μ to an uncertainty of about 7 parts per million (ppm), and thus tested electron-muon universality and the existence of a hadronic contribution of about 60 ppm to a_μ. Electroweak processes are expected to contribute to a_μ at the 1 ppm level, a contribution that is yet to be measured. Many speculative theories beyond the Standard Model predict contributions similar in size.

The muon $g-2$ experiment at Brookhaven National Laboratory (BNL) has determined a_μ^+ to an uncertainty of 1.3 ppm from data collected in 1999 and aims for a further improvement of the experimental uncertainty to 0.4 ppm. In the following, we give a brief overview of the experiment, and describe in some detail the data analysis and results. The theory is reviewed by Nath in these proceedings [4].

EXPERIMENT

The concept of the experiment at BNL is identical to that of the last CERN experiment [3] and involves the study of the orbital and spin motions of polarized muons in a magnetic storage ring.

The present experiment is situated at the Alternating Gradient Synchrotron (AGS), which in 1999 delivered 40×10^{12} protons in six 50 ns (FWHM) bunches, separated by 33 ms, over its 2.5 s cycle. The 24 GeV protons are directed onto a nickel target. Pions emitted from the target with energies of 3.1 GeV are captured into a 72 m straight section of focusing-defocusing magnetic quadrupoles, to transport the parent beam and naturally polarized muons from forward pion decays. Muons are momentum-selected at the end of the straight section, and injected into the magnetic storage ring through a field-free inflector [5] region. A pulsed magnetic kicker located at about one quarter turn from the inflector region gives a 10 mrad deflection which places the muons onto stored orbits. Electrostatic quadrupoles are used for vertical focusing, and in combination with collimators to shape the beam immediately following its injection. Typically, 50×10^3 muons were stored per AGS cycle.

The superferric storage ring [6] of 14.2 m diameter has an effective circular aperture 9 cm in diameter. Its homogeneous magnetic field of about 1.45 T is measured with a 17 probe NMR trolley that travels through the beam storage region over the full azimuthal range and with 360 NMR probes in the top and bottom walls of the vacuum chamber [7, 8]. The system is calibrated against a spherical water probe [9], which has a known relation to the free proton NMR frequency.

The muon spin rotation with frequency ω_s and momentum precession with frequency ω_c is measured as the difference frequency,

$$\omega_a = \omega_s - \omega_c = \frac{e}{mc} a_\mu \vec{B}, \tag{3}$$

via the measurement of the muon decay positrons with 24 calorimeters [10] on the open, inner side of the C-shaped ring-magnet. The above expression receives small corrections in $\beta \cdot B$ for out of plane oscillations and in $\beta \times E$ when γ is not exactly equal to its 'magic' value [3], corresponding to a momentum of 3.1 GeV/c.

DATA ANALYSIS

The analysis of a_μ follows, naturally, the separation of the measurement in the frequencies ω_p related to the magnetic field and the frequency ω_a related to the muon spin precession,

$$a_\mu = \frac{R}{\lambda - R}, \tag{4}$$

where $R = \omega_a/\omega_p$ and $\lambda = 3.183\,345\,39(10)$ (30 ppb) [11]. The analysis of ω_p and ω_a have both been performed by several largely independent groups within the muon $g-2$ collaboration. Only after each of the analyses had been finalized were the results for ω_p and ω_a from the various analyses combined, and was the value of a_μ evaluated.

The analysis of ω_p

The analysis of the 1999 magnetic field data starts with the calibration of the NMR probes in the field trolley from measurements taken at the beginning and end of the running period. In these measurements, the field in the storage region is tuned to very good

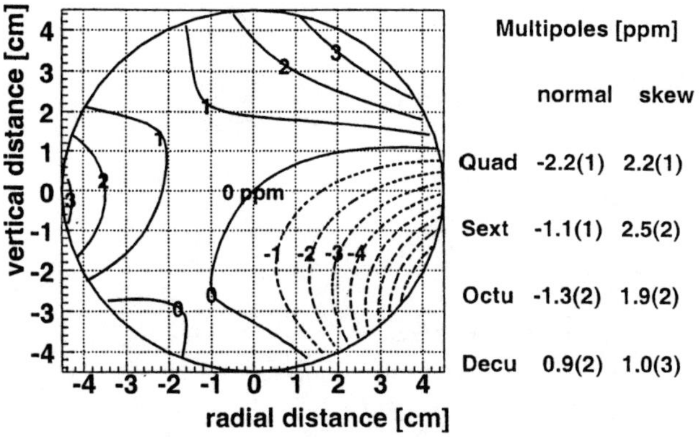

FIGURE 1. A 2-dimensional multipole expansion of the field averaged over azimuth from one out of seventeen trolley measurements in 1999. One ppm contours are shown with respect to a central azimuthal average field of 1.451 266 T. The circle indicates the muon beam storage region. The multipole amplitudes are given at the beam aperture radius of 4.5 cm.

homogeneity at specific calibration locations. Its value is then measured with the 17 NMR probes mounted in the trolley shell, and with a single probe plunged into the storage vacuum to measure the field values at the same locations. Drifts of the field during the calibration are determined by remeasuring the field with the trolley after completing the measurements with the plunging probe, and also by interpolation of the readings from nearby NMR probes in the outer top and bottom walls of the vacuum chamber. The difference of the trolley probe and plunging probe readings forms a calibration of the trolley probes with respect to the plunging probe, and hence with respect to each other. The plunging probe, as well as a subset of the trolley probes, are thereafter calibrated with respect to a standard probe [9] in similar measurements in the storage region, which is opened to air for that purpose. The leading calibration uncertainties result from the residual inhomogeneity of the field at the calibration locations and from position uncertainties in the active volumes of the NMR probes, and amount to 0.2 ppm as indicated in Table 1. The dependencies of the trolley NMR readings on the supply voltage and on other parameters were measured to be small in the range of operation. Their upper bound is included as a systematic error contribution ("other"), which also includes the effects from the measured transient kicker field caused by eddy currents.

The magnetic field along the muon orbit was measured 17 times during the data collection from January to March 1999 by pulling the field trolley through the storage vacuum chamber. Fig. 1 shows a multipole expansion of the azimuthal averages of the NMR readings from a representative measurement.

The uncertainties result from the trolley position measurement and from a region of about 1° near the inflector, where — as the result of an imperfection in the flux-trapping, superconducting shield of the inflector — the field in the storage region was inhomogeneous by several hundreds of ppm and had to be measured in separate scans

TABLE 1. Systematic errors for the ω_p analysis

Source of errors	Size [ppm]
1. Standard probe absolute calibration	0.05
2. Calibration of B_0 against standard probe	0.20
3. B_{aver} from trolley probes due to position uncertainty	0.10
4. Inflector fringe field	0.20
5. Tracking by fixed probes	0.15
6. Average over muon distribution	0.12
7. Others †	0.15
Total systematic error on ω_p	0.4

† higher multipoles, trolley temperature and its power supply voltage response, and eddy currents from the kicker.

with different operating settings for the field trolley.

The measurements with the field trolley serve, in addition, as a calibration of the 360 NMR probes in the outer top and bottom walls of the storage vacuum chamber, which are used to track the field when the field trolley is 'parked' in the storage vacuum just outside the beam region, and muons circulate in the storage ring. The calibration may change in time, for example when the magnet is ramped, and is thus repeated typically two or three times per week. Redundant trolley measurements allows the estimation of the tracking uncertainty with the fixed NMR probes.

The field frequency ω_p in Eq. 4 is the free proton NMR frequency averaged over the muon distribution and analyzed data sample. The field integral encountered by the muon beam has been studied by tracking 4000 muons for 100 turns through a measured field map. The value of the field integral over the beam trajectory is found identical to within 0.05 ppm to the azimuthally averaged field value, taken at the beam center. The beam center along the radial axis is determined from the debunching of the beam pulses, and found to be 3.7(1.0) mm outside of the central orbit. The vertical beam center is determined to be 2(2) mm above the center plane from measurements with segmented scintillation counters mounted on the front face of several of the positron calorimeters, and with scintillating fiber monitors that can be plunged into the beam region at two locations in the ring.

The resulting value for the field frequency ω_p is,

$$\omega_p = 61\,719\,256(25)\ \text{Hz}\ \ (0.4\text{ppm}), \tag{5}$$

where the uncertainty has leading contributions from the calibration of the trolley probes and the inflector fringe field, and is thus predominantly systematic. The result from a second, largely independent, analysis agrees with the above value to within 0.03 ppm.

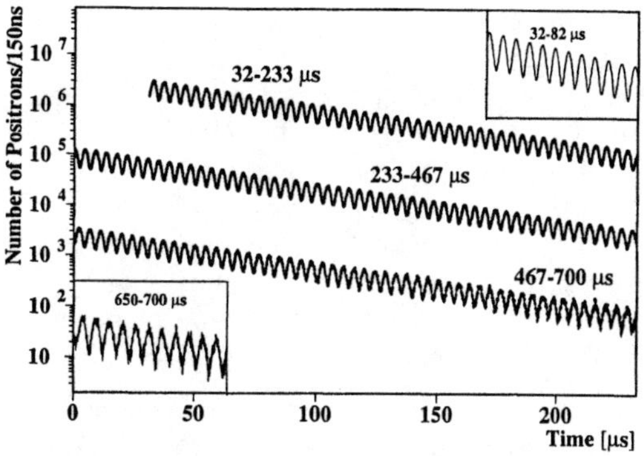

FIGURE 2. Positron time spectrum overlaid with the fitted 10 parameter function (χ^2/dof = 3818/3799). The total event sample of $1.0 \cdot 10^9$ e^+ with energies larger than 2 GeV is shown.

The analysis of ω_a

The determination of the frequency ω_a from the recorded calorimeter WFD traces starts with the identification of positron pulses and the reconstruction of their arrival times and energies. Conceptually, a sample of about 10^4 pulses in the energy range 1–3 GeV is selected for each calorimeter individually to form an average pulseshape, which is then used in fits to all WFD traces. The selection is made so as to ensure that transient effects have faded away and the traces consist of responses to single positrons. A fraction of several percent of the recordings is found to contain multiple positron pulses per WFD trace. Extensive studies of the pulse finding and fitting algorithm show that in such cases each of the pulses is identified and measured correctly, provided that the pulse separation exceeds 3 ns and the pulse energy is larger than 0.4 GeV. For intervals smaller than 3 ns, the pulses are reconstructed as a single pulse, whereas pulses with energies below 0.4 GeV escape reconstruction.

Thereafter, the data collected to study systematic effects are separated from the data with regular running conditions. Data for which the settings of the apparatus are known to be different from their nominal values are rejected, as are data for which the correspondence could not be established. The consistency of the remaining data is verified by comparing the spectra for each run with a reference spectrum created by randomly sampled events from all runs.

The time spectrum of the remaining positrons is shown in Fig. 2 in the time range 32–700 μs and for positron energies larger than 2 GeV. Corrections were applied to mask the bunched structure of the beam injection and to eliminate a small distortion from overlapping pulses, so called pileup.

The key characteristics of the spectrum arise from muon decay and spin precession, and are described by

$$N(t) = N_0(E) \exp\left(\frac{-t}{\gamma\tau}\right) [1 + A(E) \sin(\omega_a t + \phi(E))] \qquad (6)$$

in which $N_0(E)$ is the normalization, $\gamma\tau$ the dilated muon lifetime, $A(E)$ the asymmetry, and $\phi(E)$ the phase. The statistical uncertainty in the fitted frequency ω_a is inversely proportional to $A\sqrt{N_E}$ with N_E the total number of positrons in the spectrum with energies larger than E, and has its optimal value for $E \simeq 2\,\text{GeV}$. The optimal value amounts to 5 ppm for the $84 \cdot 10^6$ analyzed positrons from the 1998 data [12], which are described adequately with the function $f(t) = N(t)$.

The tenfold larger positron sample collected in 1999 requires consideration of additional effects, in particular coherent betatron oscillations and muon losses. These effects are discussed below, following the corrections to the data for beam debunching and pileup.

- The injection of the beam in narrow bunches into the storage ring results in a strong modulation of the initial part of the decay positron time spectrum with the cyclotron period of 149.185 ns. The modulation fades with a characteristic time of about 24 μs due to the 0.5% momentum spread in the beam. An effective correction for this effect is found to consist in randomization of the arrival times of all decay positrons from a single beam pulse over the cyclotron period, which in addition is chosen to be the bin-width of the fitted spectra.
- The number of overlapping pulses in the reconstructed data is proportional to the instantaneous counting rate squared $(dN/dt)^2$ and to the minimum pulse separation time of the pulse reconstruction algorithm. This fraction is about 1% of the event sample at 32 μs when the fits of ω_a to the data are started, and vanishes exponentially with a time constant equal to half the dilated muon lifetime. Pileup distorts the positron time spectrum because of miscounting of the number of pulses and misidentification of the energies and times. Since the phase ϕ in Eq. 6 depends on the positron energy and correlates strongly with the frequency ω_a in fits, pileup potentially causes a sizable error in the fitted value of ω_a. It is thus advantageous to apply a correction to the data prior to the fitting, which consists in a subtraction of a pileup spectrum that is constructed using indiscriminately recorded pulses as follows. Positron pulses found within a window at a fixed, short time after the positron pulse that triggered the WFD module are treated as if they overlap with the trigger pulse, that is, the times of both pulses are averaged and the energies added to form a single pulse. The width of the window is taken equal to the minimum pulse separation time of 3 ns. For data with energies above 2 GeV — twice the hardware threshold of about 1 GeV — the difference of the thus found spectrum and the uncorrected spectrum determines the pileup spectrum. Signals with energies below about 0.4 GeV are too small to be reconstructed with the pulse finding algorithm and are thus not accounted for in the subtraction procedure described above. These small signals distort the pulse reconstruction but do not, on average, affect the energy. They do cause small time dependent shifts in the asymmetry $A(E)$, and to a smaller extent in the phase $\phi(E)$. The observed stability of the asymmetry $A(E)$ with time is used to set a limit on the shift of the fitted value of ω_a that would result through its correlation with $\phi(E)$.
  Expressed in terms of a shift in ω_a, the size of the correction for pileup amounts to 0.3 ppm. Its uncertainty is estimated to be about twice smaller.
- Coherent betatron oscillations: The storage ring is a weak focusing spectrometer

with field index $n = 0.137$, with an aperture that is large compared to the inflector aperture of $18(\mathrm{w}) \times 57(\mathrm{h})\ \mathrm{mm}^2$. Therefore, the phase space for the betatron oscillations defined by the acceptance of the storage ring is not filled when muons are injected into the storage ring. In combination with imperfect injection angles and an imperfect horizontal kick to store the muons onto stable orbits, this results in betatron oscillations of the beam as a whole – coherent betatron oscillations (CBO). The CBO modulate the positron time spectra, since the calorimeter acceptances depend on the muon decay positions. The dominant effect is caused by the horizontal oscillations, which decay with a time constant of $\sim 120\ \mu$s. An adequate parametrization of the effect is found to be

$$b(t) = 1 + A_b \exp\left(-\frac{-t^2}{\tau_b^2}\right) \cos(\omega_b t + \phi_b) \tag{7}$$

and is included multiplicatively in the function fit to the data, $f(t) = N(t) \cdot b(t)$. The frequency $\omega_b = \omega_c(1 - \sqrt{1-n}) \simeq 470.0$ kHz, with $n \simeq 0.137$ the field focusing index and ω_c the cyclotron frequency, is determined from the Fourier spectrum of residuals of a fit to the 1999 data taking into account only muon decay and spin precession, and was measured directly during the running period using scintillating fibers that can be plunged into the storage region. The amplitude A_b is determined from the fit and found to be about $A_b \simeq 1 \times 10^{-2}$, as is the characteristic time $\tau_b \simeq 120\mu$s with which CBO vanish, owing partly to the momentum spread in the beam. The phase ϕ_b varies by detector, from 0 to 2π.

- Muon Losses: A small fraction of the stored beam is lost during the muon storage time, despite controlled scraping of the injected muon beam with the electrostatic quadrupoles for about 15 μs immediately following injection. An adequate parametrization of the residual losses is found to be

$$l(t) = 1 + n_l \exp\left(-\frac{t}{\tau_l}\right) \tag{8}$$

and is included multiplicatively in the function fit to the data, $f(t) = N(t) \cdot b(t) \cdot l(t)$. The decay time τ_l is found to be $\tau_l \simeq 20\ \mu$s, and n_l is typically less than 1%. An independent study of the losses of stored beam was made using coincident signals from three adjacent layers of scintillation fingers mounted on the front faces of five of the positron calorimeters, in the absence of energy deposits in the calorimeters.

The function fitted to the 1999 data sample is thus given by

$$f(t) = N(t) \cdot b(t) \cdot l(t), \tag{9}$$

in which $N(t)$ describes muon decay and spin precession (cf. Eq. 6), and the perturbations $b(t)$ and $l(t)$ are given in Eqs. 7 and 8. The function $f(t)$ has ten parameters that are fitted to the data, in the sense of minimizing χ^2. (The frequency ω_b is determined from a Fourier analysis, as mentioned before.)

The results from the analysis described above and from three alternative analyses [13, 14] of ω_a are found to agree to within the statistical variation of 0.4 ppm expected from

TABLE 2. Systematic errors for the ω_a analysis.

Source of errors	Size [ppm]
Pileup	0.13
AGS background	0.10
Lost muons	0.10
Timing shifts	0.10
E field and vertical betatron oscillation	0.08
Binning and fitting procedure	0.07
Coherent betatron oscillation	0.05
Beam debunching/randomization	0.04
Gain changes	0.02
Total systematic error on ω_a	0.3

slightly different data reconstructions and selections. The combined result is

$$\omega_a/2\pi = 229072.8(0.3)\ \text{Hz}\ (1.3\ \text{ppm}), \tag{10}$$

which includes a correction of +0.81(8) ppm for the small disproportionality of the observed frequency and the measured field caused by vertical betatron oscillations and electric fields in the storage ring. The stated error of 1.3 ppm reflects the total uncertainty, accounts for the strong correlations between the individual fit results, and is dominated by the statistical contribution. Table 2 lists the systematic contributions, which include background from the AGS and measured changes in the detector gains and timing, in addition to the effects described above.

RESULTS

Only after the analyses of the magnetic field data and of the spin precession data had been finalized, separately and independently, was the anomalous magnetic moment a_μ^+ evaluated,

$$a_\mu^+(\text{expt}) = 11659202(14)(6) \times 10^{-10}\ (1.3\ \text{ppm}) \tag{11}$$

The result [15] is in good agreement with previous measurements of a_μ^+ and a_μ^- [3, 16, 12] and improves the combined uncertainty by a factor of about three. When the new combined experimental result,

$$a_\mu(\text{expt}) = 11659203(15) \times 10^{-10}\ (1.3\ \text{ppm}) \tag{12}$$

was published in the spring of 1999, the most recent, published, and commonly accepted compilation [17] of evaluations from Standard Model theory was

$$a_\mu(\text{theory}) = 11659159.6(6.7) \times 10^{-10}\ (0.6\ \text{ppm}), \tag{13}$$

which differs from the experimental result by about 2.5 standard deviations.

Since the time of publication, the prediction has been reexamined and the result for the leading order hadronic contribution $a_\mu(\text{had}, 1) = 673.9(6.7) \times 10^{-10}$ [18] has been confirmed [19]. A mistake of sign, however, has recently been revealed [20] in the evaluations of the higher order contribution $a_\mu(\text{had,lbl}) = -8.5(2.5) \times 10^{-10}$ [21, 22] from hadronic light-by-light scattering.

Hence, by simply reversing the erroneous sign, one finds an updated value for the theoretical prediction,

$$a_\mu(\text{theory}) = 11\,659\,176.6(6.7) \times 10^{-10} \;\; (0.6\text{ ppm}), \tag{14}$$

and for the difference,

$$a_\mu(\text{expt}) - a_\mu(\text{theory}) = 26(16) \times 10^{-10}. \tag{15}$$

The analysis of data collected in 2000 and 2001 is well underway and, when combined with data from a requested, final run in the fall of 2002 and winter of 2003, is expected to reduce the experimental uncertainty on a_μ to 0.4 ppm. The Standard Model prediction is bound the improve with the inclusion of more and more precise data on hadron production in e^+e^- collisions in the evaluations of the hadronic contribution to a_μ.

ACKNOWLEDGMENTS

The research described in this work is supported in part by the U.S. Department of Energy, the U.S. National Science Foundation, the German Bundesminister für Bildung und Forschung, the Russian Ministry of Science, and the U.S.–Japan Collaboration in High Energy Physics.

REFERENCES

1. Charpak, G., et al., *Phys. Lett.*, **1**, 16 (1962).
2. Bailey, J., et al., *Nuovo Cimento*, **A9**, 369 (1972).
3. Bailey, J., et al., *Nucl. Phys.*, **B510**, 1 (1979).
4. Nath, P., *these proceedings* (2002).
5. Krienen, F., Loomba, D., and Meng, W., *Nucl. Instrum. Meth.*, **A283**, 5 (1989).
6. Danby, G., et al., *Nucl. Instrum. Meth.*, **A457**, 151 (2001).
7. Prigl, R., et al., *Nucl. Instrum. Methods.*, **A374**, 118 (1996).
8. Grossmann, A., *Ph.D. Thesis, Universität Heidelberg* (1998).
9. Fei, X., Hughes, V., and Prigl, R., *Nucl. Instrum. Methods.*, **A394**, 349 (1997).
10. Sedykh, S., et al., *Nucl. Instrum. Meth.*, **A455**, 346 (2000).
11. Groom, D., et al., *The European Physical Journal*, **C15**, 1 (2000).
12. Brown, H., et al., *Phys. Rev.*, **D62**, 091101 (2000).
13. Duong, L., *Ph.D. Thesis, University of Minnesota* (2001).
14. Trofimov, A., *Ph.D. Thesis, Boston University* (2001).
15. Brown, H., et al., *Phys. Rev. Lett.*, **86**, 2227 (2001).
16. Carey, R., et al., *Phys. Rev. Lett.*, **82**, 1132 (1999).
17. Czarnecki, A., and Marciano, W., *Nucl. Phys. (Proc. Suppl.)*, **B76**, 245 (1999).
18. Davier, M., and Höcker, A., *Phys. Lett.*, **B435**, 427 (1998).
19. Narison, S., *Phys. Lett.*, **B513**, 53 (2001).
20. Knecht, M., et al., *Phys. Rev. Lett.*, **88**, 071802 (2002).
21. Hayakawa, M., and Kinoshita, T., *Phys. Rev.*, **D57**, 465 (1998).
22. Bijnens, J., Pallante, E., and Prades, J., *Nucl. Phys.*, **B474**, 379 (1996).

Fermion generations birth effect in the two measures theory

E.Guendelman * and A.Kaganovich *

*Physics Department, Ben Gurion University of the Negev, Beer Sheva 84105, Israel

Abstract. A spontaneously broken $SU(2)\times U(1)$ gauge theory with just one "primordial" generation of fermions is formulated in the context of a generally covariant theory which contains two measures of integration in the action: the standard $\sqrt{-g}d^4x$ and a new one Φd^4x, where Φ is a density built out of degrees of freedom independent of the metric. Such type of models are known to produce a satisfactory answer to the cosmological constant problem. Global scale invariance is implemented. After SSB of scale invariance and gauge symmetry it is found that with the conditions appropriate to laboratory particle physics experiments, to each primordial fermion field corresponds three physical fermionic states. Two of them correspond to particles with constant masses and they are identified with the first two generations of the electro-weak theory. In space-time regions where the regular fermionic matter has the typical laboratory particle physics density, the dilaton is decoupled from the regular fermionic matter. This provides a resolution of the long-range force problem. The third fermionic states at the classical level get non-polynomial interactions which indicate the existence of fermionic condensate and fermionic mass generation. The possible role of the measure Φ in some quantum gravity effects is discussed.

INTRODUCTION

One of the most perplexing questions that have arisen in the theory of elementary particles is the origin of the families (generations) of elementary fermions: electrons and quarks. Indeed, each fermion is replicated three times: instead of having one electron, we observe in addition the muon and the tau lepton; instead of one quark doublet we have three doublets of quarks. All these replications exhibit the same charge, spin, etc. but they differ in their masses.

In this paper we will follow a geometric approach to the family problem of particle physics. Basic ideas and methods of this approach have been developed in previous papers[1, 2, 3, 4, 5, 6, 7, 8, 9] where the emphasis was on cosmological questions, in special the question of the cosmological constant problem. It was noticed however[10] that a natural solution to the family problem could be given along these lines as well. Here we generalize the results of the toy model[10] to the $SU(2)\times U(1)$ gauge theory.

The geometric approach of Refs.[1, 2, 3, 4, 5, 6, 7, 8, 9, 10] consists of using an alternative volume element Φd^4x, in addition to the standard one $\sqrt{-g}d^4x$. So a general action of the form

$$S=\int L_1\Phi d^4x+\int L_2\sqrt{-g}d^4x \tag{1}$$

is considered. In order that Φd^4x be an invariant volume element, it is necessary that Φ transforms as a density, i.e. just like $\sqrt{-g}$. This can be realized if we choose Φ to be the

CP624, *Cosmology and Elementary Particle Physics*, edited by B. N. Kursunoglu et al.

composite of 4 scalars φ_a $(a = 1,2,3,4)$

$$\Phi = \varepsilon^{\mu\nu\alpha\beta}\varepsilon_{abcd}\partial_\mu\varphi_a\partial_\nu\varphi_b\partial_\alpha\varphi_c\partial_\beta\varphi_d. \tag{2}$$

Since Φ is a total derivative, a shift of L_1 by a constant, $L_1 \rightarrow L_1 + const$, has the effect of adding to S the integral of a total derivative , which does not change equations of motion. This is why the introduction of a new volume element has consequences on the way we think about the cosmological constant problem[1, 2, 3, 4, 5].

In Eq. (1), L_1 and L_2 are Lagrangian which are functions of the matter fields, the metric, the connection (or spin-connection) but not of the "measure fields" φ_a. In such a case the action (1) has the infinite dimensional symmetry[5]: $\varphi_a \rightarrow \varphi_a + f_a(L_1)$, where $f_a(L_1)$ is an arbitrary function of L_1.

It may appear at first sight strange to think that geometry (measure, connections, metric) are relevant to particle physics. This is because we are used to think that these geometrical objects can be only related to gravity. However, as we will see, the consistency condition of equations of motion determines the ratio of two measures

$$\zeta \equiv \frac{\Phi}{\sqrt{-g}} \tag{3}$$

as a function of matter fields. The surprising feature of the theory is that neither Newton constant nor curvature appears in this constraint which means that the *geometrical scalar field* $\zeta(x)$ is determined by the matter fields configuration locally and straightforward (that is without gravitational interaction).

The use of the measure Φ for solving the cosmological constant problem was started in Refs[1, 2, 3, 4, 5]. There it was found that if Lagrangians L_1 and L_2 in Eq.(1) are φ_a independent, then the ground state of the theory is generically at zero energy density, without fine tuning. This is an approach to a solution of the so called "old cosmological constant problem".

It was found also that new realizations of scale invariance are possible in these kind of models[6, 7, 8, 9, 10] if a dilaton field is introduced. After spontaneous breaking of scale invariance[6, 7, 8, 10] or with a small explicit breaking[9], cosmological applications of these models were studied. For example, a see-saw cosmological effect was found[6, 7, 8] that allows a very small cosmological constant to be obtained from a large scale(see [11, 12, 13]). These models can also give rise to interesting quintessential scenarios without the long-range force problem[10].

As we will see, $\zeta(x)$ has a decisive influence in the determination of particle masses and in the "families birth effect". Therefore "Geometry" will be of importance, beyond what was known so far, i. e. that the geometrical objects which enter into the field theory are restricted by the metric associated to the gravitational field.

THE MODEL

To see how the theory works, let us consider a model containing the $SU(2) \times U(1)$ gauge structure (the color SU(3) can be added without changing our results), as in the standard model with standard content of the bosonic sector (gauge vector fields $\vec{A}_\mu$ and B_μ and Higgs doublet H). But in contrast to the standard model, in our model *we start from only one family of the so called "primordial" fermionic fields*: the primordial up and down quarks U and D and the primordial electron E and neutrino N. Similar to the standard model, we will proceed with the following independent fermionic degrees of freedom:

a) one primordial left quark $SU(2)$ doublet Q_L

$$Q_L = \begin{pmatrix} U_L \\ D_L \end{pmatrix}$$

and right primordial singlets U_R and D_R;

b) one primordial left lepton SU(2) doublet L_L:

$$L_L = \begin{pmatrix} N_L \\ E_L \end{pmatrix}$$

and right primordial singlet E_R.

A dilaton field ϕ is needed in order to achieve global scale invariance[6, 7, 8, 9].

According to the general prescriptions of the two measures theory, we have to start from studying the selfconsistent system of gravity and matter fields proceeding in the first order formalism. In the model including fermions in curved space-time, this means that the independent dynamical degrees of freedom are: all matter fields, vierbein e^μ_a, spin-connection ω^{ab}_μ and the measure Φ degrees of freedom, i.e. four scalar fields φ_a. We postulate that in addition to $SU(2)\times U(1)$ gauge symmetry, the theory is invariant under the global scale transformations:

$$e^a_\mu \to e^{\theta/2} e^a_\mu, \quad \omega^\mu_{ab} \to \omega^\mu_{ab}, \quad \varphi_a \to \lambda_a \varphi_a \quad where \quad \Pi\lambda_a = e^{2\theta},$$
$$\phi \to \phi - \frac{M_P}{\alpha}\theta, \quad H \to H, \quad \Psi \to e^{-\theta/4}\Psi, \quad \overline{\Psi} \to e^{-\theta/4}\overline{\Psi}; \quad \theta = const. \tag{4}$$

The global scale invariance is important for cosmological applications [6, 7, 8, 9].

The action of the model has the general structure given by Eq. (1) which is convenient to represent in the following form:

$$S = \int d^4x e^{\frac{\alpha\phi}{M_P}} (\Phi + b\sqrt{-g}) \left[\frac{1}{2} g^{\mu\nu}\phi_{,\mu}\phi_{,\nu} + \frac{1}{2} g^{\mu\nu}(D_\mu H)^\dagger D_\nu H - \frac{1}{\kappa} R(\omega, e)\right]$$
$$- \int d^4x \sqrt{-g}\left(\frac{1}{4} g^{\alpha\beta} g^{\mu\nu} B_{\alpha\mu} B_{\beta\nu} + \frac{1}{2} g^{\alpha\beta} g^{\mu\nu} Tr A_{\alpha\mu} A_{\beta\nu}\right)$$
$$- \int d^4x e^{2\alpha\phi/M_P}[\Phi V_1(H) + \sqrt{-g} V_2(H)] + \int d^4x e^{\alpha\phi/M_P}(\Phi + k\sqrt{-g}) L_{fk}$$
$$- \int d^4x e^{\frac{3}{2}\alpha\phi/M_P}\left[(\Phi + h_U\sqrt{-g}) f_U \overline{Q_L}\tilde{H} U_R + (\Phi + h_D\sqrt{-g}) f_D \overline{Q_L} H D_R\right.$$
$$\left. + (\Phi + h_E\sqrt{-g}) f_E \overline{L_L} H E_R + H.c.\right] \tag{5}$$

The notations in (5) are the following: $g^{\mu\nu} = e^\mu_a e^\nu_b \eta^{ab}$; the scalar curvature is $R(\omega, V) = V^{a\mu} V^{b\nu} R_{\mu\nu ab}(\omega)$ where $R_{\mu\nu ab}(\omega) = \partial_\mu \omega_{\nu ab} + \omega^c_{\mu a}\omega_{\nu cb} - (\mu \leftrightarrow \nu)$;

$L_{fk} = \frac{i}{2}[\overline{L}_L \not{D}^{(L)} L_L + \overline{E}_R \not{D}^{(R)} E_R + \overline{Q}_L \not{D}^{(L)} Q_L + \overline{U}_R \not{D}^{(R)} U_R + \overline{D}_R \not{D}^{(R)} D_R];$
$D_\mu H \equiv \left(\partial_\mu - \frac{i}{2} g \vec{\tau} \cdot \vec{A}_\mu - \frac{i}{2} g' B_\mu\right) H;$
$D^{(R)} \equiv e^\mu_a \gamma^a \left(\vec{\partial}_\mu + \frac{1}{2}\omega^{cd}_\mu \sigma_{cd} + ig' B_\mu\right) - \left(\overleftarrow{\partial}_\mu - \frac{1}{2}\omega^{cd}_\mu \sigma_{cd} - ig' B_\mu\right)\gamma^a e^\mu_a;$

$$\begin{aligned} D^{(L)} \equiv\; & e^\mu_a \gamma^a \left(\vec{\partial}_\mu + \frac{1}{2}\omega^{cd}_\mu \sigma_{cd} I - \frac{i}{2} g \vec{\tau} \cdot \vec{A}_\mu + \frac{i}{2} g' B_\mu\right) \\ & - \left(\overleftarrow{\partial}_\mu - \frac{1}{2}\omega^{cd}_\mu \sigma_{cd} I + \frac{i}{2} g \vec{\tau} \cdot \vec{A}_\mu - \frac{i}{2} g' B_\mu\right)\gamma^a e^\mu_a; \end{aligned} \tag{6}$$

and finally $B_{\mu\nu} \equiv \partial_\mu B_\nu - \partial_\nu B_\mu$, $A_{\mu\nu} \equiv \partial_\mu A_\nu - \partial_\nu A_\mu - ig[A_\mu A_\nu - A_\nu A_\mu]$ where $A_\mu = \frac{1}{2}\vec{A}_\mu \cdot \vec{\tau}$; I is 2×2 unit matrix in the isospin space.

A few explanations concerning our choice of the action (5) are in necessary:

1) In order to avoid a possibility of negative energy contribution from the space-time derivatives of the dilaton ϕ and Higgs H fields (see Ref.[10]) we have chosen the coefficient b in front of $\sqrt{-g}$ in the first integral of (5) to be a common factor of the gravitational term $-\frac{1}{\kappa}R(\omega, e)$ and of the kinetic terms for ϕ and H. This guarantees that this item can not be an origin of ghosts in quantum theory.

2) For the same reasons we choose the kinetic terms of the gauge bosons in the conformal invariant form which is possible only if these terms are coupled to the measure $\sqrt{-g}$. Introducing the coupling of these terms to the measure Φ would lead to the nonlinear equations and non positivity of the energy.

3) For simplicity, we have taken the coupling of the kinetic terms of the fermions to the measures to be universal (see the forth integral in Eq.(5)).

Except for these three items, Eq.(5) describes the most general action of the two measures theory satisfying the formulated above symmetries.

CLASSICAL EQUATIONS OF MOTION

After SSB of scale and gauge symmetries, proceeding in the unitary gauge, the Higgs field can be represented in the standard form

$$H = \begin{pmatrix} 0 \\ \frac{1}{\sqrt{2}}(\upsilon + \chi) \end{pmatrix}$$

Varying the measure fields φ_a, we get $A^\mu_a \partial_\mu L_1 = 0$ where L_1 is defined, according to Eq. (1), as the part of the integrand of the action (5) coupled to the measure Φ and $A^\mu_a = \varepsilon^{\mu\nu\alpha\beta}\varepsilon_{abcd}\partial_\nu\varphi_b\partial_\alpha\varphi_c\partial_\beta\varphi_d$.

Since $Det(A^\mu_a) = \frac{4^{-4}}{4!}\Phi^3$ it follows that if $\Phi \neq 0$,

$$L_1 = sM^4 = const \tag{7}$$

Here $s = \pm 1$ and the appearance of a nonzero integration constant M^4 of the dimension of mass spontaneously breaks the scale invariance (4).

Complete system of equations corresponding to the action (5) is very bulky. Variation of S with respect to vierbein e^{μ}_{a} yields the gravitational equation linear both in the curvature and in the scalar field ζ, defined by Eq. (3). Contracting this equation with e^{μ}_{a}, solving for the curvature scalar R and replacing in Eq. (7) we obtain the following consistency condition of the theory:

$$(\zeta-b)\left[sM^4e^{-\alpha\phi/M_p}+V_1e^{\alpha\phi/M_p}-L_{fk}+\frac{\upsilon+\chi}{\sqrt{2}}e^{\frac{1}{2}\alpha\phi/M_p}\sum_i f_i\overline{\Psi}_i\Psi_i\right]+$$
$$+2V_2e^{-\alpha\phi/M_p}+\frac{1}{2}(\zeta-3k)L_{fk}+\sqrt{2}(\upsilon+\chi)e^{\frac{1}{2}\alpha\phi/M_p}\sum_i f_ih_i\overline{\Psi}_i\Psi_i=0 \tag{8}$$

where $i=E,U,D$, Ψ_i labels E, U and D fermion fields and L_{fk} was defined in Sec.2. Using equations of motion for all the fermion fields, it is easy to check that the following relation is true

$$\begin{aligned} L_{fk} &= \frac{e^{\frac{1}{2}\alpha\phi/M_p}}{\zeta+k}\left[(\zeta+h_E)f_E\overline{L_L}HE_R+(\zeta+h_U)f_U\overline{Q_L}\tilde{H}U_R\right.\\ &+ \left.(\zeta+h_D)f_D\overline{Q_L}HD_R+H.c.\right] \end{aligned} \tag{9}$$

Due to this relation, the consistency condition (8) becomes a constraint having a fundamental role for the theory.

In order to get the physical content of the theory it is required to express it in terms of variables where all equations of motion acquire a canonical form in an Einstein-Cartan space-time (for detail see Ref.[2]). This is possible after performing the following redefinitions of the vierbein (and metric) and all fermion fields:

$$\tilde{g}_{\mu\nu}=e^{\frac{\alpha\phi}{M_p}}(\zeta+b)g_{\mu\nu};\;\tilde{e}_{a\mu}=e^{\frac{\alpha\phi}{2M_p}}(\zeta+b)^{1/2}e_{a\mu};\;\Psi'=e^{-\frac{\alpha\phi}{4M_p}}\frac{(\zeta+k)^{1/2}}{(\zeta+b)^{3/4}}\Psi. \tag{10}$$

With these variables, the spin-connections become those of the Einstein-Cartan space-time. Since $\tilde{e}_{a\mu}$ and Ψ' are invariant under the scale transformations (4), spontaneous breaking of the scale symmetry (4) (by means of Eq. (7)) is reduced in the new variables to the spontaneous breaking of the shift symmetry $\phi\to\phi+const$ for the dilaton field.

One can check that equations of motion for the gauge fields in the new variables are canonical and after the Higgs develops VEV, the gauge bosons mass generation is standard, that is exactly the same as it is in the Weinberg-Salam electroweak theory: photon, $W^{\pm}$ and Z bosons as well as the Weinberg angle appear as the result of the standard procedure of the Weinberg-Salam theory.

The gravitational equations of motion in the new variables take the form

$$G_{\mu\nu}(\tilde{g}_{\alpha\beta})=\frac{\kappa}{2}T^{eff}_{\mu\nu} \tag{11}$$

$$\begin{aligned} T^{eff}_{\mu\nu} &= \phi_{,\mu}\phi_{,\nu}-K_{\phi}\tilde{g}_{\mu\nu}+\chi_{,\mu}\chi_{,\nu}-K_{\chi}\tilde{g}_{\mu\nu}+\tilde{g}_{\mu\nu}V_{eff}\\ &+T^{(gauge,can)}_{\mu\nu}+T^{(ferm,can)}_{\mu\nu}-\tilde{g}_{\mu\nu}\sum_i F_i(\zeta,\upsilon+\chi)\overline{\Psi}'_i\Psi'_i, \end{aligned} \tag{12}$$

Here $G_{\mu\nu}(\tilde{g}_{\alpha\beta})$ is the Einstein tensor in the Riemannian (or, more exactly, Einstein-Cartan) space-time with metric $\tilde{g}_{\mu\nu}$; $K_\phi \equiv \frac{1}{2}\tilde{g}^{\alpha\beta}\phi_{,\alpha}\phi_{,\beta}$;
$K_\chi \equiv \frac{1}{2}\tilde{g}^{\alpha\beta}\chi_{,\alpha}\chi_{,\beta}$; $T_{\mu\nu}^{(gauge,can)}$ is the canonical energy momentum tensor for gauge bosons, including mass terms of $W^\pm$ and Z bosons. $T_{\mu\nu}^{(ferm,can)}$ is the canonical energy momentum tensor for (primordial) fermions in curved space-time including also their standard electromagnetic and weak interactions with gauge bosons. Functions V_{eff} and $F_i(\zeta, \upsilon+\chi)$ $(i=E',U',D')$ are defined by equations

$$V_{eff} = \frac{b\left(sM^4e^{-2\alpha\phi/M_p} + V_1\right) - V_2}{(\zeta+b)^2} \tag{13}$$

$$F_i \equiv \frac{(\upsilon+\chi)f_i}{2\sqrt{2}(\zeta+k)^2(\zeta+b)^{1/2}}[\zeta^2 + (3h_i - k)\zeta + 2b(h_i-k) + kh_i];\; i = E',U',D' \tag{14}$$

The scalar field ζ in the above equations is defined by the constraint determined by means of Eqs. (8) and (9). In the new variables (10) this constraint takes the form

$$(\zeta - b)\left[sM^4e^{-\frac{2\alpha\phi}{M_p}} + V_1(\upsilon+\chi)\right] + 2V_2(\upsilon+\chi) + (\zeta+b)^2\sum_i F_i(\zeta,\upsilon+\chi)\overline{\Psi}_i'\Psi_i' = 0 \tag{15}$$

The dilaton ϕ and Higgs χ field equations in the new variables are the following

$$\frac{1}{\sqrt{-\tilde{g}}}\partial_\mu(\sqrt{-\tilde{g}}\tilde{g}^{\mu\nu}\partial_\nu\phi) - \frac{\alpha}{M_p(\zeta+b)}\left[sM^4e^{-2\alpha\phi/M_p} - \frac{(\zeta-b)V_1 + 2V_2}{\zeta+b}\right] = -\frac{\alpha}{M_p}\sum_i F_i\overline{\Psi}_i'\Psi_i', \tag{16}$$

$$\frac{1}{\sqrt{-\tilde{g}}}\partial_\mu(\sqrt{-\tilde{g}}\tilde{g}^{\mu\nu}\partial_\nu\chi) + \frac{\zeta V_1' + V_2'}{(\zeta+b)^2} = -\frac{\upsilon+\chi}{\sqrt{2}(\zeta+b)^{1/2}(\zeta+k)}\sum_i(\zeta+h_i)f_i\overline{\Psi}_i'\Psi_i'. \tag{17}$$

Equations for the primordial fermions in terms of the variables (10) take the standard form of fermionic equations in the Einstein-Cartan space-time where the standard interactions to the gauge fields present also. All the novelty consists of the form of the ζ depending "masses" $m_i(\zeta)$ of the massive primordial fermions:

$$m_i(\zeta) = \frac{f_i\upsilon(\zeta+h_i)}{\sqrt{2}(\zeta+k)(\zeta+b)^{1/2}} \qquad i = E',U',D'. \tag{18}$$

VACUUM AND FAMILIES BIRTH EFFECT

Let us consider the following two limiting cases:

(i) **In the absence of massive fermions**, solving ζ from the constraint (15)

$$\frac{1}{\zeta+b} = \frac{sM^4e^{-2\alpha\phi/M_p} + V_1}{2\left[b\left(sM^4e^{-2\alpha\phi/M_p} + V_1(\upsilon+\chi)\right) - V_2(\upsilon+\chi)\right]} \tag{19}$$

one can check that in this case the dilaton and Higgs fields equations (16) and (17) take the form of the canonical scalar fields equations with the effective potential

$$V_{eff}(\phi,\upsilon+\chi)=\frac{\left[sM^4e^{-2\alpha\phi/M_p}+V_1(\upsilon+\chi))\right]^2}{4\left[b\left(sM^4e^{-2\alpha\phi/M_p}+V_1(\upsilon+\chi)\right)-V_2(\upsilon+\chi)\right]} \tag{20}$$

From this we immediately conclude that the stable vacuum of the scalar fields ($<\phi>\equiv\bar{\phi}$ and υ) is realized as a manifold determined by the equation

$$sM^4e^{-2\alpha\bar{\phi}/M_p}+V_1(\upsilon)=0 \tag{21}$$

provided that $V_2(\upsilon)<0$ in this degenerate vacuum. The masses of the dilaton and Higgs fields excitations above this degenerate vacuum are respectively

$$m^2_{dilat}=\frac{\alpha^2M^8}{M_p^2|V_2(\upsilon)|}e^{-4\alpha\bar{\phi}/M_p}; \qquad m^2_{higgs}=\frac{(V_1'(\upsilon))^2}{|V_2(\upsilon)|}. \tag{22}$$

Notice that we did not assume any specific properties of V_1 and V_2 so far. If we wish to provide conditions for a big Higgs mass we see from the second equation in (22) that there is no need for big "pre-potentials" $V_1(\upsilon)$ and $V_2(\upsilon)$ but rather they both can be small as compared to a typical energy scale of particle physics, however $V_2(\upsilon)$ must be very small.

An important feature of the degenerate vacuum (21) is that the effective vacuum energy density of the scalar fields is equal to zero without any sort of fine tuning regardless of the detailed shape of the potentials V_1 and V_2 as well as of the initial conditions. This fact has been very extensively explored as a way to solve the cosmological constant problem[5]. In this paper, however, we will concentrate our attention on the applications of the theory to particle physics.

Notice that according to Eq. (19), $\zeta=\infty$ in the degenerate vacuum (21). However, in the presence of any small "contamination" by massive fermions, it follows from the constraint (15) that ζ is large but finite. Therefore we must return to the general form of the effective potential (13) which will be small but non zero. This means that zero vacuum energy is practically unachievable, and there must be a correlation between the fermion content of the universe and the vacuum energy. This correlation might be a possible mechanism for the explanation of the "cosmic coincidence" problem[14].

(ii) **Case where fermion densities are of the typical laboratory particle physics scales.** Assuming as it was done before that $M^4e^{-2\alpha\phi/M_p}$, V_1 and V_2 are small as compared to the typical particle physics energy densities of fermions, we see from the constraint (15) that now there are no reasons for ζ to be large. On the contrary, it has to be of the same order as the dimensionless parameters of the theory (b, k and h_i) which we assume are of order one. So, for the case when fermion densities are of the typical laboratory particle physics scales, ζ has to satisfy the simplified form of the constraint(15):

$$(\zeta+b)[F_E(\zeta)\overline{E'}E'+F_U\overline{U'}U'+F_D\overline{D'}D']=0. \tag{23}$$

To see the meaning of the constraint in this case, let us take one single primordial fermionic state: or E', or U', or D'. Then we have three solutions for each of $\zeta^{(i)}$,

($i=E',U',D'$): two constant solutions are defined by the condition $F_i(\zeta)=0$

$$\zeta_{1,2}^{(i)}=\frac{1}{2}\left[k-3h_i\pm\sqrt{(k-3h_i)^2+8b(k-h_i)-4kh_i}\right],\qquad i=E',U',D' \tag{24}$$

and the third solution $\zeta+b=0$.

The first two solutions correspond to two different states of the $i's$ primordial fermion with different constant masses determined by Eq.(18) where we have to substitute $\zeta_{1,2}^{(i)}$ instead of ζ. These two states can be identified with *the first two generations of the physical leptons and quarks.*

Surprisingly that the same combination that we see in the l.h.s. of the constraint (23) appears in the last terms of Eqs. (16) and (12) (we assume here that $\zeta+b\neq 0$). Therefore, in the regime where the regular fermionic matter (i.e. u and d quarks, e^- and ν_e) dominates, the last terms of Eqs. (16) and (12) *automatically* vanish. In Eq. (16), this means that *the fermion densities are not a source for the dilaton and thus the long-range forces disappear automatically.* Notice that there is no need to require no interactions of the dilaton with fermionic matter at all to have agreement with observations but it is rather enough that these interactions vanish in the appropriate regime where regular fermionic matter has the typical laboratory particle physics density. In Eq. (12), the condition (23) means that in the region where the regular fermionic matter dominates, the fermion energy-momentum tensor becomes equal to the canonical energy-momentum tensor of fermion fields in GR (see also Ref.[10]).

The third solution $\zeta+b=0$ is singular one as we see from equations of motion. This means that one can not neglect the first two terms in the constraint (15). Then instead of $\zeta+b=0$ we have to take the solution $\zeta+b\approx 0$ by solving $\zeta+b$ in terms of the dilaton and Higgs fields and the primordial fermionic fields themselves.

$$\frac{1}{\sqrt{\zeta_3+b}}\approx\left[\frac{\upsilon\left[f_E(b-h_E)\overline{E'}E'+f_U(b-h_U)\overline{U'}U'+f_D(b-h_D)\overline{D'}D'\right]}{4\sqrt{2}(b-k)\left[b\left(sM^4e^{-2\alpha\phi/M_p}+V_1\right)-V_2\right]}\right]^{1/3} \tag{25}$$

This leads to non-polynomial fermion interactions. A full treatment of the third family requires the study of quantum corrections and fermion condensates which will give the third family appropriate masses. Interestingly enough that the effective coupling constants of the non-polynomial interactions are dimensionless, which suggests that the quantum corrections of this theory may be meaningful.

DISCUSSION AND CONCLUSIONS

For lack of space we have discussed here only some results concerning the particle physics in the two measures theory although the theory has a lot to say in the context of cosmology, in particular concerning the cosmological constant problem[1, 2, 3, 4, 5, 6, 7, 8, 9]. In the field theory/particle physics applications we have focused mostly on the families problem. When fermion densities are of the typical laboratory particle physics scales, three families have been obtained from the three different solutions (of the fundamental constraint of the theory) for the value of the scalar field $\zeta \equiv \frac{\Phi}{\sqrt{-g}}$, the ratio of the two measures.

It is important to point out that these three possible values of ζ **do not** represent three fifferent "vacuua" in the normal sense, like when one discusses the different expectation values of a normal scalar field. This is because ζ is **not** a normal scalar field: describing classical dynamics in terms of variables defined by Eq. (10) we have seen that space-time derivatives of ζ do not appear in the constraint and in the equations of motion. The scalar field ζ is determined by a constraint (15) and it can change from one value to the other very quickly in time and have very big gradients in space without this costing any energy. The three families can indeed coexist in the same space-time region just for this reason.

The quantization of this theory requires further study. Among the most interesting aspects of the quantum theory which should be studied are the quantization of the measure fields φ_a. In fact, we expect the "families birth effect" to be closely related to the functional integration over those measures fields. There we expect that functional integration will be restricted by the configurations dictated by the constraint (15). The integration over the φ_a fields should contain an integration over ζ and integrations over volume preserving variables (i.e. those that preserve the value of Φ). At each point, the integration over ζ selects then the values where the constraint is satisfied and for the fermion densities corresponding to laboratory conditions, three possible values of ζ are then selected. We hope to give more details concerning these quantum aspects of the theory in a future publication.

Finally, it is important to notice that the theory explained here allows for transitions from a certain family to another. One can indeed notice from the constraint itself that the three distinct values of ζ (again, for the fermion densities corresponding to laboratory conditions) can coincide when several types of fermions are present at the same space-time point. Once one reaches these "unification points", it is clear that transitions from family to family are possible. The calculation of the amplitudes of these transitions appear to be technically complicated but are in principle calculable. Therefore the parameters of the Kobayashi-Maskawa mass matrix should be indeed calculable as a function of the parameters of the theory.

In this report we have ignored the question of a possible neutrino mass. but there is no problem to incorporate it in our formalism. In this case the physics of neutrino will have some resemblance to the situation with quarks. It would be very important to see how the phenomenon of neutrino oscillations could appear in the context of this theory.

All the discussion so far in this paper has been in the context of situation where $\Phi \neq 0$ and finite, i.e. there is a well defined mapping from the x^μ space to the φ_a space. It could be however that configurations where the mapping between the x^μ and φ_a spaces is singular play a fundamental role in quantum gravity. An indication in this direction is the calculations of Farhi, Guth and Guven (FGG)[15] for the creation of a universe in the laboratory via tunneling when starting from a false vacuum bubble (which classically would collapse). They find out that in the tunneling solution, the bubble trajectory covers certain points more then once. To make sense of this solution, FGG are forced to introduce a "covering space" ξ^μ, different from the x^μ space, and allow the volume of integration to change sign. As FGG correctly point out, this corresponds to introducing a non Riemannian structure. This construction is very close to the introduction of the modified measure we have been studying here. The role of the "covering space" being similar to that of our φ_a scalars.

ACKNOWLEDGMENTS

We want to thank Arthur Chernin and Phillip Mannheim for usefull discussions.

REFERENCES

1. Guendelman, E., and Kaganovich, A., *Phys. Rev. D*, **53**, 7020–7025 (1996).
2. Guendelman, E., and Kaganovich, A., *Phys. Rev. D*, **55**, 5970–5980 (1997).
3. Guendelman, E., and Kaganovich, A., *Phys. Rev. D*, **56**, 3548–3554 (1997).
4. Guendelman, E., and Kaganovich, A., *Phys. Rev. D*, **57**, 7200–7203 (1998).
5. Guendelman, E., and Kaganovich, A., *Phys. Rev. D*, **60**, 065004–065028 (1999).
6. Guendelman, E., *Mod.Phys.Lett. A*, **14**, 1043–1052 (1999).
7. Guendelman, E., *Class.Quant.Grav.*, **17**, 361–372 (2000).
8. Guendelman, E., *Mod.Phys.Lett.A*, **14**, 1397–1408 (1999).
9. Kaganovich, A., *Phys.Rev. D*, **63**, 025022–025041 (2001).
10. Guendelman, E., and Kaganovich, A. (2001), URL `hep-th/0110040`.
11. Antoniadis, I., S.Dimopoulos, and Dvali, G., *Nucl. Phys. B*, **516**, 70–85 (1998).
12. Arkani-Hamed, N., Hall, L., Kolda, C., and Murayama, H., *Phys. Rev. Lett.*, **85**, 4434–4438 (2000).
13. Chernin, A. (2001), URL `astro-ph/0110003`.
14. Steinhardt, P., Wang, L., and Zlatev, I., *Phys.Rev. D*, **59**, 123504–123524 (1999).
15. Farhi, E., Guth, A., and Guven, J., *Nuclear Physics B*, **339**, 417–490 (1990).

Theoretical Status of Muon (g-2)

Utpal Chattopadhyay*, Achille Corsetti† and Pran Nath†

*Department of Theoretical Physics, Tata Institute of Fundamental Research, Homi Bhaba Road, Mumbai 400005, India
†Department of Physics, Northeastern University, Boston, Massachusetts 02115, USA

Abstract. The theoretical status of the muon anomaly is reviewed including the recent change in the light by light hadronic correction. Specific attention is given to the implications of the shift in the difference between the BNL experiment result and the standard model prediction for sparticle mass limits. The implication of the BNL data for Yukawa unification is discussed and the role of gaugino mass nonuniversalities in the satisfaction of Yukawa unification is explored. An analysis of the BNL constraint for the satisfaction of the relic density constraint and for the search for dark matter is also given.

INTRODUCTION

In this talk we discuss the current status of theory vs experiment for $a_\mu = (g_\mu - 2)/2$ and the implications for new physics. Recently a reevaluation of the light by light hadronic contribution to a_μ has resulted in a change in the sign of this contribution reducing the difference between the BNL experimental result and the standard model prediction from 2.6σ to 1.6σ. In view of this change we reconsider the implications for supersymmetry. We carry out the analysis using a 1σ and a 1.5σ error corridor around the central value of the difference between experiment and theory. For the 1σ analysis we find that the upper limits on sparticle masses remain unchanged from those predicted with the 2.6σ difference between experiment and the standard model result with a 2σ error corridor. For the 1.5σ analysis we find that the upper limits are substantially increased from the old analysis and the upper limits of the sparticle masses may sometimes lie on the borderline of what is accessible at the Large Hadron Collider. An important result that arises from the Brookhaven experiment is that the sign of the μ parameter is determined to be positive for a broad class of supersymmetric models. However, it is known that Yukawa coupling unification typically prefers a negative μ. We discuss a possible way out of this problem using nonuniversality of gaugino masses. Finally we consider the implications of the Brookhaven result for neutralino relic density and for the direct detection of supersymmetric dark matter in dark matter detectors.

$G_\mu - 2$: EXPERIMENT VS STANDARD MODEL

Over the last three months the theoretical prediction of a_μ in the standard model has undergone a significant revision because of the change in sign of the light by light (LbL) hadronic correction to a_μ. Thus the previous average for $a_\mu^{had}(LbL)$ was[1, 2]

CP624, *Cosmology and Elementary Particle Physics*, edited by B. N. Kursunoglu et al.

$a_\mu^{had}(LbL) = -8.5(2.5) \times 10^{-10}$. However, recent reevaluations[3, 4, 5, 6, 7] give a $a_\mu^{had}(LbL)$ opposite in sign to the previous evaluations. The reevaluations are summarized in Table 1.

Table 1: light by light hadronic correction

authors	$a_\mu^{had}(LbL)$
Knecht *et. al.*[3]	$8.3(2.5) \times 10^{-10}$
Hayakawa & Kinoshita[4]	$8.9(1.5) \times 10^{-10}$
Bijnens *et. al.*[6]	$(8.3 \pm 3.2) \times 10^{-10}$
Blockland *et. al.*[5] (π^0 *pole*)	5.6×10^{-10}

Now with the old value of the LbL hadronic correction the total standard model prediction of $a_\mu^{SM} = a_\mu^{QED} + a_\mu^{EW} + a_\mu^{had}$ was $a_\mu^{SM} = 11659159.7(6.7) \times 10^{-10}$. Using the BNL experimental result[8] of $a_\mu^{exp} = 11659203(15) \times 10^{-10}$ one finds $a_\mu^{exp} - a_\mu^{SM} = 43(16) \times 10^{-10}$ which gives the old 2.6σ deviation between experiment and the standard model. However, taking an average of the top three entries in Table 1 for $a_\mu^{had}(LbL)$ the revised difference between experiment and the standard model is

$$a_\mu^{exp} - a_\mu^{SM} = 26(16) \times 10^{-10} \tag{1}$$

which is now only a 1.6σ deviation between experiment and the standard model prediction. [More recently another evaluation of $a_\mu^{had}(LbL)$ based on chiral perturbation theory has been given in Ref.[7] which gives $a_\mu^{had}(LbL) = (5.5^{+5}_{-6} + 3.1\hat{C}) \times 10^{-10}$ where $\hat{C}$ stands for unknown low energy constants arising from subleading contributions. The authors of Ref.[7] view a $\hat{C}$ range of -3 to 3 or even larger as not unreasonable. The result of Eq.(1) corresponds essentially to a $\hat{C} = 1$ and a much larger value will significantly affect Eq.(1) and the conclusions resulting from it.] Aside from the issue of LbL hadronic correction, the remaining part of the hadronic correction contains α^2 and α^3 vacuum polarization corrections. In the deduction of Eq.(1) we used the evaluation of Ref.[9] for the α^2 correction. However, there is considerable amount of controversy regarding these corrections and this issue is still under scrutiny[10].

SUPERSYMMETRIC CORRECTION TO $G_\mu - 2$

If indeed there is discrepancy between experiment and the standard model prediction of a_μ then it would have important implications for new physics. Such new physics could be supersymmetry, compact extra dimensions, muon compositeness,techni-color, anomalous W couplings, new gauge bosons, lepto-quarks, or radiative muon masses[11]. We focus here on supersymmetric models and specifically on supergravity models[12] which arise from gravity mediated breaking of supersymmetry. The soft SUSY breaking sector of the minimal supergravity model (mSUGRA) is defined by four parameters: these consist of the universal scalar mass m_0, the universal gaugino mass $m_{1/2}$, the universal trilinear coupling A_0 and $\tan\beta = \langle H_2 \rangle / \langle H_1 \rangle$ where H_2 gives mass to the up quark and H_1 gives mass to the down quarks and the leptons. We will use SUGRA models as a benchmark and similar analyses can be carried out in other models such as those based on gauge mediation and anomaly mediation breaking mechanisms of supersymmetry. We begin by discussing the basic one loop contribution to $g_\mu - 2$ in supersymmetry[13]. Here the basic contributions are from the chargino $\tilde{W}$ and neutralino χ_i (i=1,.,4) exchange. For the CP conserving case the chargino contribution is the largest and here one has

$$a_\mu^{\tilde{W}} = \frac{m_\mu^2}{48\pi^2}\frac{A_R^{(a)2}}{m_{\tilde{W}_a}^2}F_1(\left(\frac{m_{\tilde{\nu}}}{m_{\tilde{W}_a}}\right)^2) + \frac{m_\mu}{8\pi^2}\frac{A_R^{(a)}A_L^{(a)}}{m_{\tilde{W}_a}}F_2(\left(\frac{m_{\tilde{\nu}}}{m_{\tilde{W}_a}}\right)^2) \tag{2}$$

where $A_L(A_R)$ are the left(right) chiral amplitudes and are defined by

$$A_R^{(1)} = -\frac{e}{\sqrt{2}\sin\theta_W}\cos\gamma_1; \quad A_L^{(1)} = (-1)^\theta \frac{em_\mu \cos\gamma_2}{2M_W \sin\theta_W \cos\beta}$$

$$A_R^{(2)} = -\frac{e}{\sqrt{2}\sin\theta_W}\sin\gamma_1; \quad A_L^{(2)} = -\frac{em_\mu \sin\gamma_2}{2M_W \sin\theta_W \cos\beta} \tag{3}$$

and where γ_i are the mixing angles and F_1, F_2 are form factors. Recently, the absolute signs of the supersymmetric contribution was checked by taking the supersymmetric limit[14]. There are some interesting features of the SUSY contribution. One finds that since $A_L \sim 1/\cos\beta$ one has[15, 16] $a_\mu^{SUSY} \sim \tan\beta$. Further, it is easy to show that the sign of a_μ^{SUSY} is correlated with the sign of μ[15, 16]. Thus one finds that $a_\mu^{SUSY} > 0$ for $\mu > 0$ and $a_\mu^{SUSY} < 0$ for $\mu < 0$ where we use the sign convention of Ref.[17].

IMPLICATIONS OF DATA

Upper limits on sparticle masses. Soon after the BNL result became available[8] a large number of analyses appeared in the literatures exploring the implications of the results for new physics[18]. These analyses were based on the result $a_\mu^{exp} - a_\mu^{SM} = 43(16) \times 10^{-10}$ which as is now realized is based on using the wrong sign of the light by light hadronic correction. Using the above result and a the 2σ error corridor so that $10.6 \times 10^{-10} < a_\mu^{SUSY} < 76.2 \times 10^{-10}$ the BNL data leads to the following sparticle mass limits in mSUGRA[19]: $m_{\tilde{W}} \leq 650$ GeV, $m_{\tilde{\nu}} \leq 1.5$ TeV ($\tan\beta \leq 55$) and $m_{1/2} \leq 800$ GeV, $m_0 \leq 1.5$ TeV ($\tan\beta \leq 55$). Since the LHC can explore squarks/gluinos up to 2 TeV the BNL result implies that sparticles should become visible at the LHC[20]. Next we assess the situation as a consequence of the change in the sign of the light by light hadronic correction which results in Eq.(1). In this case a 2σ error corridor would not lead to upper limits for the sparticle masses. However, one can get interesting constraints if one imposes a 1σ or a 1.5σ constraint. A 1σ constraint actually yields exactly the same upper limits as before so in this case the analysis of Ref.[19] remains valid as far as the upper limits are concerned. The case of 1.5σ was analyzed in Ref.[21] and it was found, as expected, that the upper limits go up considerably. In Fig.(1) results

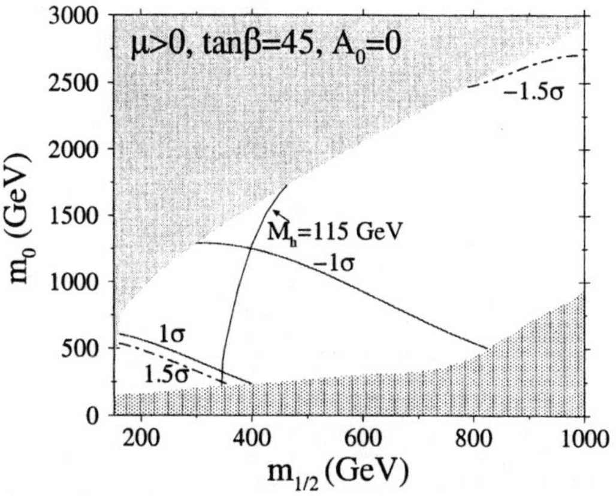

FIGURE 1. Upper and lower limits in $m_0 - m_{1/2}$ plane corresponding to 1σ and 1.5σ constraints from a_μ^{SUSY} for $\tan\beta = 45$. The top left gray region does not satisfy the radiative electroweak symmetry breaking requirement while the bottom patterned region near the higher $m_{1/2}$ side and on the border of the white allowed regions is discarded either because of stau or the CP-odd Higgs boson turning tachyonic at the tree level (from Ref.[21]).

are presented for the case of $\tan\beta = 45$. Here one finds that the upper limits on m_0 and $m_{1/2}$ are considerably larger than for the analysis of Ref.[19]. Specifically one finds from Fig.(1) that the upper limit on m_0 in the range of the parameter space exhibited already exceeds 2.5 TeV which is on the borderline of the reach of the LHC. The upper limits for other values of $\tan\beta$ are sharply dependent on the value of $\tan\beta$. A more complete analysis of the constraint for the 1.5σ case can be found in Ref.[21].

Another interesting implication of the BNL result is that under the assumption of CP conservation and setting $a_\mu^{SUSY} = a_\mu^{exp} - a_\mu^{SM}$ the BNL data determines $sign(\mu) = +1$ (see, e.g., Ref.[16]). It known that $\mu > 0$ is favored by the $b \to s + \gamma$ constraint[22, 23] and also favored for dark matter searches. The implications of $\mu > 0$ for dark matter will be discussed in further detail below. One issue of concern relates to the possibility that the supersymmetric effects may be masked by effects arising from low lying extra dimensions. This possibility was examined in Ref.[24] in a model with one large extra dimension compactified on S^1/Z_2 with radius R ($M_R = 1/R = O($ TeV$)$). The extra dimension contributes to the Fermi constant and by a comparison of the standard model prediction with the experimental value of the Fermi constant[24] one can place a limit on the extra dimension of about $M_R > 3$ TeV. With this size value of M_R one finds that the contribution of the extra dimension to a_μ is negligible[24] compared to the supersymmetric contribution. For the case of strong gravity the effect on a_μ from the Kaluza-Klein excitations of the graviton in the case d=2 is also small[25] since here the fundamental Planck scale M_* is found to have a lower limit of $M_* > 3.5$ TeV from the recent gravity experiment[26]. The above exhibits the fact that $g_\mu - 2$ is not an efficient probe of extra dimensions. Other techniques such as energetic dileptonic signals at LHC would be more efficient signals for the exploration of extra dimensions[27]. An

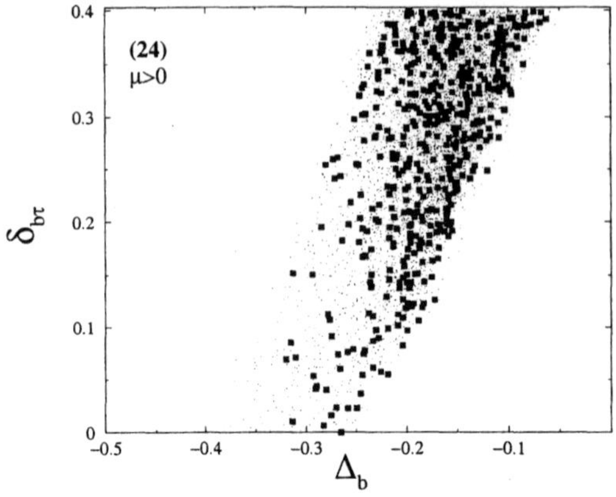

FIGURE 2. $\delta_{b\tau}$ vs Δ_b, the SUSY correction to m_b, for the 24-plet case, when $\tan\beta < 55$, $0 < m_0 < 2$ TeV, -1 TeV $< C_{24}m_{1/2} < 1$ TeV, -6 TeV $< A_0 < 6$ TeV and $\mu > 0$, where C_{24} is as defined in Ref.[35]. The dots refer to $b-\tau$ unification at the shown level and filled squares additionally represent points which satisfy both the $b \to s+\gamma$ and the muon $g-2$ constraints (from Ref.[35]).

important effect that can modify the supersymmetric contribution is the effect of CP violating phases. This topic has been analyzed in several works[14]. Specifically it is found that the BNL data can be used to constrain the CP phases and strong constraint on the phases are found to exist[28].

Positivity of μ and Yukawa Unification. We discuss now another aspect of the $g_\mu - 2$ constraint and this concerns Yukawa unification in supersymmetric models. It has been known for some time that $b-\tau$ Yukawa coupling unification typically prefers a negative μ[29, 30]. Thus the positively of the μ sign implied by the BNL data appears a priori to pose a problem for Yukawa unification. Now the reason why Yukawa unification typically prefers a negative μ is easily understood from the fact that such unification requires a negative supersymmetric correction to the b quark mass and a negative correction to the b quark mass is easily obtained when μ is negative. To illustrate this phenomenon we consider the gluino and chargino exchanges which contribute the largest supersymmetric correction to the b quark mass. Thus one has[31] $\Delta_b^{\tilde{g}} = \frac{2\alpha_3\mu M_{\tilde{g}}}{3\pi}\tan\beta I(m^2_{\tilde{b}_1}, m^2_{\tilde{b}_2}, M^2_{\tilde{g}})$ and $\Delta_b^{\tilde{\chi}^+} = \frac{Y_t\mu A_t}{4\pi}\tan\beta I(m^2_{\tilde{t}_1}, m^2_{\tilde{t}_2}, \mu^2)$ where $Y_t = \lambda_t^2/4\pi$ where $I > 0$. Generally the gluino exchange contribution tends to be the larger one and here one finds that for a positive $M_{\tilde{g}}$ which is typically the case a negative μ indeed leads to a negative correction to the b quark mass which in turn leads to the usual result that $b-\tau$ unification prefers a negative μ. There are several solutions discussed recently to overcome this problem[32, 33, 34, 35]. One simple possibility discussed in Refs.[34, 35] is that of nonuniversal gaugino masses where the sign of the gluino mass is negative relative to the mass of the SU(2) gaugino mass. In this case one can obtain a negative contribution to the b quark mass while maintaining a positive μ. The basic idea is that with nonuniversalities one can have the sign of $SU(2)$ and $SU(3)$

gauginos to be opposite. A positive positive μ and a positive $\tilde{m}_2$ are consistent with the BNL data, while a positive μ and a negative $\tilde{m}_3$ gives a negative correction to the b qaurk mass and leads to $b-\tau$ unification. We considered two classes of models, one based on SU(5) and the other on SO(10). For the SU(5) case one has that the the gaugino mass matrix can transform like the symmetric product of $(24\times24)_{sym}$ which has the expansion of $1+24+75+200$. In this case the gaugino masses arising from the 24 plet has the opposite sign between the SU(2) and SU(3) gauginos[36]. Thus one can choose μ positive and the gluino mass to be negative which gives a negative contribution to the b quark mass and allows for the satisfaction of the $b-\tau$ unification constraints. The degree of unification defined by $\delta_{b\tau}=(|\lambda_b-\lambda_\tau|)/\lambda_\tau$ vs the correction to the b quark mass is exhibited in Fig.(2) for the 24 plet case with a positive μ sign. One finds that $b-\tau$ unification can be satisfied to a high degree of accuracy with an appropriate negative correction Δ_b to the b quark mass. A similar analysis holds for the SO(10) case. Here the gaugino mass matrix can transform like symmetric product of $(45\times45)_{sym}$ which has the expansion of $1+54+210+770$. Here for the case when the symmetry breaking occurs pattern is of the form $SO(10)\rightarrow SU(4)\times SU(2)\times SU(2)$ $\rightarrow SU(3)\times SU(2)\times U(1)$ one finds that the $SU(3), SU(2), U(1)$ gaugino masses arising from the 54 plet are in the ratio[37] $M_3:M_2:M_1=1:-3/2:-1$. However, for the symmetry pattern $SO(10)\rightarrow SU(2)\times SO(7)\rightarrow SU(3)\times SU(2)\times U(1)$ one finds that the $SU(3), SU(2)$, and $U(1)$ gaugino masses arising from the 54 plet are in the ratio[37] $M_3:M_2:M_1=1:-7/3:1$. We will call this case $54'$. An analysis similar to that for the 24 plet case can be carried out for the 54 and $54'$ cases and one finds that $b-\tau$ unification occurs once again for these cases.

Implications for Relic Density and Dark Matter Search. As noted earlier the BNL data indicates a positive value of μ for mSUGRA. This result has important implications for dark matter. As already indicated a positive μ is preferred by the constraint imposed by the flavor changing neutral current process $b\rightarrow s+\gamma$[22] in that the experimental value for this branching ratio imposes severe constraints on the SUSY parameter space for the negative sign of μ but imposes much less stringent constraints on the parameter space for a positive value of μ. Thus a positive μ is more favorable for supersymmetric dark matter analysis in that it allows for a large amount of the parameter space of the model where relic density constraints can be satisfied along with satisfying the $b\rightarrow s+\gamma$ constraint. Furthermore, it also turns out that detection rates for a positive μ are generally larger than for a negative μ. Thus after the BNL data became available it was immediately realized that the positive μ sign indicated by the BNL data was favorable for dark matter searches[19, 18]. More detailed analyses were done in several subsequent works and the parameter space of mSUGRA was further constrained from the relic density constraints. Now another way that the BNL data constrains dark matter is through the Yukawa unification conditions. Here we discuss the implications of this constraint on dark matter. As discussed in the section on $b-\tau$ unification above, one finds that this unification can be achieved with a positive μ in a variety of ways. One possibility discussed above arose from nonuniversal gaugino masses. We discussed two main scenarios for nonuniversalities corresponding to the SU(5) and SO(10) cases. For SU(5) the gaugino mass nonuniversalities arising from the 24 plet case allows a negative contribution to the b quark mass with a positive μ and leads to regions of the parameter space where $b-\tau$ unification occurs. Analysis in this region exhibits that all of the spectrum lies within the usual naturalness limits[38]. It is interesting to investigate if this region of the parameter space also leads to a satisfaction of the relic density constraint. We consider a rather liberal corridor here corresponding to the range $0.02 \leq \Omega h^2 \leq 0.3$. The analysis shows that significant regions of the parameter space exist where these constraints are satisfied. An analysis of the detection rates in the direct detection of dark matter was also given[21]. One finds[21] that the detection rates lie in a range that can be fully explored in the new generation of experiments currently underway and those which are planned in the future (see, e.g., Ref.[39]). A similar analysis can be carried out for the SO(10) case. The sparticle masses consistent with the BNL 1σ constraint as given by Eq.(1), consistant with the $b \to s+\gamma$ constraint and consistent with the relic constraint are given in Table 2 for the 24 plet case of SU(5) and for the 54 and $54'$ cases of SO(10).

Table 2: Sparticle mass ranges for 24, 54, and $54'$ cases from Ref.[21]

Particle	**24** (GeV)	**54** (GeV)	$54'$ (GeV)
χ_1^0	32.3 - 75.2	32.3 - 81.0	32.3 - 33.4
$\chi_1^{\pm}$	86.9 - 422.6	94.6 - 240.8	145.8 - 153.9
$\tilde{g}$	479.5 - 1077.2	232.5 - 580.3	229.8 - 237.4
$\tilde{\mu}_1$	299.7 - 1295.9	480.5 - 1536.8	813.1 - 1196.3
$\tilde{\tau}_1$	203.5 - 1045.1	294.2 - 1172.6	579.4 - 863.7
$\tilde{u}_1$	533.6 - 1407.2	566.7 - 1506.4	822.9 - 1199.8
$\tilde{d}_1$	535.1 - 1407.5	580.3 - 1546.2	845.1 - 1232.5
$\tilde{t}_1$	369.9 - 975.2	271.5 - 999.6	513.7 - 819.9
$\tilde{b}_1$	488.2 - 1152.8	158.1 - 1042.0	453.2 - 749.9
h	104.3 - 114.3	103.8 - 113.3	108.1 - 110.9

Here one finds some interesting features in the spectrum. Thus in these scenarios the neturalino mass lies below 81 GeV and the higgs boson mass lies below 115 GeV in the three scenarios considered in Table 2. The Higgs mass ranges of Table 2 are consistent with the current Higgs mass limits from LEP[40] taking into account the $\tan\beta$ dependence[41]. Further these mass ranges can be fully explored in RUNII of the Tevatron. Similarly the mass ranges of the other sparticle masses of Table 2 can be explored in RUNII of the Tevatron via the trileptonic signal[42] and other techniques[17] while the full range for most of the spectrum of Table 2 can be explored at the LHC[20].

CONCLUSION

There is a significant amount of more data from the 2000 runs and BNL eventually hopes to measure a_μ to an accuracy of 4×10^{-10}. On the other hand, reanalyses of the hadronic correction are still underway to pin down further the size of these corrections. If the deviation between the central value of experiment and the standard model prediction persists at the current level but the error is significantly reduced one could still see a possibility of approaching the discovery limit. Needless to say the implications of a sizable deviation between experiment and theory are enormous as forseen in early works[13] and elucidated further in several subsequent works[15, 16, 18, 19]. Specifically, the light Higgs boson should show up in RUNII of the Tevatron and most of the sparticles ($\tilde{g}, \tilde{q}, \tilde{W}, \chi^0$ etc) should become visible at the LHC. Further, a positive μ sign implied by the BNL data along with a low lying sparticle spectrum is very encouraging for the search for supersymmetric dark matter.

ACKNOWLEDGMENTS

This work was supported in part by NSF grant PHY-9901057.

REFERENCES

1. H. Hayakawa, T. Kinoshita and A. Sanda, Phys. Rev. Lett. **75**, 790(1995); Phys. Rev. **D54**, 3137(1996); M. Hayakwa and T. Kinoshita, Phys. Rev. **D57**, 465(1998).
2. J. Bijnens, E. Pallante and J. Prades, Phys. Rev. Lett. **75**, 3781(1995); E. Nucl. Phys. **B474**, 379(1996). See also: Ref. [6].
3. M. Knecht and A. Nyffeler, arXiv:hep-ph/0111058; M. Knecht, A. Nyffeler, M. Perrottet and E. de Rafael, Phys. Rev. Lett. **88**, 071802 (2002).
4. M. Hayakawa and T. Kinoshita, arXiv:hep-ph/0112102.
5. I. Blokland, A. Czarnecki and K. Melnikov, Phys. Rev. Lett. **88**, 071803 (2002).
6. J. Bijnens, E. Pallante and J. Prades, arXiv:hep-ph/0112255.
7. M. Ramsey-Musolf and M. B. Wise, theory," arXiv:hep-ph/0201297.
8. H.N. Brown et al., Muon $(g-2)$ Collaboration, Phys. Rev. Lett. **86**, 2227 (2001).
9. M. Davier and A. Höcker, *Phys. Lett.* B **435**, 427 (1998) .
10. For other assessments of the hadronic error see, F.J. Yndurain, hep-ph/0102312; J.F. De Troconiz and F.J. Yndurain, arXiv:hep-ph/0106025; S. Narison, Phys. Lett. B **513**, 53 (2001); K. Melnikov, Int. Jour. of Mod. Phys. **A16**, 4591, (2001) [arXiv:hep-ph/0105267]; G. Cvetic, T. Lee and I. Schmidt, Phys. Lett. B **520**, 222 (2001). For a review of status of the hadronic error see, W.J. Marciano and B.L. Roberts, "Status of the hadronic contribution to the muon $g-2$ value", arXiv:hep-ph/0105056; J. Prades, "The Standard Model Prediction for Muon $g-2$", arXiv:hep-ph/0108192
11. A. Czarnecki and W.J. Marciano, *Nucl. Phys. (Proc. Suppl.)* **B76**, 245(1999).
12. A.H. Chamseddine, R. Arnowitt and P. Nath, Phys. Rev. Lett. **49**, 970 (1982) ; R. Barbieri, S. Ferrara and C.A. Savoy, *Phys. Lett.* B **119**, 343 (1982) ; L. Hall, J. Lykken, and S. Weinberg, *Phys. Rev.* D **27**, 2359 (1983) : P. Nath, R. Arnowitt and A.H. Chamseddine, *Nucl. Phys.* B **227**, 121 (1983) . For reviews, see P. Nath, R. Arnowitt and A.H. Chamseddine, "Applied N=1 Supergravity", world scientific, 1984; H.P. Nilles, Phys. Rep. **110**, 1(1984).
13. T. C. Yuan, R. Arnowitt, A. H. Chamseddine and P. Nath, *Z. Phys.* C **26**, 407 (1984) ; D.A. Kosower, L.M. Krauss, N. Sakai, *Phys. Lett.* B **133**, 305 (1983) ;
14. T. Ibrahim and P. Nath, Phys. Rev. bf D61,095008(2000); Phys. Rev. **D62**, 015004(2000).
15. J.L. Lopez, D.V. Nanopoulos, X. Wang, *Phys. Rev.* D **49**, 366 (1994) .
16. U. Chattopadhyay and P. Nath, *Phys. Rev.* D **53**, 1648 (1996) ; T. Moroi, *Phys. Rev.* D **53**, 6565 (1996) ; M. Carena, M. Giudice and C.E.M. Wagner, *Phys. Lett.* B **390**, 234 (1997) ; E. Gabrielli and U. Sarid, Phys. Rev. Lett. **79**, 4752 (1997) ; T. Ibrahim and P. Nath, Phys. Rev. bf D61,095008(2000); Phys. Rev. **D62**, 015004(2000); K.T. Mahanthappa and S. Oh, *Phys. Rev.* D **62**, 015012 (2000) ; T. Blazek, arXiv:hep-ph/9912460; U.Chattopadhyay , D. K. Ghosh and S. Roy, *Phys. Rev.* D **62**, 115001 (2000) .
17. SUGRA Working Group Collaboration (S. Abel et. al.), arXiv:hep-ph/0003154.

18. L. L. Everett, G. L. Kane, S. Rigolin and L. Wang, Phys. Rev. Lett. **86**, 3484 (2001); J. L. Feng and K. T. Matchev, Phys. Rev. Lett. **86**, 3480 (2001); E. A. Baltz and P. Gondolo, Phys. Rev. Lett. **86**, 5004 (2001); U. Chattopadhyay and P. Nath, Phys. Rev. Lett. **86**, 5854 (2001); S. Komine, T. Moroi, and M. Yamaguchi, Phys. Lett. B **506**, 93 (2001); Phys. Lett. B **507**, 224 (2001); T. Ibrahim, U. Chattopadhyay and P. Nath, Phys. Rev. **D64**, 016010(2001); J. Ellis, D.V. Nanopoulos, K. A. Olive, Phys. Lett. B **508**, 65 (2001); R. Arnowitt, B. Dutta, B. Hu, Y. Santoso, Phys. Lett. B **505**, 177 (2001); S. P. Martin, J. D. Wells, Phys. Rev. D **64**, 035003 (2001); H. Baer, C. Balazs, J. Ferrandis, X. Tata, Phys.Rev.**D64**: 035004, (2001); M. Byrne, C. Kolda, J.E. Lennon, arXiv:hep-ph/0108122. For a more complete set of references see, U. Chattopadhyay and P. Nath, arXiv:hep-ph/0108250.
19. U. Chattopadhyay and P. Nath, in Ref.[18].
20. CMS Collaboration, Technical Proposal: CERN/LHCC 94-38(1994); ATLAS Collaboration, Technical Proposal, CERN/LHCC 94-43(1944); H. Baer, C-H. Chen, F. Paige and X. Tata, Phys. Rev. **D52**, 2746(1995); Phys. Rev. **D53**, 6241(1996).
21. U. Chattopadhyay, A. Corsetti and P. Nath, arXiv:hep-ph/0201001.
22. P. Nath and R. Arnowitt, *Phys. Lett.* B **336**, 395 (1994) ; Phys. Rev. Lett. **74**, 4592 (1995) ; F. Borzumati, M. Drees and M. Nojiri, *Phys. Rev.* D **51**, 341 (1995) ; H. Baer, M. Brhlik, D. Castano and X. Tata, *Phys. Rev.* D **58**, 015007 (1998) .
23. M. Carena, D. Garcia, U. Nierste, C.E.M. Wagner, Phys. Lett. **B499** 141 (2001); G. Degrassi, P. Gambino, G.F. Giudice, JHEP 0012, 009 (2000) and references therein; W. de Boer, M. Huber, A.V. Gladyshev, D.I. Kazakov, Eur. Phys. J. C **20**, 689 (2001).
24. P. Nath and M. Yamaguchi, Phys. Rev. D **60**, 116004 (1999); Phys. Rev. D **60**, 116006 (1999). For a review see, P. Nath, arXiv:hep-ph/0011177
25. M. L. Graesser, Phys. Rev. D **61**, 074019 (2000) [arXiv:hep-ph/9902310].
26. C. D. Hoyle, U. Schmidt, B. R. Heckel, E. G. Adelberger, J. H. Gundlach, D. J. Kapner and H. E. Swanson, Phys. Rev. Lett. **86**, 1418 (2001) [arXiv:hep-ph/0011014].
27. P. Nath, Y. Yamada and M. Yamaguchi, Phys. Lett. B **466**, 100 (1999) [arXiv:hep-ph/9905415]; I. Antoniadis, K. Benakli and M. Quiros, Phys. Lett. B **460**, 176 (1999) [arXiv:hep-ph/9905311]; T.G. Rizzo, Phys. Rev. **D61**,055005(2000).
28. T. Ibrahim, U. Chattopadhyay and P. Nath, Phys. Rev. D **64**, 016010 (2001) [arXiv:hep-ph/0102324].
29. D. Pierce, J. Bagger, K. Matchev and R. Zhang, Nucl. Phys. **B491**, 3(1997); H. Baer, H. Diaz, J. Ferrandis and X. Tata, Phys. Rev. **D61**, 111701(2000).
30. W. de Boer, M. Huber, A.V. Gladyshev, D.I. Kazakov, Eur. Phys. J. C **20**, 689 (2001); W. de Boer, M. Huber, C. Sander, and D.I. Kazakov, arXiv:hep-ph/0106311;
31. L.J. Hall, R. Rattazzi and U. Sarid, Phys. Rev **D50**, 7048 (1994); R. Hempfling, Phys. Rev **D49**, 6168 (1994); M. Carena, M. Olechowski, S. Pokorski and C. Wagner, Nucl. Phys. **B426**, 269 (1994); D. Pierce, J. Bagger, K. Matchev and R. Zhang, Nucl. Phys. **B491**, 3 (1997).
32. H. Baer and J. Ferrandis, Phys. Rev. Lett.**87**, 211803 (2001).
33. T. Blazek, R. Dermisek and S. Raby, arXiv:hep-ph/0107097; R. Dermisek, arXiv:hep-ph/0108249; S. Raby, arXiv:hep-ph/0110203.
34. S. Komine and M. Yamaguchi, arXiv:hep-ph/0110032
35. U. Chattopadhyay and P. Nath, arXiv:hep-ph/0110341 (to be published in Phys. Rev. D).
36. G. Anderson, C.H. Chen, J.F. Gunion, J. Lykken, T. Moroi, and Y. Yamada, arXiv:hep-ph/9609457; G. Anderson, H. Baer, C-H Chen and X. Tata, Phys. Rev. D **61**, 095005 (2000).
37. N. Chamoun, C-S Huang, C Liu, and X-H Wu, Nucl. Phys. **B624**, 81 (2002).
38. K.L. Chan, U. Chattopadhyay and P. Nath, *Phys. Rev.* D **58**, 096004 (1998) .
39. H.V. Klapor-Kleingrothaus, et.al., "GENIUS, A Supersensitive Germanium Detector System for Rare Events: Proposal", MPI-H-V26-1999, arXiv:hep-ph/9910205.
40. [LEP Higgs Working Group Collaboration], "Searches for the neutral Higgs bosons of the MSSM: Preliminary combined results using LEP data collected at energies up to 209-GeV," arXiv:hep-ex/0107030.
41. A. Sopczak, arXiv:hep-ph/0112086.
42. P. Nath and R. Arnowitt, Mod. Phys.Lett.**A2**, 331(1987); H. Baer and X. Tata, Phys. Rev.**D47**, 2739(1993); V. Barger and C. Kao, Phys. Rev. **D60**, 115015(1999).

Simulation of Models with adjoint Bosons in 1+1 Dimensions

S. Pinsky and I. Filippov

Department of Physics
Ohio State University, Columbus, OH 43210

Abstract. We consider the N=(1,1) SYM theory that is obtain by dimensionally reducing SYM theory in 2+1 dimensions to 1+1 dimensions and discuss soft supersymmetry breaking. We discuss the numerical simulation of this theory using SDLCQ when the fermion has a larger mass. We compare our results to the pure adjoint boson DLCQ calculation of Klebanov, Demeterfi and Bhanot. When a large fermion mass is added to the theory we find that it is not necessary to add operators to obtain a sensible theory. This theory of adjoint boson is a theory that has stringy bound states similar to the full SYM theory. We also discuss another theory of adjoint bosons with a spectrum similar to that of Klebanov, Demeterfi and Bhanot.

INTRODUCTION

In recent years we worked extensively on a numerical method [1] for solving exactly supersymmetric field theories in $1+1$ and $2+1$ dimensions. We call this method supersymmetric discrete light-cone quantization (SDLCQ) and we have successfully applied it to many theories and addressed a number of interesting issues[2, 3, 4, 5, 6, 7, 8, 9]. The world is however not exactly supersymmetric and it is therefore important to learn how to generalize SDLCQ to solve none supersymmetric theories.

With this objective in mind we focus in this work on the interrelation of the numerical simulations of two theories. The objective is to learn something about soft supersymmetry breaking within the context of these numerical simulations. We want to know if softly broken theories make sense in the context of SDLCQ and how the broken theories is related to non-supersymmetric theories of adjoint bosons that have been discussed in the literature.

The non-supersymmtric simulations is that of Klebanov, Demeterfi and Bhanot[10] (KDB) a calculation of a gauged adjoint boson in $1+1$ dimensions. The supersymmetric theory is the N=(1,1) SYM that one obtains by dimensionally reducing SYM in 2+1 dimensions to 1+1 dimensions[1, 2]. Henceforth we will refer to this theory as SYM. Starting with this theory we consider the theory that is obtained by adding a large mass for the fermion fields. The large mass freezes out that field leaving the low mass bound state to have primarily only constituents of the boson field.

The numerical method that we use to simulate these theories is discrete light-cone quantization (DLCQ). When this method is applied to the continuum hamiltonian of a theory it is referred to just as DLCQ and it produces a finite dimensional hamiltonian. When this method is applied to the continuum supercharge of a supersymmetric theory it

CP624, *Cosmology and Elementary Particle Physics*, edited by B. N. Kursunoglu et al.

produces a finite dimensional supercharge which is then used to calculate a finite dimensional hamiltonian, $P^- = (Q^-)^2/\sqrt{2}$. We refer to this approximation as supersymmetric DLCQ or SDLCQ. To discretize the hamiltonian or supercharge, we introduce discrete longitudinal momenta k^+ as fractions nP^+/K of the total longitudinal momentum P^+, where K is an integer that determines the resolution of the discretization and is known in DLCQ as the harmonic resolution [11]. We then convert the mass eigenvalue problem $2P^+P^-|M\rangle = M^2|M\rangle$ $(P^\pm = (P^z \pm P^0)/\sqrt{2})$ to a matrix eigenvalue problem by introducing a basis where P^+ is diagonal. Because light-cone longitudinal momenta are always positive, K and each n are positive integers; the number of constituents is then bounded by K. The continuum limit is then recovered by taking the limit $K \to \infty$.

Of course we can write the continuum hamiltonian for a supersymmetric theory and apply DLCQ to it. This yields a different finite dimensional approximation to the hamiltonian than SDLCQ. Recently we have developed a technique for writing down a finite dimensional DLCQ hamiltonians directly, that are identical to the hamiltonians obtained in SDLCQ[9].

THE SUPERSYMMETRIC THEORY

SYM is difined by the supercharge given in [1] which in turn defines the hamiltionian by virtue of the anticommutation relation $\{Q^-, Q^-\} = 2\sqrt{2}P^-$. Through out this paper we will write expression in a continuum form for notational convenience, however it is to be understood that all the calculation are discrete in momentum space. The numerical method SDLCQ simply means that we apply DLCQ to the supercharge and then square the finite dimensional representation of the supercharge to get the Hamiltonian. Recently we found the hamiltonian that in the DLCQ approximation reproduces the SDLCQ hamiltonian,

$$\begin{aligned}
P^- &= \frac{g^2 N_c}{4\pi}\int_0^\infty dk_1 \frac{\mu^2(k_1)}{k_1}(a^\dagger a + b^\dagger b) + \frac{g^2}{4\pi}\int_0^\infty dk_1 dk_2 dk_3 dk_4 [\\
&+ A_1 b^\dagger b^\dagger bb + A_2(b^\dagger bbb - b^\dagger b^\dagger b^\dagger b) + B_1 a^\dagger a^\dagger aa + B_2(a^\dagger aaa + a^\dagger a^\dagger a^\dagger a) \\
&+ C_1 b^\dagger b^\dagger aa + C_2 a^\dagger a^\dagger bb + C_3 b^\dagger a^\dagger ba + C_4 a^\dagger b^\dagger ab + C_5 b^\dagger a^\dagger ab + C_6 a^\dagger b^\dagger ba \\
&+ D_1(a^\dagger abb - a^\dagger b^\dagger b^\dagger a) + D_2(a^\dagger bab - b^\dagger a^\dagger b^\dagger a) + D_3(a^\dagger bba - b^\dagger b^\dagger a^\dagger a) \\
&+ D_4(b^\dagger baa + b^\dagger a^\dagger a^\dagger b) + D_5(b^\dagger aba + a^\dagger b^\dagger a^\dagger b) + D_6(b^\dagger aab + a^\dagger a^\dagger b^\dagger b)],
\end{aligned} \quad (1)$$

where the coefficient in front of the dynamic mass term is given by [9],

$$\mu^2(k_1) = \int_0^{k_1} dk_2 \frac{(k_1+k_2)^2}{k_2(k_1-k_2)^2} = \int_0^{k_1} dk_2 \left(\frac{4k_1}{(k_2-k_1)^2} + \frac{1}{k_2}\right). \quad (2)$$

It is convenient to define the instantaneous mass contribution to the Hamiltonian, $P^{-}_{Imass}(boson)$,

$$P^{-}_{Imass}(boson) = \frac{g^2N_c}{\pi}\int_0^\infty dk_1 a(k_1)^\dagger a(k_1)\int_0^{k_1}\frac{dk_2}{(k_2-k_1)^2} \tag{3}$$

which is part of μ defined above. The coefficients of the pure boson terms are,

$$\begin{aligned} B_1 &= \frac{1}{\sqrt{4k_1k_2k_3k_4}}\left(\frac{(k_1-k_2)(k_3-k_4)}{(k_1+k_2)^2} - PV\frac{(k_1+k_3)(k_2+k_4)}{(k_4-k_2)^2}\right) \\ B_2 &= \frac{1}{\sqrt{4k_1k_2k_3k_4}}\left(\frac{(k_3-k_2)(k_1+k_4)}{(k_3+k_2)^2} + \frac{(k_1-k_2)(k_3+k_4)}{(k_1+k_2)^2}\right). \end{aligned} \tag{4}$$

These term arises from normal order the square of the discrete supercharges rather than normal ordering the continuum formulation of the supercharge and then discretizing this result. A detail discussion of the origin of these terms can be found in [9].

The singularities in P_{Imass} cancel the singularities in B_1. This cancellation is commonly seen in light-cone calculation and is not related to supersymmetry. We will see that KDB in their theory of adjoint bosons treat these singularities differently than they are treated in SDLCQ.

The $1/k_2$ term in μ is a real logarithmic mass divergence. In a non-supersymmetric theory this requires a mass renormalization. Here in a supersymmetric theory the bound states are such that this term is finite.

PURE ADJOINT BOSON THEORIES

The pure boson theories that we want to compare are KDB and the SYM theory with a large mass for the adjoint fermion. Physically the boson are the transverse gluons of the 2+1 dimensional parent theory, therefore these states are effectively 1+1 dimensional glueballs.

First we consider the result of simply adding a large fermion mass to the SYM theory. This leaves us with the pure boson theory which contains the Imass term and the logarithmic divergent mass term. Surprisingly the low mass states of the boson theory, seen in Fig.[1a], are linear as a function of $1/K$ and therefore the continuum spectrum obtained by extrapolating to $K=\infty$ appears to remains finite. We have calculated one state out to resolution twenty to check for any logarithmic dependence and found none. A partial explanation for this has to lie in the stringy nature of these bound states. In the spectrum of the pure SYM we found that as we increased the resolution new bound states appeared with more partons and with a lower mass. This abundance of low mass states with many partons is what we refer to as the stringy nature of the theory. Interestingly we see this property for this pure glue theory here in Fig.[1a]. The fact that the logarithmic divergent mass term does not give rise to a divergent spectrum as a function of the resolution is apparently related to the stingy nature of the bound states.

We now want to contrast this stringy theory with the model considered by KDB [10]. There are two main differences between these two models. First KDB renormalizes away

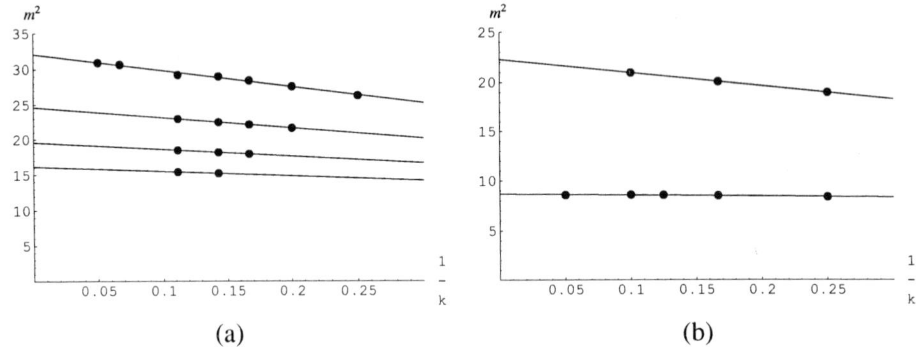

FIGURE 1. $\frac{\pi M^2}{g^2 N_c}$ vs. $\frac{1}{K}$, where K is the resolution (a) state of SYM with a very large fermion mass (b)SDLCQ with a heavy fermion mass add, the logarithmic divergent boson mass subtracted and a bare mass of $m^2 = 2g^2N_c/\pi$ added

the logarithmic divergent mass term. Secondly they use a different approach to treat the singularity in P^-_{Imass} term and the singular term in B_1. To regularize the singularity in B_1 they add and subtract a term. The remaining finite part of P_{Imass} has the form a mass term with $m^2 = -2g^2N/\pi$. They lump this term together with the logarithmic divergent mass term and a bare mass term to form a renormalized mass term. In principle this appears to be simply a different way of renormalizing the mass and a different way of making the singularity finite in the principle value prescription.

Numerically KDB find a very different spectrum than we found above. Their spectrum is very QCD like with the lowest mass bound states having primarily two gluon and a mass of about $4g^2N_c/\pi$. The higher mass states have dominant components with a small number of particles. KDB did the calculation using anti-periodic boundary conditions and we have repeated the calculation using both periodic and anti-periodic boundary conditions. We get the same results by both methods and our anti-periodic calculation agrees exactly with KDB. The convergence is similar in both methods when one takes into account the fact the you have to go to twice the resolution with anti-periodic boundary conditions to get the same number of data points.

It is interesting to make a direct comparison of the KDB approach with SDLCQ. To make this comparison we will repeat the SDLCQ calculation but now we will drop the divergent mass term and add a bare mass term as KDB do and then calculate the spectrum for various values of the bare mass.

With the bare mass equal to zero the theory is unstable and generates negative mass for the bound states. For $m^2 = g^2N_c/\pi$ the mass of the lowest bound state does not converge well as a function of $1/K$. It The fit is,

$$M^2 = \frac{g^2N_c}{\pi}(6.00 - 2.80\frac{1}{K} + 1.29078Log[K]). \tag{5}$$

The lowest mass bound state for $m^2 = 2g^2N_c/\pi$ is shown in Fig.[1b] and we see that the bound state is well behaved and is fit nicely by a linear plot in $1/K$. In fact the entire spectrum for this theory is well behaved. Any inspection of the wavefunctions of this theory shows that it has a valence structure similar to KDB but not identical.

DISCUSSION

We have consider the effect of adding a fermion mass to SYM and compared the results with known pure bosonic theories that exist in the literature [10].

When we add a large fermion mass to the SYM theory the resulting bosonic theory does appear to be a sensible theory. The masses are linear in $1/K$ and have a well defined continuum limit. The spectrum is very stringy like the spectrum of the original SYM theory. As we increase the resolution we find states with lower masses and more partons. In comparing it to the work of KDB we see that they renormalized away the logarithmic divergent mass term, treat the coulomb singularity differently than SDCLQ and adds a bare mass. The spectrum that they find also converges well in $1/K$ and the spectrum is very QCD like. The low mass states have a few valence partons and as one increase the resolution one finds higher mass states. When we renormalize away the logarithmic divergence mass and add a bare mass but use the SDLCQ treatment of the coulomb singularity we find that for some values of the bare mass that we also find a QCD like spectrum. If the bare mass is too small however we find that the spectrum is badly behaved.

ACKNOWLEDGMENTS

This work was supported in part by the US Department of Energy. One of the authors (S.P) would like to acknowledge the hospitality of the Aspen Center of Physics where part of the work was completed. The authors would like to acknowledge conversations with U. Trittmann, O. Lunin and J. Hiller.

REFERENCES

1. Y. Matsumura, N. Sakai, and T. Sakai, *Phys. Rev.* **D52** (1995) 2446, hep-th/9504150.
2. F. Antonuccio, O. Lunin, and S. Pinsky, *Phys. Lett.* **B429** (1998) 327, hep-th/9803027; *Phys. Rev.* **D58** (1998) 085009, hep-th/9803170; *Phys. Lett.* **B442** (1998) 173, hep-th/9809165; *Phys. Rev.* **D59** (1999) 085001, hep-th/9811083; *Phys. Lett.* **B459** (1999) 159, hep-th/9904027.
3. F. Antonuccio, H. C. Pauli, S. Pinsky, and S. Tsujimaru, *Phys. Rev.* **D58** (1998) 125006, hep-th/9808120.
4. F. Antonuccio, O. Lunin, S. Pinsky, H. C. Pauli, and S. Tsujimaru, *Phys. Rev.* **D58** (1998) 105024, hep-th/9806133.
5. F. Antonuccio, O. Lunin, S. Pinsky, and A. Hashimoto, *JHEP* **07** (1999) 029.
6. J.R. Hiller, O. Lunin, S. Pinsky, and U. Trittmann, *Phys. Lett.* **B482** (2000) 409, hep-th/0003249.
7. P. Haney, J.R. Hiller, O. Lunin, S. Pinsky, and U. Trittmann *Phys. Rev.* **D62** (2000) 075002, hep-th/9911243.
8. J.R. Hiller, S. Pinsky, and U. Trittmann, *Phys. Rev.* **D63**(2001) 105017 hep-th/0101120; *Phys. Rev.* **D64**(2001) 105027 hep-th/0101120; "Simulation of Dimensionally Reduced SYM-Chern-Simons theory", submitted to PRD hep-th/0112151.

9. I. Filippov, S. Pinsky, J.R. Hiller, "Renormalizing DLCQ using supersymmetry" *Phys. Lett.* **B506** (2001) 221, hep-th/0011106.
10. K. Demeterfi, I.R. Klebanov, and G. Bhanot, *Nucl. Phys.* **B418** (1994) 15.
11. S.J. Brodsky, H.C. Pauli, and S.S. Pinsky, *Phys. Rep.* **301** (1998) 299, hep-ph/9705477.

PROGRESS IN M-THEORY*

M. J. DUFF†
Michigan Center for Theoretical Physics
Randall Laboratory, Department of Physics, University of Michigan
Ann Arbor, MI 48109–1120, USA

After reviewing how M-theory subsumes string theory, we report on some new and interesting developments, focusing on the "brane-world".

1. M-theory and dualities

Not so long ago it was widely believed that there were five different superstring theories each competing for the title of "Theory of everything," that all-embracing theory that describes all physical phenomena. See Table 1.

Moreover, on the (d, D) "branescan" of supersymmetric extended objects with d worldvolume dimensions moving in a spacetime of D dimensions, all these theories occupied the same $(d = 2, D = 10)$ slot. See table 2. The orthodox wisdom was that while $(d = 2, D = 10)$ was the Theory of Everything, the other branes on the scan were Theories of Nothing. All that has now changed. We now know that there are not five different theories at all but, together with $D = 11$ supergravity, they form merely six different corners of a deeper, unique and more profound theory called "M-theory" where M stand for Magic, Mystery or Membrane. M-theory involves all of the other branes on the branescan, in particular the eleven-dimensional membrane $(d = 3, D = 11)$ and eleven-dimensional fivebrane $(d = 6, D = 11)$, thus resolving the mystery of why strings stop at ten dimensions while supersymmetry allows eleven[2].

Although we can glimpse various corners of M-theory, the big picture still eludes us. Uncompactified M-theory has no dimensionless parameters, which is good from the uniqueness point of view but makes ordinary perturbation theory impossible since there are no small coupling constants to provide the expansion parameters. A low energy, E, expansion is possible in powers of E/M_P, with M_P the Planck mass, and leads to the familiar $D = 11$ supergravity plus corrections of higher powers in the curvature. Figuring out what governs these corrections would go a long way in pinning down what M-theory really is.

Why, therefore, do we place so much trust in a theory we cannot even define? First we know that its equations (though not in general its vacua) have the maximal number of 32 supersymmetry charges. This is a powerful constraint and provides many "What else can it be?" arguments in guessing what the theory looks like when

*Research supported in part by DOE Grant DE-FG02-95ER40899.
†mduff@umich.edu

CP624, *Cosmology and Elementary Particle Physics*, edited by B. N. Kursunoglu et al.

	Gauge Group	Chiral?	Supersymmetry charges
Type I	$SO(32)$	yes	16
Type IIA	$U(1)$	no	32
Type IIB	–	yes	32
Heterotic	$E_8 \times E_8$	yes	16
Heterotic	$SO(32)$	yes	16

Table 1: The Five Superstring Theories

compactified to $D < 11$ dimensions. For example, when M-theory is compactified on a circle S^1 of radius R_{11}, it leads to the Type IIA string, with string coupling constant g_s given by

$$g_s = {R_{11}}^{3/2} \tag{1}$$

We recover the weak coupling regime only when $R_{11} \to 0$, which explains the earlier illusion that the theory is defined in $D = 10$. Similarly, if we compactify on a line segment (known technically as S^1/Z_2) we recover the $E_8 \times E_8$ heterotic string. Moreover, although the corners of M-theory we understand best correspond to the weakly coupled, perturbative, regimes where the theory can be approximated by a string theory, they are related to one another by a web of dualities, some of which are rigorously established and some of which are still conjectural but eminently plausible. For example, if we further compactify Type IIA string on a circle of radius R, we can show rigorously that it is equivalent to the Type IIB string compactified on a circle or radius $1/R$. If we do the same thing for the $E_8 \times E_8$ heterotic string we recover the $SO(32)$ heterotic string. These well-established relationships which remain within the weak coupling regimes are called *T-dualities.* The name *S-dualities* refers to the less well-established strong/weak coupling relationships. For example, the $SO(32)$ heterotic string is believed to be S-dual to the $SO(32)$ Type I string, and the Type IIB string to be self-S-dual. If we compactify more dimensions, other dualities can appear. For example, the heterotic string compactified on a six-dimensional torus T^6 is also believed to be self-S-dual. There is also the phenomenon of *duality of dualities* by which the T-duality of one theory is the S-duality of another. When M-theory is compactified on T^n, these S and T dualities are combined into what are termed U-dualities. All the consistency checks we have been able to think of (and after 5 years there dozens of them) have worked and convinced us that all these dualities are in fact valid. Of course we can compactify M-theory on more complicated manifolds such as the four-dimensional $K3$ or the six-dimensional Calabi-Yau spaces and these lead to a bewildering array of other dualities. For example: the heterotic string on T^4 is dual to the Type II string on $K3$; the heterotic string on T^6 is dual to the the Type II string on Calabi-Yau; the Type IIA string on a Calabi-Yau manifold is dual to the Type IIB string on the mirror Calabi-Yau manifold. These more complicated compactifications lead to many more parameters in the theory, known to the mathematicians as *moduli*, but

$D \uparrow$												
11	.			S			T					
10	.	V	S/V	V	V	V	S/V	V	V	V	V	
9	.	S				S						
8	.				S							
7	.			S			T					
6	.	V	S/V	V	S/V	V	V					
5	.	S		S								
4	.	V	S/V	S/V	V							
3	.	S/V	S/V	V								
2	.	S										
1	.											
0	.	.	.	.	.	.	.	.	.	.	.	.
	0	1	2	3	4	5	6	7	8	9	10	11 $d \rightarrow$

Table 2: The branescan, where S, V and T denote scalar, vector and antisymmetric tensor multiplets.

in physical uncompactified spacetime have the interpretation as expectation values of scalar fields. Within string perturbation theory, these scalar fields have flat potentials and their expectation values are arbitrary. So deciding which topology Nature actually chooses and the values of the moduli within that topology is known as the *vacuum degeneracy problem.*

2. Branes

In the previous section we outlined how M-theory makes contact with and relates the previously known superstring theories, but as its name suggest, M-theory also relies heavily on membranes or more generally p-branes, extended objects with $p = d - 1$ spatial dimensions (so a particle is a 0-brane, a string is a 1-brane, a membrane is a 2-brane and so on). In $D = 4$, a charged 0-brane couples naturally to an Maxwell vector potential A_μ, with field strength $F_{\mu\nu}$ and carries an electric charge

$$Q \sim \int_{S^2} * F_2 \tag{2}$$

and magnetic charge

$$P \sim \int_{S^2} F_2 \tag{3}$$

where F_2 is the Maxwell 2-form field strength, $*F_2$ is its 2-form dual and S^2 is a 2-sphere surrounding the charge. This idea may be generalized to p-branes in D dimensions. A p-brane couples to $(p+1)$-form potential $A_{\mu_1 \mu_2 \dots \mu_{p+1}}$ with $(p+2)$-form field strength $F_{\mu_1 \mu_2 \dots \mu_{p+2}}$ and carries an "electric" charge per unit p-volume

$$Q \sim \int_{S^{D-p-2}} * F_{D-p-2} \tag{4}$$

and "magnetic" charge per unit p-volume

$$P \sim \int_{S^{p+2}} F_{p+2} \tag{5}$$

where F_{p+2} is the $(p+2)$-form field strength, $*F_{D-p-2}$ its $(D-p-2)$-form dual and S^n is an n-sphere surrounding the brane. A special role is played in M-theory by the BPS (Bogomolnyi-Prasad-Sommerfield) branes whose mass per unit p-volume, or tension T, is equal to the charge per unit p-volume

$$T \sim Q \tag{6}$$

This formula may be generalized to the cases where the branes carry several electric and magnetic charges. The supersymmetric branes shown on the branescan are always BPS, although the converse is not true. M-theory also makes use of non-BPS and non-supersymmetric branes not shown on the branescan, but the supersymmetric ones do play a special role because they are guaranteed to be stable.

The letters S, V, and T on the branescan refer to scalar, vector and antisymmetric tensor supermultiplets of fields that propagate on the worldvolume of the brane. Historically, these points on the branescan were discovered in three different ways. The S branes were classified by writing down spacetime supersymmetric worldvolume actions that generalize the Green-Schwarz actions on the superstring worldsheet[5]. By contrast, the V and T branes were shown to arise as soliton solutions[a] of the underlying supergravity theories[6]. However, the solitonic V branes found this way were bound by $p \leq 7$. The 8-brane and 9-brane slots were included on the scan only because they were allowed by supersymmetry[6]. Subsequently, all the V-branes were given a new interpretation as Dirichlet p-branes, called D-branes, surfaces of dimension p on which open strings can end and which carry R-R (Ramond-Ramond) charge[9]. The IIA theory has D-branes with $p = 0, 2, 4, 6, 8$ and the IIB theory has D-branes with $p = 1, 3, 5, 7, 9$. They are related to one another by T-duality. In terms of how their tensions depend on the string coupling g_s, the D-branes are mid-way between the fundamental (F) strings and the solitonic (S) fivebranes:

$$T_{F1} \sim {m_s}^2, \qquad T_{Dp} \sim \frac{{m_s}^{p+1}}{g_s}, \qquad T_{S5} \sim \frac{{m_s}^6}{{g_s}^2} \tag{7}$$

Since they are BPS, there is a no-force condition between the branes that allows us to have many branes of the same charge parallel to one another. The gauge group on a single D-brane is an abelian $U(1)$. If we stack N such branes on top of one another, the gauge group is the non-abelian $U(N)$. As we separate them this decomposes into its subgroups, so in fact there is a Higgs mechanism at work whereby the vacuum expectation values of the Higgs fields are related to the separation of the branes. For example the theory that lives on a stack of N Type IIB $D3$ branes is a four-dimensional $U(N)$ $n = 4$ super Yang-Mills theory. In the limit of large N

[a]The 3-brane soliton of Type IIB supergravity was an early candidate for a 'brane-world', firstly because of its dimensionality[7,8] and secondly because gauge fields propagate on its worldvolume[8].

the geometry of this configuration tends to the product of five dimensional anti-de Sitter space and a five dimensional sphere, $AdS_5 \times S^5$.

In $D = 11$, M-theory has two BPS branes, an electric 2-brane and its magnetic dual which is a 5-brane. Their tensions are related to each other and the Planck mass by

$$T_2^3 \sim T_5 \sim M_P^6 \tag{8}$$

if we stack N such branes on top of one another, the M2-brane geometry tends in the large N limit to $AdS_4 \times S^7$ and the M5-brane geometry to $AdS_7 \times S^4$. In addition there are two other objects in $D = 11$, the plane wave and the Kaluza-Klein monopole, which though not branes are still BPS. When spacetime is compactified a p-brane may remain a p-brane or else become a $(p-k)$-brane if it wraps around k of the compactified directions. For example, the Type IIA fundamental string emerges by wrapping the M2-brane around S^1 and shrinking its radius to zero, and the Type IIA 4-brane emerges in a similar way from the $M5$-brane.

3. Spin-offs of M-theory

What do we now know with M-theory that we did not know with old-fashioned string theory? Here are a few examples, references to which may be found in Ref. 2.

1) Electric-magnetic (strong/weak coupling) duality in $D = 4$ is a consequence of string/string duality in $D = 6$ which in turn is a consequence of membrane/fivebrane duality in $D = 11$.

2) *Exact* electric-magnetic duality, first proposed for the maximally supersymmetric conformally invariant $n = 4$ super Yang-Mills theory, has been extended to *effective* duality by Seiberg and Witten to non-conformal $n = 2$ theories: the so-called Seiberg-Witten theory. This has been very successful in providing the first proofs of quark confinement (albeit in the as-yet-unphysical super QCD) and in generating new pure mathematics on the topology of four-manifolds. Seiberg-Witten theory and other $n = 1$ dualities of Seiberg may, in their turn, be derived from M-theory.

3) Indeed, it seems likely that all supersymmetric quantum field theories with any gauge group, and their spontaneous symmetry breaking, admit a geometrical interpretation within M-theory as the worldvolume fields that propagate on the common intersection of stacks of p-branes wrapped around various cycles of the compactified dimensions, with the Higgs expectation values given by the brane separations.

4) In perturbative string theory, the vacuum degeneracy problems arises because there are billions of Calabi-Yau vacua which are distinct according to classical topology. Like higher-dimensional Swiss cheeses, each can have different number of p-dimensional holes. This results in many different kinds of four-dimensional gauge theories with different gauge groups, numbers of families and different choices of quark and lepton representations. Moreover, M-theory introduces new non-perturbative effects which allow many more possibilities, making the degeneracy problem apparently even worse. However, most (if not all) of these manifolds are

in fact smoothly connected in M-theory by shrinking the p-branes that can wrap around the p-dimensional holes in the manifold and which appear as black holes in spacetime. As the wrapped-brane volume shrinks to zero, the black holes become massless and effect a smooth transition from one Calabi-Yau manifold to another. Although this does not yet cure the vacuum degeneracy problem, it puts it in a different light. The question is no longer why we live in one topology rather than another but why we live in one particular corner of the unique topology. This may well have a dynamical explanation.

5) Ever since the 1970's, when Hawking used macroscopic arguments to predict that black holes have an entropy equal to one quarter the area of their event horizon, a microscopic explanation has been lacking. But treating black holes as wrapped p-branes, together with the realization that Type II branes have a dual interpretation as Dirichlet branes, allows the first microscopic prediction in complete agreement with Hawking. The fact that M-theory is clearing up some long standing problems in quantum gravity gives us confidence that we are on the right track.

6) It is known that the strengths of the four forces change with energy. In supersymmetric extensions of the standard model, one finds that the fine structure constants $\alpha_3, \alpha_2, \alpha_1$ associated with the $SU(3) \times SU(2) \times U(1)$ all meet at about 10^{16} GeV, entirely consistent with the idea of grand unification. The strength of the dimensionless number $\alpha_G = GE^2$, where G is Newton's constant and E is the energy, also almost meets the other three, but not quite. This near miss has been a source of great interest, but also frustration. However, in a universe of the kind envisioned by Witten, spacetime is approximately a narrow five dimensional layer bounded by four-dimensional walls. The particles of the standard model live on the walls but gravity lives in the five-dimensional bulk. As a result, it is possible to choose the size of this fifth dimension so that all four forces meet at this common scale. Note that this is much less than the Planck scale of 10^{19} GeV, so gravitational effects may be much closer in energy than we previously thought; a result that would have all kinds of cosmological consequences.

So what is M-theory?

There is still no definitive answer to this question, although several different proposals have been made. By far the most popular is M(atrix) theory[10]. The matrix models of M-theory are $U(N)$ supersymmetric gauge quantum mechanical models with 16 supersymmetries. Such models are also interpretable as the effective action of N coincident Dirichlet 0-branes.

The theory begins by compactifying the eleventh dimension on a circle of radius R, so that the longitudinal momentum is quantized in units of $1/R$ with total P_L N/R with $N \to \infty$. The theory is *holographic* in that it contains only degrees of freedom which carry the smallest unit of longitudinal momentum, other states being composites of these fundamental states. This is, of course entirely consistent with their identification with the Kaluza-Klein modes. It is convenient to describe these N degrees of freedom as $N \times N$ matrices. When these matrices commute, their simultaneous eigenvalues are the positions of the 0-branes in the conventional sense. That they will in general be non-commuting, however, suggests that to properly understand M-theory, we must entertain the idea of a fuzzy spacetime in which spacetime coordinates are described by non-commuting matrices. In any event, this matrix approach has had success in reproducing many of the expected properties of M-theory such as $D = 11$ Lorentz covariance, $D = 11$ supergravity as the low-energy limit, and the existence of membranes and fivebranes.

It was further proposed that when compactified on T^{d-1}, the quantum mechanical model should be replaced by an d-dimensional $U(N)$ Yang-Mills field theory defined on the dual torus $\tilde{T}^{d-1}$. Another test of this M(atrix) approach, then, is that it should explain the U-dualities. For $d = 4$, for example, this group is $SL(3, Z) \times SL(2, Z)$. The $SL(3, Z)$ just comes from the modular group of T^3 whereas the $SL(2, Z)$ is the electric/magnetic duality group of four-dimensional $n = 4$ Yang-Mills. For $d > 4$, however, this picture looks suspicious because the corresponding gauge theory becomes non-renormalizable and the full U-duality group has still escaped explanation. There have been speculations on what compactified M-theory might be, including a revival of the old proposal that it is really M(embrane)theory. In other words, perhaps $D = 11$ supergravity together with its BPS configurations: plane wave, membrane, fivebrane, KK monopole and the $D = 11$ embedding of the Type IIA eightbrane, are all there is to M-theory and that we need look no further for new degrees of freedom, but only for a new non-perturbative quantization scheme. At the time of writing this is still being hotly debated.

What seems certain, however, is that M-theory is not a string theory. It can be approximated by a string theory only in certain peculiar corners of parameter space. So "string phenomenology" will become an oxymoron unless, for some as yet unknown reason, our universe happens to occupy one of these corners.

4. AdS/CFT and the brane-world

The year 1998 marked a renaissance in anti de-Sitter space (AdS) brought about by Maldacena's conjectured duality between physics in the bulk of AdS and a conformal field theory on its boundary[11]. For example, M-theory on $AdS_4 \times S^7$ is dual to a non-abelian ($n = 8, d = 3$) superconformal theory, Type IIB string theory on $AdS_5 \times S^5$ is dual to a ($n = 4, d = 4$) $U(N)$ super Yang-Mills theory and M-theory on $AdS_7 \times S^4$ is dual to a non-abelian ($(n_+, n_-) = (2, 0), d = 6$) conformal theory. In particular, as has been spelled out most clearly in the $d = 4$ $U(N)$ Yang-Mills case, there is seen to be a correspondence between the Kaluza-Klein mass spectrum in the bulk and the conformal dimension of operators on the boundary[12,13]. We note that, by choosing Poincaré coordinates on AdS_5, the metric may be written as

$$ds^2 = e^{-2y/L}(dx^\mu)^2 + dy^2, \tag{9}$$

where x^μ, ($\mu = 0, 1, 2, 3$), are the four-dimensional brane coordinates. In this case the superconformal Yang-Mills theory is taken to reside at the boundary $y \to -\infty$.

The AdS length scale L is given by

$$L^4 = 4\pi\alpha'^2(g_{YM}^2 N) \tag{10}$$

The string coupling g_s and the Yang-Mills coupling g_{YM} are related by

$$g_s = {g_{YM}}^2 \tag{11}$$

The full quantum string theory on this spacetime is difficult to deal with, but we can approximate it by classical Type IIB supergravity provided

$$L^2 >> \alpha' \tag{12}$$

so that stringy correction to supergravity are small, and that $g_s << 1$ or

$$N \to \infty \tag{13}$$

so that loop corrections can be neglected. There is now overwhelming evidence in favor of this correspondence and it allows us to calculate previously uncalculable strong coupling effects in the gauge theory starting from classical supergravity. Models of this kind, where a bulk theory with gravity is equivalent to a boundary theory without gravity, have also been advocated by 't Hooft[14] and by Susskind[15] who call them *holographic* theories. Many theorists are understandably excited about the AdS/CFT correspondence because of what it can teach us about non-perturbative QCD. In my opinion, however, this is, in a sense, a diversion from the really fundamental question: What is M-theory? So my hope is that this will be a two-way process and that superconformal field theories will also teach us more about M-theory.

The Randall-Sundrum mechanism[16] also involves AdS but was originally motivated, not via the decoupling of gravity from D3-branes, but rather as a possible mechanism for evading Kaluza-Klein compactification by localizing gravity in the presence of an uncompactified extra dimension. This was accomplished by inserting a positive tension 3-brane (representing our spacetime) into AdS_5. In terms of the Poincaré patch of AdS_5 given above, this corresponds to removing the region $y < 0$, and either joining on a second partial copy of AdS_5, or leaving the brane at the end of a single patch of AdS_5. In either case the resulting Randall-Sundrum metric is given by

$$ds^2 = e^{-2|y|/L}(dx^\mu)^2 + dy^2, \tag{14}$$

where $y \in (-\infty, \infty)$ or $y \in [0, \infty)$ for a 'two-sided' or 'one-sided' Randall-Sundrum brane respectively.

The similarity of these two scenarios led to the notion that they are in fact closely tied together. To make this connection clear, consider the one-sided Randall-Sundrum brane. By introducing a boundary in AdS_5 at $y = 0$, this model is conjectured to be dual to a cutoff CFT coupled to gravity, with $y = 0$, the location of the Randall-Sundrum brane, providing the UV cutoff. This extended version of the Maldacena conjecture[17] then reduces to the standard AdS/CFT duality as the boundary is pushed off to $y \to -\infty$, whereupon the cutoff is removed and gravity becomes completely decoupled. Note in particular that this connection involves a single CFT at the boundary of a single patch of AdS_5. For the case of a brane sitting between two patches of AdS_5, one would instead require two copies of the CFT, one for each of the patches. A crucial test of this assumed complementarity of the Maldacena and Randall-Sundrum pictures is that both should yield the same corrections to Newton's law [18].

A third development in the brane-world has been the idea that the extra dimensions are compact but much larger than the conventional Planck sized dimensions in traditional Kaluza-Klein theories[19]. This is possible if the standard model fields are confined to the $d = 4$ brane with only gravity propagating in the $d > 4$ bulk[19]. We shall not pursue this possibility here.

References

1. J. H. Schwarz, *Superstrings. The first fifteen years of superstring theory*, (World Scientific, 1985).
2. M. J. Duff, *The world in eleven dimensions: supergravity, supermembranes and M-theory*, (IOP Publishing, 1999).
3. J. H. Schwarz, *Recent progress in superstring theory*, hep-th/0007130.
4. J. H. Schwarz, *Does superstring theory have a conformally invariant limit?*, hep-th/0008009.
5. A. Achucarro, J. Evans, P. Townsend and D. Wiltshire, *Super p-branes*, *Phys. Lett.* **B198**, 441 (1987).
6. M. J. Duff, R. R. Khuri and J. X. Lu, *String solitons*, *Phys. Rep.* **259**, 213 (1995), [hep-th/9412184].
7. G. T. Horowitz and A. Strominger, *Black strings and p-branes*, *Nucl. Phys.* **B360** 197 (1991).
8. M. J. Duff and J. X. Lu, *The self-dual Type IIB superthreebrane*, *Phys. Lett.* **B273** 409 (1991).
9. J. Polchinski, *Dirichlet-branes and Ramond-Ramond charges*, *Phys. Rev. Lett.* **75**, 4724 (1995).
10. T. Banks, W. Fischler, S. H. Shenker and L. Susskind, *M-theory as a matrix model: a conjecture*, *Phys. Rev.* **D55**, 5112 (1997).
11. J. Maldacena, *The large N limit of superconformal field theories and supergravity*, *Adv. Theor. Math. Phys.* **2**, 231 (1998) [hep-th/9711200].
12. S. S. Gubser, I. R. Klebanov and A. M. Polyakov, *Gauge theory correlators from non-critical string theory*, *Phys. Lett.* **B428**, 105 (1998) [hep-th/9802109].
13. E. Witten, *Anti-de Sitter space and holography*, *Adv. Theor. Math. Phys.* **2**, 253 (1998) [hep-th/9802150].
14. G. 't Hooft, *Dimensional reduction in quantum gravity*, gr-qc/9310026.
15. L. Susskind, *The world as a hologram*, *J. Math. Phys.* **36**, 6377 (1995) [hep-th/9409089].
16. L. Randall and R. Sundrum, *An alternative to compactification*, *Phys. Rev. Lett.* **83**, 4690 (1999) [hep-th/9906064].
17. L. Susskind and E. Witten, *The Holographic Bound in Anti-de Sitter Space*, hep-th/9805114.
18. M. J. Duff and James T. Liu, *Complementarity of the Maldacena and Randall-Sundrum*

pictures, *Phys. Rev. Lett.* **85**, 2052 (2000) [hep-th/0003237].
19. N. Arkani-Hamed, S. Dimopoulos and G. Dvali, *The universe's unseen dimensions*, Scientific American, August 2000, 62.

Areal Theory

Thomas Curtright
University of Miami, Coral Gables, Florida 33124

Abstract

New features are described for models with multi-particle area-dependent potentials, in any number of dimensions. The corresponding many-body field theories are investigated for classical configurations. Some explicit solutions are given, and some conjectures are made about chaos in such field theories.

Area-dependent potentials, $V(A = \mathbf{r}_1 \wedge \mathbf{r}_2)$, or their mathematical equivalents, appear in several physical problems of contemporary interest. Notable among these problems are Yang-Mills theory (especially for spatially homogeneous configurations, with only time dependence) [3], extended supersymmetric field theories with (pseudo)scalar self-interactions [17, 12, 19], and more recently, membrane models [10, 15]. In the last century, Feynman [13] even assigned an exercise involving such potentials to students: Show the energy spectrum for a quantized areal potential model is discrete and quantum particles cannot escape from such a potential, even though the spectrum for the classical model is continuous and classical particles can escape along special trajectories for which $A = 0$. Many of us pondered this problem over the intervening years, especially in the field theory context [4]. Meanwhile, Simon worked out five or six solutions to Feynman's exercise and published them [24].

Classically, such potential models are interesting insofar as they may provide simple examples of chaotic systems [23, 2]. While there are special trajectories for which the motion is quite regular, including those for which particles can escape (such as straight line, free particle motion), for most trajectories this is not the case. The current consensus is such models are not integrable in the Liouville sense, and do not admit the construction of a Lax pair. Is this really so?

Quantum mechanically, there are obviously more interesting questions to ask about such potential models than simply whether the energy spectrum is discrete. In particular, just what *is* that discrete spectrum? Is the quantized model, or some simple variant of it, completely integrable [18], even if the classical is not? Do quantum effects sufficiently ameliorate any chaotic classical trajectories [16] to

CP624, *Cosmology and Elementary Particle Physics*, edited by B. N. Kursunoglu et al.

permit closed-form solutions for the wave functions or useful forms for the propagator?

With these questions in mind, I will describe in this talk[1] some new features of models with area-dependent potentials. As non-relativistic many-body field theories, I believe that area-dependent models are very interesting. I will argue this beginning with the classical field versions. I explicitly solve these for some special configurations, and then investigate general situations. The classical field theory problem always reduces to solving the linear Schrödinger equation in a self-consistent effective potential. In general, the effective potential for the $V = A$ model is anisotropic and linear, while for the $V = A^2$ model it is anisotropic and quadratic. In both cases the effective potential is determined from the initial data by a closed set of coupled time-dependent equations. I will analyze these equations, especially for the case of spherically symmetric particle density, and I will solve them in that special situation. However, I have not completely solved the area-dependent models for general, anisotropic initial data. Much more work is needed for the $V = A^2$ model. Additional work is also needed to construct the quantum versions of these field theories. These particular models seem to be ripe for exploration using deformation quantization [1, 6, 9] and ideas from non-commutative geometry [14].

To begin, we construct a many-body field theory on the plane with a three-body potential that is just the *signed* area of the three-body triangle (*not* the absolute value of the area). We take a collection of *three types of non-identical particles*, i.e. three particle "species", each represented by its own local field ψ_a , $a = 1, 2, 3$. In this case, the Hamiltonian has no lower bound for configurations with large but negative A, so the model is reminiscent of a linear potential single-particle QM. (Actually there is more than reminiscing going on here as we shall soon see.) Explicitly, we define a multi-particle configuration by the field theory Lagrangian

$$L = \sum_{a=1,2,3} i \int (d\mathbf{r})\, \psi_a^* (\mathbf{r}) \frac{\partial}{\partial t} \psi_a (\mathbf{r}) - H \, .$$

It should be understood that all the fields also depend on a common time, t, but usually we will not explicitly indicate this. We then use the obvious symmetrical

[1]In the actual talk, I began with a simple model of N distinguishable point particles on the plane, with $V = A$. This is based on work in collaboration with Alexios Polychronakos and Cosmas Zachos [8]. We have completely understood this model's classical and quantum properties. Since the potential is quadratic, this is not very difficult to do, in principle, but nevertheless it was necessary to develop an efficient formalism to handle an arbitrary number of particles. We have done this using so-called cyclotomic coordinates that have previously been used to analyze toroidal membrane dynamics [11]. I view the analysis of this $V = A$ model as a precursor to point particle models with $V = A^2$. These latter models are much more interesting, but are not so easy to solve. While I have made a modicum of progress in an unfinished attempt to construct a Lax pair for certain variants of $V = A^2$ models, in the point particle approach [22], I will not discuss this here. Instead, I will shift from point particle models to field theories.

Hamiltonian

$$H = \int (d\mathbf{r}) \sum_{a=1,2,3} \psi_a^* (\mathbf{r}) \left[\tfrac{-\nabla^2}{2m}\right] \psi_a (\mathbf{r})$$
$$+ k \iiint (d\mathbf{r}_1) (d\mathbf{r}_2) (d\mathbf{r}_3)\ A (\mathbf{r}_1, \mathbf{r}_2, \mathbf{r}_3)\ |\psi_1 (\mathbf{r}_1) \psi_2 (\mathbf{r}_2) \psi_3 (\mathbf{r}_3)|^2$$

involving A, (twice) the area of the triangle formed by the three particles.

$$A (\mathbf{r}_1, \mathbf{r}_2, \mathbf{r}_3) = \mathbf{r}_1 \wedge \mathbf{r}_2 + \mathbf{r}_2 \wedge \mathbf{r}_3 + \mathbf{r}_3 \wedge \mathbf{r}_1 \,.$$

This may be rewritten (exactly) as

$$H = \sum_{a=1,2,3} \int (d\mathbf{r}) \, \psi_a^* (\mathbf{r}) \left[\tfrac{-\nabla^2}{2m}\right] \psi_a (\mathbf{r}) + \tfrac{1}{3} \sum_{a=1,2,3} \int (d\mathbf{r}) \left(S_a [\psi] + \mathbf{V}_a [\psi] \cdot \mathbf{r}\right) |\psi_a (\mathbf{r})|^2$$

where we have introduced the combinations $S_a [\psi] + \mathbf{V}_a [\psi] \cdot \mathbf{r}$ for $a = 1, 2, 3$ to serve as *effective potentials* for the three fields. These effective potentials are defined in terms of the fields as

$$V_a^i [\psi] = \tfrac{1}{2} k \varepsilon^{ij} \sum_{b,c=1,2,3} \varepsilon^{abc} \iint (d\mathbf{r}_2) (d\mathbf{r}_3)\ (\mathbf{r}_2 - \mathbf{r}_3)^j\ |\psi_b (\mathbf{r}_2) \psi_c (\mathbf{r}_3)|^2$$
$$S_a [\psi] = \tfrac{1}{2} k \varepsilon^{abc} \iint (d\mathbf{r}_2) (d\mathbf{r}_3)\ (\mathbf{r}_2 \wedge \mathbf{r}_3)\ |\psi_b (\mathbf{r}_2) \psi_c (\mathbf{r}_3)|^2$$

Note that S_a and $\mathbf{V}_a$ do *not* depend on ψ_a but *do* depend on the other two field configurations, $\psi_{b \neq a}$. Further observe that $S_a [\psi] = 0$ if *either* of the two ψ's involved are of definite parity (either even or odd functions of $\mathbf{r}$), and $V_a^i [\psi] = 0$ if *both* of the two ψ's involved are of definite parity. Thus the special configurations where all three species have definite parity reduce to free fields! Again, this is a model on the plane, so each species coordinate $\mathbf{r}_a$ is a two-vector. To construct a potential linear in the signed area in higher dimensions, we would need to take an inner product of $A (\mathbf{r}_1, \mathbf{r}_2, \mathbf{r}_3)$ with some constant 2-form.

Like the point-particle theory in [8], this classical field theory is always *solvable*, even in the general situation when all the fields do *not* have definite parity. We work in real coordinates here. The field equations in terms of the effective potentials are:

$$i \frac{\partial}{\partial t} \psi_a (\mathbf{r}, t) + \frac{1}{2m} \nabla^2 \psi_a (\mathbf{r}, t) = \left(S_a [\psi] + \mathbf{V}_a [\psi] \cdot \mathbf{r}\right) \psi_a (\mathbf{r}, t) \qquad \text{no sum } a.$$

For each species we have just Schrödinger's equation with a linear (in r^i) potential, albeit a non-isotropic linear potential, in general, with non-constant (in t) configuration-dependent coefficients. From the above definitions we obtain

$$S_a [\psi] = \tfrac{1}{2} k \sum_{\substack{b,c=1,2,3 \\ i,j=1,2}} \varepsilon^{ij} \varepsilon^{abc} N_b R_b^i N_c R_c^j \,, \quad V_a^i [\psi] = k \sum_{\substack{b,c=1,2,3 \\ j=1,2}} \varepsilon^{ij} \varepsilon^{abc} N_b R_b^j N_c$$

$$N_a R_a^j \equiv \int (d\mathbf{r})\; r^j\; |\psi_a(\mathbf{r},t)|^2 \;, \quad N_a P_a^j \equiv -i \int (d\mathbf{r})\; \psi_a^*(\mathbf{r},t)\; \overleftrightarrow{\nabla}^j \psi_a(\mathbf{r},t) \;,$$
$$N_a \equiv \int (d\mathbf{r})\; |\psi_a(\mathbf{r},t)|^2 \,,$$

with no sum over a in any of these. While these are indeed configuration-dependent coefficients, the implicit field dependence of the coefficients is completely tractable. From the field equations, the coefficients in the effective potential obey simple first order time-derivative equations:

$$\frac{d}{dt} N_a = 0 \,, \quad \frac{d}{dt} R_a^k = \frac{1}{2m} P_a^k \,,$$
$$\frac{d}{dt} P_a^j = -2V_a^j = -2k\varepsilon^{ij} \sum_{b,c=1,2,3} \varepsilon^{abc} N_b N_c R_b^j \,, \quad \frac{d}{dt}\left(\sum_{a=1,2,3} N_a P_a^j \right) = 0 \,.$$

The last equation represents conservation of the system's total momentum, while the next to last equation shows that the individual particle species momenta are not separately conserved, in general.

These first order equations combine to yield linear second order equations with *constant* coefficients for any given initial data (hence easily solved).

$$m \frac{d^2}{dt^2} R_a^i = k \sum_{j,b} M_{(ia)(jb)} R_b^j \,, \qquad M_{(ia)(jb)} = \varepsilon^{ij} \sum_{c=1,2,3} \varepsilon^{acb} N_c N_b \qquad (\text{no sum } b) \,.$$

The time-*in*dependent eigenvalues of M are 0 and $\pm\sqrt{N_1 N_2 N_3}\sqrt{N_1 + N_2 + N_3}$, with each of these three possibilities occurring twice. (Note how the three point-particle mechanics eigenvalues in [8] are obtained by setting all the N's equal to one.) Therefore the R's, and hence all the terms in the effective potentials, can be solved for in terms of the initial data. The three independent classes of solutions for the R's, corresponding to the three eigenvalues of M, are functions linear in t, real exponentials, and oscillations. The time dependencies of the terms in the effective potential then follow from their expression in terms of the R's. Finally, to complete the solution, the linear Schrödinger equations for the individual fields must be solved using the solutions for the effective potentials.

I believe, and assert without proof, that this last step can always be carried out, although there is some possibility of unusual behavior for the fields because the potential terms in the Schrödinger equation are combinations of polynomials and complex exponentials in t. Technically, I cannot rigorously rule out some sort of chaotic behavior in the phases of the fields, yet, but since the behavior of all the field bilinears is so deterministic, including those involving spatial derivatives, my current opinion is that chaotic behavior in the fields, if any, must be limited to only time-dependent phases and is completely innocuous. (I would devote more time to this issue, but once again, I view this model as just a warm-up exercise for the A^2 model presented below.) I leave it to the reader to compare the above system of equations to those of the standard isotropic oscillator [20, 21].

For consistency with the treatment of the point-particle case in [8], and to warm the hearts of conformal field theorists, the above analysis should perhaps be re-done using complex coordinates. Also, a generalization to a polygon area potential, involving N types of particles and $A(\mathbf{r}_1, \mathbf{r}_2, \cdots, \mathbf{r}_N) = \mathbf{r}_1 \wedge \mathbf{r}_2 + \mathbf{r}_2 \wedge \mathbf{r}_3 + \cdots + \mathbf{r}_N \wedge \mathbf{r}_1$, is straightforward. An elegant touch would be to introduce a tensor $E^{a_1 \cdots a_N}$ on the particle type such that the combination $A(\mathbf{r}_1, \mathbf{r}_2, \cdots, \mathbf{r}_N) E^{a_1 \cdots a_N}$ is symmetric under any pair interchange $\mathbf{r}_i, a_i \leftrightarrow \mathbf{r}_j, a_j$. Again the most streamlined way to do this on the plane is to work with complex coordinates, as was the case in the point-particle situation.

We next construct an area-squared field theory, where the potential is A^2, a quartic function of field coordinates. This can be incorporated into a many-body field theory with only a single complex field ψ in any number of spatial dimensions n. The Hamiltonian for the model is

$$\begin{aligned} H[\psi] &= \frac{1}{2m} \int d^n s \, \partial_i \bar{\psi} \partial_i \psi \\ &+ \tfrac{1}{3} k \int d^n s_1 \int d^n s_2 \int d^n s_3 \, |\psi(s_1)|^2 |\psi(s_2)|^2 |\psi(s_3)|^2 \left[s_{12}^2 s_{23}^2 - (s_{12} \cdot s_{23})^2 \right] \end{aligned}$$

where $\mathbf{s}_{ab} \equiv \mathbf{s}_a - \mathbf{s}_b$ for each ab pair of the identical particles' coordinate vectors. The equations of motion from $L[\psi] = \int d^n s \, i \bar{\psi} \partial_t \psi - H[\psi]$ are

$$i \partial_t \psi(r) + \frac{1}{2m} \nabla^2 \psi(r) = V[\psi](r) \, \psi(r)$$

In this case the effective potential corresponds to a non-isotropic oscillator

$$\begin{aligned} V[\psi](r) &= k \int d^n s_2 \int d^n s_3 \, |\psi(s_2)|^2 |\psi(s_3)|^2 \left[s_{12}^2 s_{23}^2 - (s_{12} \cdot s_{23})^2 \right] \Big|_{s_1 = r} \\ &= K_{ij} r^i r^j + B_i r^i + Z = K_{ij} \left(r^i r^j - 2 R^i r^j \right) + Z \\ &= K_{ij} \left(r^i - R^i \right) \left(r^j - R^j \right) + Z - K_{ij} R^i R^j \end{aligned}$$

with the definitions and relations

$$\begin{aligned} K_{ij} &\equiv k \int d^n s_2 \int d^n s_3 \, |\psi(s_2)|^2 |\psi(s_3)|^2 \left(\delta^{ij} s_{23}^2 - s_{23}^i s_{23}^j \right) \\ &= 2kN^2 \left(\left(S - R^2 \right) \delta_{ij} - \left(S_{ij} - R_i R_j \right) \right) \end{aligned}$$

$$\begin{aligned} B_i &\equiv -2k \int d^n s_2 \int d^n s_3 \, |\psi(s_2)|^2 |\psi(s_3)|^2 \left(s_2^i \, s_{23}^2 - s_{23}^i \, s_2 \cdot s_{23} \right) \\ &= 4kN^2 \left(S_{ij} R_j - S R_i \right) = -2 K_{ij} R_j \end{aligned}$$

$$Z \equiv k \int d^n s_2 \int d^n s_3 \, |\psi(s_2)|^2 |\psi(s_3)|^2 \left(s_2^2 s_3^2 - (s_2 \cdot s_3)^2 \right) = kN^2 \left(S^2 - S_{ij} S_{ji} \right)$$

$$N \equiv \int d^n s \; |\psi(s)|^2 \;, \quad R_i \equiv \frac{1}{N}\int d^n s \; s_i \, |\psi(s)|^2 \;, \quad P_i[\psi] \equiv \frac{-i}{N}\int d^n s \left(\bar{\psi}\overleftrightarrow{\partial_i}\psi\right) \;,$$

$$S_{ij} \equiv \frac{1}{N}\int d^n s \; s_i s_j \, |\psi(s)|^2 \;, \quad S = \sum_{j=1}^{n} S_{jj} \;, \quad K_{ij} = 2kN^2 \left(S\delta_{ij} - S_{ij}\right) \;.$$

Note that everything in $V[\psi]$ factorizes into integrals of bilinear densities. Actually, this is not too surprising given the polynomial character and factorized form of the terms of the three-body potential. So the effective potential for the area^2 model becomes

$$\begin{aligned} V[\psi](r) = {} & 2kN^2 \left[S(r-R)^2 - S_{ij}\left(r^i - R^i\right)\left(r^j - R^j\right)\right] + 2kN^2\left(S_{ij}R^iR^j - SR^2\right) \\ & + kN^2\left(S^2 - S_{ij}S_{ij}\right) - 2kN^2\left(R^2r^2 - (r\cdot R)^2\right) \end{aligned}$$

As already noted, this is a non-isotropic quadratic potential. Note the presence of the last term $R^2r^2 - (r\cdot R)^2 = (\mathbf{r}\wedge\mathbf{R})^2$ which is the squared area of the parallelogram formed from r and R. The negative coefficient of this term and the potential instability it would produce for large $r \perp R$ are completely offset by the additional presence of $2kN^2Sr^2$ in the effective potential, along with the standard inequality $S \geq R^2$. Also note that V is non-zero only through the field spreading out around the center of mass location. Were $|\psi(s)|^2$ a delta-function at R, all coefficients would vanish in V.

Also note the absence of *nontrivial* "zero modes" for the quadratic part of V (i.e. those zero-modes that would persist even in the CM frame, and not those associated with the free translation of the CM). This is straightforward to see, even though for the field coordinates alone $\det(s_is_j) = 0 = \det(s^2\delta_{ij} - s_is_j)$, and so these coordinate matrices have zero eigenvalues. However, $\langle \det S\rangle \neq \det\langle S\rangle$ just as $\langle x^2\rangle \neq \langle x\rangle^2$. In fact $\det(S\delta_{ij} - S_{ij}) = \frac{1}{N^n}\det\int d^n s\;(s^2\delta_{ij} - s_is_j)\,|\psi(s)|^2 \geq 0$. Moreover, in the CM $\det K > 0$ *and all the eigenvalues of K_{ij} are greater than zero.* Or equivalently, all the eigenvalues of $S\delta_{ij} - S_{ij}$ are positive. Proof: For any real vector v, we have $v_i(S\delta_{ij} - S_{ij})v_j = \frac{1}{N}\int d^n s\;\left(s^2v^2 - (s\cdot v)^2\right)|\psi(s)|^2$. But the integrand is non-negative by the Schwarz inequality $s^2v^2 - (s\cdot v)^2 \geq 0$. Thus any eigenvalue λ of K_{ij} is non-negative, $\lambda \geq 0$. Now to rule out the case where any eigenvalue vanishes in the CM, we note the Schwarz inequality is only an equality $s^2v^2 - (s\cdot v)^2 = 0$ if $s_i \propto v_i$. Therefore the only way to have a zero eigenvalue is to have the full integrand $\left(s^2v^2 - (s\cdot v)^2\right)|\psi(s)|^2$ restricted to a line where $s_i \propto v_i$. Forget that! There is always some transverse spread in $|\psi(s)|^2$. A proper wave function (or classical field) in $n \geq 2$ dimensions does not have support just on a line. So $V[\psi](r)$ always has a quadratic r^2 part and therefore the field is always localized around the CM. Or, perhaps more physically, all the particles are bound by the effective potential to the CM.

For solutions the energy is

$$H[\psi] = \frac{1}{8m}NT + kN^3\left(S^2 - S_{ij}S_{ij}\right)$$

using $\mathcal{S}_{ij} = S_{ij} - R_i R_j$, $\mathcal{S} = \mathcal{S}_{jj}$, $T = T_{jj}$, with the stress-tensor defined as

$$T_{ij}[\psi] \equiv \frac{-1}{N}\int d^n s\, \left(\bar{\psi}\overleftrightarrow{\partial_i}\overleftrightarrow{\partial_j}\psi\right) = \frac{-4}{N}\int d^n s\, \left(\bar{\psi}\partial_i\partial_j\psi\right)$$

Now what is the time-dependence of the three effective potential coefficients K_{ij}, B_i, and Z? Or rather, what is the time-dependence of the quantities that compose them? Obviously, N, R_i, and P_i obey the equations

$$\frac{d}{dt}N = 0\,, \quad \frac{d}{dt}P_i = 0\,, \quad \frac{d}{dt}R_i = \frac{1}{2m}P_i$$

with the solutions

$$N(t) = N_0 \equiv N\,, \quad P_i(t) = P_{0i} \equiv P_i\,, \quad R_i(t) = R_{0i} + \frac{1}{2m}P_i t$$

But what about S_{ij}? With the definition (note that $Q_{ij} \neq Q_{ji}$)

$$Q_{ij}[\psi] \equiv \frac{-i}{N}\int d^n s\, s_i\left(\bar{\psi}\overleftrightarrow{\partial_j}\psi\right) = \frac{-2i}{N}\int d^n s\, s_i\left(\bar{\psi}\partial_j\psi\right) - i\delta_{ij}$$

we find

$$\frac{d}{dt}S_{ij} = \frac{1}{2m}\left(Q_{ij} + Q_{ji}\right)$$

So we must include Q_{ij} into the mix of differential equations. It's time derivative is

$$\frac{d}{dt}Q_{ij} = \frac{1}{2m}T_{ij} - 4S_{ik}K_{kj} - 2R_i B_j$$

where $T_{ij}[\psi]$ is the previous stress-tensor. We are one step away from closing the system. We just need the final time derivative

$$\frac{d}{dt}T_{ij} = -4\left(K_{ik}Q_{kj} + K_{jk}Q_{ki}\right) - 2P_i B_j - 2P_j B_i$$

The system of equations is now closed. The problem then is to solve these equations to obtain S_{ij} and hence the effective potential in terms of the initial field configuration . Note that only the symmetric tensor part of Q_{ij} varies with time.

Let's break down the tensors into somewhat more traditional combinations. The symmetric and antisymmetric parts of Q_{ij} are (almost) familiar from rotations and (at least when traced) coordinate rescaling.

$$J_{ij} = Q_{ij} - Q_{ji}\,, \quad D_{ij} = Q_{ij} + Q_{ji}$$

These nicely occur above with their values for (or about) the CM removed. Thus define

$$\mathcal{D}_{ij} = D_{ij} - \left(R_i P_j + R_j P_i\right)\,, \quad \mathcal{J}_{ij} = J_{ij} - \left(R_i P_j - R_j P_i\right)$$

Similarly remove from the quadrupole and stress tensors the CM coordinate and momentum dyads.

$$\mathcal{S}_{ij} = S_{ij} - R_i R_j \ , \qquad \mathcal{T}_{ij} = T_{ij} - P_i P_j$$

Incorporating momentum and angular momentum conservation, i.e. $B_i = -2K_{ij}R^j$ and $K_{ik}\mathcal{S}_{kj} - R_j K_{ik} R^k = K_{jk}\mathcal{S}_{ki} - R_i K_{jk} R^k$ (just a matrix commutator, $[\mathcal{S}, K]_{ij} = 0$), the time evolution equations are then

$$\begin{aligned}
&\frac{d}{dt}\mathcal{S}_{ij} = \frac{1}{2m}\mathcal{D}_{ij} \ , \quad \frac{d}{dt}N = 0 \ , \quad \frac{d}{dt}\mathcal{J}_{ij} = 0 \\
&\frac{d}{dt}\mathcal{D}_{ij} = \frac{1}{m}\mathcal{T}_{ij} - 4\mathcal{S}_{ik}K_{kj} - 4\mathcal{S}_{jk}K_{ki} = \frac{1}{m}\mathcal{T}_{ij} - 4\left\{\mathcal{S}, K\right\}_{ij} \qquad (1) \\
&\frac{d}{dt}\mathcal{T}_{ij} = -2K_{ik}\left(\mathcal{D}_{kj} + \mathcal{J}_{kj}\right) - 2K_{jk}\left(\mathcal{D}_{ki} + \mathcal{J}_{ki}\right) = -2\left\{K, \mathcal{D}\right\}_{ij} - 2\left[K, \mathcal{J}\right]_{ij}
\end{aligned}$$

where now $K_{ij} = 2kN^2\left(\mathcal{S}\delta_{ij} - \mathcal{S}_{ij}\right)$. Anyone skilled in the black arts of nuclear physics should feel comfortable with equations of this sort.

An invariant arises from taking traces. Thus

$$\frac{d}{dt}\left(\mathrm{tr}\,\mathcal{T}_{ij} + 8mkN^2\left(\mathcal{S}^2 - \mathrm{tr}\,\mathcal{S}_{ij}^2\right)\right) = 0 \ .$$

Now, this is essentially just the energy in the center-of-mass.

$$P^2 + \mathrm{tr}\,\mathcal{T}_{ij} + 8mkN^2\left(\mathcal{S}^2 - \mathrm{tr}\,\mathcal{S}_{ij}^2\right) = \mathrm{tr}\,T_{ij} + 8mkN^2\left(S^2 - \mathrm{tr}\,S_{ij}^2\right) = 8mH[\psi]/N$$

where

$$H[\psi] = \frac{1}{8m}NT + kN^3\left(S^2 - S_{ij}S_{ij}\right) \ .$$

As a special situation, but really without loss of generality, consider co-moving with the system in the CM, where $R_i = 0 = P_i$. Then the effective potential coefficients reduce to

$$K_{ij} = 2kN^2\left(\mathcal{S}\delta_{ij} - \mathcal{S}_{ij}\right) \ , \qquad B_i = 0 \ , \qquad Z = kN^2\left(\mathcal{S}^2 - \mathcal{S}_{ij}\mathcal{S}_{ij}\right)$$

with the $\mathcal{S}_{ij}$, $\mathcal{D}_{ij}$, and $\mathcal{T}_{ij}$ time derivatives unchanged in form. (Note in the CM $S_{ij} = \mathcal{S}_{ij}$, $D_{ij} = \mathcal{D}_{ij}$, and $T_{ij} = \mathcal{T}_{ij}$.)

For now, I cannot solve the non-isotropic equations (1). Perhaps closed-form solutions can be obtained in various limiting situations, but most probably the system is chaotic for general initial data[2]. Nevertheless, I can solve exactly a further specialization, which *is* with loss of generality. Suppose the various densities are actually isotropic about the CM. Then we have

$$\mathcal{D}_{ij} = \frac{1}{n}\mathcal{D}\delta_{ij} \ , \quad \mathcal{S}_{ij} = \frac{1}{n}\mathcal{S}\delta_{ij} \ , \quad \mathcal{T}_{ij} = \frac{1}{n}\mathcal{T}\delta_{ij} \ , \quad K_{ij} = \frac{1}{n}K\delta_{ij} = 2kN^2\left(\frac{n-1}{n}\right)\mathcal{S}\delta_{ij}$$

$$K = 2kN^2\left(n-1\right)\mathcal{S} \ , \quad Z = kN^2\left(\frac{n-1}{n}\right)\mathcal{S}^2$$

[2] Since giving this talk, extensive numerical investigations have shown that the model is indeed chaotic for general initial data [7].

and the time derivatives are $\frac{d}{dt}K_{ij} = 2kN^2\left(\frac{n-1}{n}\right)\frac{1}{2m}\mathcal{D}\delta_{ij}$, etc. These may all be traced now w.l.o.g. to yield

$$\frac{d}{dt}K = \frac{1}{m}kN^2\left(n-1\right)\mathcal{D}\,,\quad \frac{d}{dt}Z = \frac{1}{m}kN^2\left(\frac{n-1}{n}\right)\mathcal{S}\mathcal{D}\,,\quad \frac{d}{dt}\mathcal{S} = \frac{1}{2m}\mathcal{D}\,,$$

and the remaining equations

$$\frac{d}{dt}\mathcal{D} = \frac{1}{m}\mathcal{T} - 16kN^2\left(\frac{n-1}{n}\right)\mathcal{S}^2\,,\quad \frac{d}{dt}\mathcal{T} = -8kN^2\left(\frac{n-1}{n}\right)\mathcal{S}\mathcal{D}$$

Using $\frac{d}{dt}\mathcal{S} = \frac{1}{2m}\mathcal{D}$ the $\mathcal{T}$ equation becomes $\frac{d}{dt}\mathcal{T} = -8mkN^2\left(\frac{n-1}{n}\right)\frac{d}{dt}\mathcal{S}^2$ which integrates immediately to

$$\mathcal{T}\left(t\right) = \mathcal{T}_0 + 8mkN^2\left(\frac{n-1}{n}\right)\left(\mathcal{S}_0^2 - \mathcal{S}\left(t\right)^2\right)$$

Inserting this into the $\mathcal{D}$ equation and using $\frac{d}{dt}\mathcal{S} = \frac{1}{2m}\mathcal{D}$ once again, we have a second-order equation for $\mathcal{S}$.

$$2m\frac{d^2}{dt^2}\mathcal{S} = \frac{1}{m}\mathcal{T}_0 + 8kN^2\left(\frac{n-1}{n}\right)\mathcal{S}_0^2 - 24kN^2\left(\frac{n-1}{n}\right)\mathcal{S}^2$$

Thus $\mathcal{S}$ alone evolves as a nonlinear oscillator subject to a cubic self-interaction. That is, the effective potential governing the time evolution of $\mathcal{S}$ is $\mathcal{V}\left(\mathcal{S}\right) = a\mathcal{S}^3 - b\mathcal{S}$, where the constant, initial-data-dependent coefficients are $a = 8kN^2\left(\frac{n-1}{n}\right)$, $b = \frac{1}{m}\mathcal{T}_0 + 8kN^2\left(\frac{n-1}{n}\right)\mathcal{S}_0^2$, and are both positive. So the potential has a local minimum/maximum at $3a\mathcal{S}^2 = b$ or $\mathcal{S}_{\min/\max} = \pm\sqrt{\frac{b}{3a}}$, hence

$$\mathcal{S}_{\min/\max} = \pm\sqrt{\frac{1}{3}\mathcal{S}_0^2 + \frac{1}{24mkN^2}\left(\frac{n}{n-1}\right)\mathcal{T}_0}$$

Only the local minimum here is physical. Recall that we are in the CM, so that $D = \mathcal{D}$, $T = \mathcal{T}$, and most importantly the inequality $\mathcal{S} = S \equiv \frac{1}{N}\int d^n s\, s^2\, \left|\psi\left(s\right)\right|^2 \geq 0$,so we see that physically $\mathcal{S}$ is constrained to be positive (and $\mathcal{S} = 0$ only for $\left|\psi\left(s\right)\right|^2$ ultra-localized at the origin, like a delta-function $\cdots$ a collapse/condensation of the matter field).

Is it possible that we encounter singularities here, in a finite time? By it's definition we are constrained to have $S \geq 0$, but for "special" initial data, perhaps the equation evolves S to zero, the boundary of the unphysical $S < 0$ region, hence the matter field would collapse to the origin. After all, the equation for S is that of a nonlinear "cubic" oscillator. Do we have here a baby version of black-hole physics?

The answer is no, $S = 0$ is never achieved, as we now explain. The effective potential for $\mathcal{S}$ will require $\mathcal{S} > 0$ to be true *for all times* if and only if $\mathcal{S}_0 > 0$ and $\mathcal{E}(\mathcal{S}) < 0$, where $\mathcal{E}$ is the conserved *effective energy* for $\mathcal{S}$.

$$\mathcal{E}(\mathcal{S}) \equiv m\left(\frac{d}{dt}\mathcal{S}\right)^2 + \mathcal{V}(\mathcal{S}) = \mathcal{E}_0$$

$$\mathcal{E}_0 = \frac{1}{4m}\mathcal{D}_0^2 + 8kN^2\left(\tfrac{n-1}{n}\right)\mathcal{S}_0^3 - \left(\frac{1}{m}\mathcal{T}_0 + 8kN^2\left(\tfrac{n-1}{n}\right)\mathcal{S}_0^2\right)\mathcal{S}_0 = \frac{1}{4m}\left(\mathcal{D}_0^2 - 4\,\mathcal{T}_0\mathcal{S}_0\right)$$

Consider the two terms on the RHS of this last expression. The standard QM uncertainty relation (basically the Schwarz inequality) for any two hermitean operators $\mathbf{a}$ and $\mathbf{b}$, assuming $\langle\mathbf{a}\rangle = 0 = \langle\mathbf{b}\rangle$, is $\langle\mathbf{a}^2\rangle\langle\mathbf{b}^2\rangle \geq \left\langle\frac{-i}{2}(\mathbf{ab}-\mathbf{ba})\right\rangle^2 + \left\langle\frac{1}{2}(\mathbf{ab}+\mathbf{ba})\right\rangle^2$. Each of the two terms on the RHS of the inequality are positive. Usually we discard the second of these RHS terms and keep the first, especially in the case for canonically conjugate variables with $\mathbf{xp} - \mathbf{px} = i\hbar$. But we could equally well discard the first RHS term to obtain$\langle\mathbf{a}^2\rangle\langle\mathbf{b}^2\rangle - \left\langle\frac{1}{2}(\mathbf{ab}+\mathbf{ba})\right\rangle^2 \geq 0$. This is a strict inequality for conjugate variables. Applying this to the situation at hand we conclude $4\mathcal{T}\mathcal{S} - \mathcal{D}^2 > 0$. Hence the cubic oscillator system above is always bound to the stable minimum, with $\mathcal{E} < 0$. This is what we wanted to show.

We will discuss static solutions in more detail below. For now, we note that these are possible if we are at the local minimum of $\mathcal{V}(\mathcal{S})$ with $\mathcal{D} = 0$, where we require

$$\mathcal{S}_{\min} = \mathcal{S}_0 = \sqrt{\frac{1}{3}\mathcal{S}_0^2 + \frac{1}{24mkN^2}\left(\frac{n}{n-1}\right)\mathcal{T}_0}\,, \quad \text{or} \quad \mathcal{T}_0 = 16mkN^2\left(\frac{n-1}{n}\right)\mathcal{S}_0^2\,,$$

as well as $\mathcal{D}_0 = 0$, of course. The fields themselves may be worked out explicitly in this case and are nothing but Gaussians.

Even in the non-static case the conserved energy for $\mathcal{S}$ reduces it's determination to quadrature and allows us to solve for the $\mathcal{S}$ motion in textbook fashion.

$$\frac{d}{dt}\mathcal{S} = \pm\sqrt{\left(\mathcal{E}_0 - \mathcal{V}(\mathcal{S})\right)/m}$$

$$\pm\frac{t}{\sqrt{m}} = \int_{\mathcal{S}_0}^{\mathcal{S}(t)}\frac{d\mathcal{S}}{\sqrt{\mathcal{E}_0 - \mathcal{V}(\mathcal{S})}} = \frac{1}{\sqrt{a}}\int_{\mathcal{S}_0}^{\mathcal{S}(t)}\frac{d\mathcal{S}}{\sqrt{(\mathcal{S}_{hi} - \mathcal{S})(\mathcal{S} - \mathcal{S}_{mid})(\mathcal{S} - \mathcal{S}_{lo})}}$$

where for $\mathcal{E}_0 < 0$, $\{\mathcal{S}_{lo}, \mathcal{S}_{mid}, \mathcal{S}_{hi}\}$ are the three distinct zeroes of the cubic $\mathcal{E}_0 + b\mathcal{S} - a\mathcal{S}^3$ with $\mathcal{S}_{lo} < 0 < \mathcal{S}_{mid} \leq \mathcal{S}(t) \leq \mathcal{S}_{hi}$. The RHS here is an elliptic integral.

$$\int_{\mathcal{S}_{mid}}^{\mathcal{S}(t)}\frac{\sqrt{\mathcal{S}_{hi}-\mathcal{S}_{lo}}\;d\mathcal{S}}{\sqrt{4(\mathcal{S}_{hi}-\mathcal{S})(\mathcal{S}-\mathcal{S}_{mid})(\mathcal{S}-\mathcal{S}_{lo})}} = K\left(\sqrt{\tfrac{\mathcal{S}_{hi}-\mathcal{S}_{mid}}{\mathcal{S}_{hi}-\mathcal{S}_{lo}}}\right) - F\left(\sqrt{\tfrac{\mathcal{S}_{hi}-\mathcal{S}(t)}{\mathcal{S}_{hi}-\mathcal{S}_{mid}}}, \sqrt{\tfrac{\mathcal{S}_{hi}-\mathcal{S}_{mid}}{\mathcal{S}_{hi}-\mathcal{S}_{lo}}}\right)$$

K and F are the standard elliptic integrals, with $F(1,z) = K(z)$, and $F(0,z) = 0$. Thus $\mathcal{S}(t)$ is an elliptic function. The solution oscillates between the distinct turning points $\mathcal{S}_{mid} \leq \mathcal{S}(t) \leq \mathcal{S}_{hi}$ with period

$$\tfrac{1}{2}t_{period} = \sqrt{\frac{m}{a}} \int_{\mathcal{S}_{mid}}^{\mathcal{S}_{hi}} \frac{d\mathcal{S}}{\sqrt{(\mathcal{S}_{hi} - \mathcal{S})(\mathcal{S} - \mathcal{S}_{mid})(\mathcal{S} - \mathcal{S}_{lo})}} = \sqrt{\tfrac{4m}{a(\mathcal{S}_{hi}-\mathcal{S}_{lo})}} K\left(\sqrt{\tfrac{\mathcal{S}_{hi}-\mathcal{S}_{mid}}{\mathcal{S}_{hi}-\mathcal{S}_{lo}}}\right)$$

We may extend the analysis to higher dimensions for models with potentials depending on higher multi-particle coordinate forms. Assume a $(d+1)$-body potential U in $n \geq d$ dimensions which is a $(d\text{-form})^2$. That is

$$F(\mathbf{r}_1, \mathbf{r}_2, \mathbf{r}_3, \cdots, \mathbf{r}_d, \mathbf{r}_{d+1}) = (\mathbf{r}_1 - \mathbf{r}_2) \wedge (\mathbf{r}_2 - \mathbf{r}_3) \wedge \cdots \wedge (\mathbf{r}_d - \mathbf{r}_{d+1})$$

$$\begin{aligned} U(\mathbf{r}_1, \mathbf{r}_2, \cdots, \mathbf{r}_{d+1}) &= kF^2(\mathbf{r}_1, \mathbf{r}_2, \cdots, \mathbf{r}_{d+1}) \\ &= k\ \delta^{i_1 i_2 \cdots i_d}_{j_1 j_2 \cdots j_d}\ \mathbf{r}_{12}^{i_1}\mathbf{r}_{12}^{j_1}\mathbf{r}_{23}^{i_2}\mathbf{r}_{23}^{j_2}\mathbf{r}_{34}^{i_3}\mathbf{r}_{34}^{j_3} \cdots \mathbf{r}_{dd+1}^{i_d}\mathbf{r}_{dd+1}^{j_d} \end{aligned}$$

$$\begin{aligned} H = &\int (d\mathbf{r})\ \psi^*(\mathbf{r}) \left[-\frac{1}{2m}\nabla^2\right] \psi(\mathbf{r}) \\ &+ \frac{1}{d+1} \int \cdots \int (d\mathbf{r}_1) \cdots (d\mathbf{r}_{d+1})\ U(\mathbf{r}_1, \mathbf{r}_2, \cdots, \mathbf{r}_{d+1})\ |\psi(\mathbf{r}_{d+1}) \cdots \psi(\mathbf{r}_2)\psi(\mathbf{r}_1)|^2 \end{aligned}$$

with $\int \cdots \int (d\mathbf{r}_2) \cdots (d\mathbf{r}_{d+1})\ U(\mathbf{r}, \mathbf{r}_2, \cdots, \mathbf{r}_{d+1})\ |\psi(\mathbf{r}_{d+1}) \cdots \psi(\mathbf{r}_2)|^2\ \psi(\mathbf{r})$ in the field equation.

If we assume spherically symmetric fields we always obtain from this isotropic harmonic effective potentials.

$$\begin{aligned} &\int \cdots \int (d\mathbf{r}_2) \cdots (d\mathbf{r}_{d+1})\ U(\mathbf{r}, \mathbf{r}_2, \cdots, \mathbf{r}_{d+1})\ |\psi(\mathbf{r}_{d+1}) \cdots \psi(\mathbf{r}_2)|^2 \\ &= k(n-1)(n-2)\cdots(n-d+1)\left(dM_0 r^2 + M_2\right)\left(\frac{M_2}{n}\right)^{d-1} \end{aligned}$$

where $M_k \equiv \int (d\mathbf{r})(\mathbf{r}^2)^{k/2} |\psi(r)|^2 = \Omega_n \int_0^\infty |\psi|^2 r^{k+n-1} dr$ and $\Omega_n = 2\pi^{n/2}/\Gamma(n/2)$. As usual the interpretation (at least in the quantum theory) is that $N \equiv M_0$ is the total number of particles. Thus the field equation for such spherically symmetric solutions becomes

$$i\frac{\partial}{\partial t}\psi(r) + \frac{1}{2m}\left(\frac{d^2}{dr^2}\psi(r) + \frac{n-1}{r}\frac{d}{dr}\psi(r)\right) = \frac{M_2^{d-1}(n-1)!}{n^{d-1}(n-d)!}\left(dM_0 r^2 + M_2\right) k\psi(r)\ .$$

For a Gaussian ansatz, $\psi(\mathbf{r}, t) = C\exp\left(-\tfrac{1}{2}c\mathbf{r}^2\right)\exp(-iEt)$, this field equation reduces to

$$E + \frac{1}{2m}\left(c^2 r^2 - nc\right) = k\frac{(n-1)!}{(n-d)!}\left(\frac{N}{2c}\right)^{d-1}\left(dNr^2 + \frac{nN}{2c}\right)$$

and so we have a solution provided

$$\frac{1}{2m}c^{d+1} = k\frac{(n-1)!}{(n-d)!}N^d\frac{d}{2^{d-1}}\,, \qquad E = \frac{nc}{2m}\left(1+\frac{1}{2d}\right)$$

The E above is almost, but not quite, the energy per particle. That is, the value of H, the total energy, is almost but not quite EN for the spherically symmetric solution. For the Gaussian ansatz

$$H = \frac{ncN}{4m}\left(1+\frac{1}{d}\right)$$

The above suggests what to do also in the situation where ψ is not necessarily spherically symmetric. For any $U(\mathbf{r},\mathbf{r}_2,\mathbf{r}_3\cdots,\mathbf{r}_d,\mathbf{r}_{d+1})$ which is a quadratic form in the components of $\mathbf{r}$ we may write quite generally

$$Z+\mathbf{r}^iB_i+\mathbf{r}^i\mathbf{r}^jK_{ij} = \int\cdots\int(d\mathbf{r}_2)\cdots(d\mathbf{r}_{d+1})\;U(\mathbf{r},\mathbf{r}_2,\cdots,\mathbf{r}_{d+1})\;|\psi(\mathbf{r}_{d+1})\cdots\psi(\mathbf{r}_2)|^2\,.$$

This is true for the case at hand. $U(\mathbf{r},\mathbf{r}_2,\mathbf{r}_3\cdots,\mathbf{r}_d,\mathbf{r}_{d+1})$ is a quadratic form in the components of $\mathbf{r}$ (as well as quadratic in all the other individual $\mathbf{r}_a$) even though it is of order $2d$ altogether. Hence

$$K_{ij} = k\delta^{ii_2\cdots i_d}_{jj_2\cdots j_d}\int\cdots\int(d\mathbf{r}_2)\cdots(d\mathbf{r}_{d+1})\;\mathbf{r}^{i_2}_{23}\mathbf{r}^{j_2}_{23}\cdots\mathbf{r}^{i_d}_{dd+1}\mathbf{r}^{j_d}_{dd+1}\;|\psi(\mathbf{r}_{d+1})\cdots\psi(\mathbf{r}_2)|^2$$

$$B_i = -2k\delta^{ii_2\cdots i_d}_{j_1j_2\cdots j_d}\int\cdots\int(d\mathbf{r}_2)\cdots(d\mathbf{r}_{d+1})\;\mathbf{r}^{j_1}_2\;\mathbf{r}^{i_2}_{23}\mathbf{r}^{j_2}_{23}\cdots\mathbf{r}^{i_d}_{dd+1}\mathbf{r}^{j_d}_{dd+1}\;|\psi(\mathbf{r}_{d+1})\cdots\psi(\mathbf{r}_2)|^2$$

$$Z = k\delta^{i_1i_2\cdots i_d}_{j_1j_2\cdots j_d}\int\cdots\int(d\mathbf{r}_2)\cdots(d\mathbf{r}_{d+1})\;\mathbf{r}^{i_1}_2\mathbf{r}^{j_1}_2\;\mathbf{r}^{i_2}_{23}\mathbf{r}^{j_2}_{23}\cdots\mathbf{r}^{i_d}_{dd+1}\mathbf{r}^{j_d}_{dd+1}\;|\psi(\mathbf{r}_{d+1})\cdots\psi(\mathbf{r}_2)|^2$$

It is now straightforward to use the field equations

$$i\frac{\partial}{\partial t}\psi(\mathbf{r})+\frac{1}{2m}\nabla^2\psi(\mathbf{r}) = \left(Z+\mathbf{r}^iB_i+\mathbf{r}^i\mathbf{r}^jK_{ij}\right)\psi(\mathbf{r})$$

to determine a closed set of time-derivative equations obeyed by the coefficients in the effective potential. Once these are solved, either in special situations or perhaps more generally, then the field equation itself is to be solved using the time-dependence determined for the coefficients, in a self-consistent way. Sounds easy, even if it is not in practice, but perhaps the resulting non-isotropic equations can always be solved in closed form in some limit, such as large n.

There was neither enough time in the talk nor enough space in this written version to discuss either the supersymmetric extensions of these models or their quantization using deformation methods [8, 5]. These subjects will be treated elsewhere.

Acknowledgments

This work was supported in part by NSF Award 0073390 and by US Department of Energy, Division of High Energy Physics, Contract W-31-109-ENG-38. I thank the Particle Theory Group at Argonne National Laboratory for their hospitality in the summer of 2001 during which a portion of this research was completed.

References

[1] F Bayen, et al., Ann Phys **111** (1978) 61; *ibid.* 111.

[2] A Carnegie and I C Percival, J Phys **A 17** (1984) 801-813.

[3] S-J Chang, Phys Rev **D 29** (1984) 259-268.

[4] T Curtright, ICTP seminar, 1989, [*http://server.physics.miami.edu/~curtright/ictp89.html*].

[5] T Curtright, unpublished.

[6] T Curtright, D Fairlie, and C Zachos, Phys Rev **D58** (1998) 025002 [*hep-th/9711183*].

[7] T Curtright and H Kocak, in preparation.

[8] T Curtright, A Polychronakos, and C Zachos, to appear in Phys Lett A [*hep-th/0111173*].

[9] T Curtright, T Uematsu, and C Zachos, J Math Phys **42** (2001) 2396-2415. [*hep-th/0011137*]

[10] B de Wit, J Hoppe, and H Nicolai, Nucl Phys **B305** (1988) 545-581; B de Wit, M Lüscher, and H Nicolai, Nucl Phys **B320** (1989) 135-159.

[11] D Fairlie and C Zachos, Phys Lett **B224** (1989) 101; E Floratos, Phys Lett **B228** (1989) 335-340.

[12] S Ferrara and A Zaffaroni, [*hep-th/9802203*].

[13] R P Feynman, Nucl Phys **B188** (1981) 479-512.

[14] J M Gracia-Bondía, J C Várilly, and H Figueroa, *Elements of Noncommutative Geometry*, Birkhäuser, 2001.

[15] G M Graf, D Hasler, J Hoppe, [*math-ph/0109032*]; J Froehlich, G M Graf, D Hasler, J Hoppe, S-T Yau, Nucl Phys **B567** (2000) 231-248 [*hep-th/9904182*].

[16] M C Gutzwiller, *Chaos in Classical and Quantum Mechanics*, Springer-Verlag, 1990.

[17] P S Howe and P C West, [hep-th/9611074].

[18] A Jaffe and E Witten, "Quantum Yang-Mills Theory", [*http://www.claymath.org/prizeproblems/yang_mills.pdf*].

[19] S Lee, S Minwalla, M Rangamani, and N Seiberg, [*hep-th/9806074*].

[20] H R Lewis, Jr., and W B Riesenfeld, J Math Phys **10** (1969) 1458-1473.

[21] G D Mahan, *Many-Particle Physics*, Kluwer Academic/Plenum Publishers 2000.

[22] A M Perelomov, *Integrable Systems of Classical Mechanics and Lie Algebras* (Volume I) Birkhäuser 1990.

[23] G K Savvidy, Nucl Phys **B246** (1982) 302 and Phys Lett **130B** (1983) 203; S G Martinyan, E B Prokhorenko, and G K Savvidy, Nucl Phys **B298** (1988) 414.

[24] B Simon, Ann Phys (NY) **146** (1983) 209-220.

IV. NEUTRINOS AND CP VIOLATION

High Energy Tau Neutrinos

S. Iyer Dutta*, M. H. Reno† and I. Sarcevic**

*Department of Physics and Astronomy, SUNY Stony Brook, Stony Brook, NY 11794
†Department of Physics and Astronomy, University of Iowa, Iowa City, IA 52242
**Department of Physics, University of Arizona, Tucson, AZ 85721

Abstract. Muon neutrino oscillation into tau neutrinos is the most likely interpretation of the SuperKamiokande atmospheric data. Tau neutrino propagation in matter and interaction in detectors are described for several energy dependences of incident fluxes. Strategies for detecting an oscillation signal are discussed.

INTRODUCTION

Since the SuperKamiokande Collaboration presented their results on the likelihood of neutrino oscillations in their atmospheric flux data [1], a discussion of tau neutrinos and their measurement in large underground detectors has become a realistic possibility. The early SuperK results confirmed earlier work by a variety of experiments showing a deficit of atmospheric muon neutrino events while at the same time measuring the predicted amount of electron neutrino events [2]. More compelling were the SuperK results showing the energy and angular dependence of the deficit. The data are consistent with $\nu_\mu \leftrightarrow \nu_\tau$ oscillations with mixing parameters $\Delta m^2 = 1.5 - 5 \times 10^{-3}$ eV^2 and $\sin^2 2\theta \geq 0.88$ [1]. Oscillations into purely sterile neutrinos are disfavored, although an admixture of tau neutrinos and sterile neutrinos is allowed.

At neutrino energies above 1 TeV, atmospheric sources of tau neutrinos are small. The flux from oscillations of muon neutrinos through the diameter of the Earth is approximately $P_{\nu_\mu \to \nu_\tau} \simeq 5 \times 10^{-2}(\mathrm{TeV}/E)$. Atmospheric fluxes of tau neutrinos through cosmic ray production of $D_s \to \tau\nu_\tau$ are small as well [3].

Given the likelihood of ν_μ oscillations into ν_τ and bimaximal mixing ($\sin^2 2\theta = 1$), galactic and extragalactic sources of muon neutrinos become, after evolution over astronomical distances, sources of tau neutrinos. The distance average of the probability for oscillations is

$$\begin{aligned} \langle P_{\nu_\mu \to \nu_\tau} \rangle &\simeq \langle \sin^2 2\theta \sin^2 \left(\frac{1.27 \Delta m^2/\mathrm{eV}^2 \, L/\mathrm{km}}{E/\mathrm{GeV}} \right) \rangle \qquad (1) \\ &\simeq \sin^2 2\theta \cdot \frac{1}{2} \simeq \frac{1}{2}, \end{aligned}$$

for L the distance to the source and E the energy of the propagating neutrino. Specifically, the sine function of L and E averages to 1/2 for $\Delta m^2 \gg 10^{-16}$ eV^2 for a distance $L = 1$ Mpc and $E = 1$ TeV. The net result of bimaximal mixing is that for a flux ratio of

CP624, *Cosmology and Elementary Particle Physics*, edited by B. N. Kursunoglu et al.

$\nu_\tau : \nu_\mu : \nu_e = 0 : 2 : 1$ at the source, oscillations result in a ratio of $1 : 1 : 1$ over astronomical distances. Unitarity of the mixing matrix and bimaximal $\nu_\mu \to \nu_\tau$ mixing makes the prediction of equal flux ratios even more robust in terms of the initial flux ratios [4, 5].

In the next section, we will describe ν_τ flux attenuation and how it differs from ν_μ flux attenuation in the Earth. This is followed by a discussion of detection strategies in the range of 1 TeV-1 PeV neutrino energies.

NEUTRINO ATTENUATION

Qualitative Behavior

The behavior of tau neutrino fluxes as they pass through the Earth is markedly different from muon neutrino fluxes, as pointed out by Halzen and Saltzberg in Ref. [6]. Charged current interactions of muon neutrinos result in muons which lose energy via electromagnetic processes at a significant rate. As a comparison, the length scale characterizing the electromagnetic energy loss $1/\beta_\mu$ and the decay length (scaled by density ρ) are

$$\frac{1}{\beta_\mu} \sim 10^6 \text{ g/cm}^2 \qquad \gamma c \tau_\mu \rho = 6.2 \times 10^5 \text{ g/cm}^2 \frac{E}{1 \text{ GeV}} \frac{\rho}{1 \ g/cm^3}, \tag{2}$$

so for high energies, electromagnetic energy loss is always important. On the other hand, for tau leptons produced by tau neutrino charged current interactions [7],

$$\frac{1}{\beta_\tau} \sim 10^6 \text{ g/cm}^2 \qquad \gamma c \tau_\tau \rho = 4.9 \times 10^{-3} \text{ g/cm}^2 \frac{E}{1 \text{ GeV}} \frac{\rho}{1 \ g/cm^3}, \tag{3}$$

so until $E \sim 10^8 - 10^9$ GeV, tau lepton energy loss is not an issue. Qualitatively, then, when tau neutrinos interact to make taus, the tau has an energy $(1 - \langle y \rangle)E_\nu = 0.7E_\nu$. The tau decays before interacting, and deposits $\sim 40\%$ of its energy in the decay neutrino, so a neutrino with energy $E_\nu \to 0.7 \times 0.4 E_\nu \sim 0.3 E_\nu$ returns to the flux. The neutrino associated with muon decay returns with a dramatically lower energy and is "lost" to the high energy flux. Because of the tau neutrino energy loss, the "pileup" of neutrino flux will be flux dependent. Since the feed-down process depends on an initial neutrino interaction, the attenuation and pileup will also be angle dependent.

Quantitative Results

Semi-analytic solution of the transport equations for neutrinos and leptons is required to evaluate the attenuation of the neutrino fluxes. The general transport equations are:

$$\frac{\partial F_\nu(E,X)}{\partial X} = -\frac{F_\nu(E,X)}{\mathscr{L}_\nu^{\text{int}}} + \int_E^\infty dE_y G^{\nu \to \nu}(E, E_y, X) \tag{4}$$

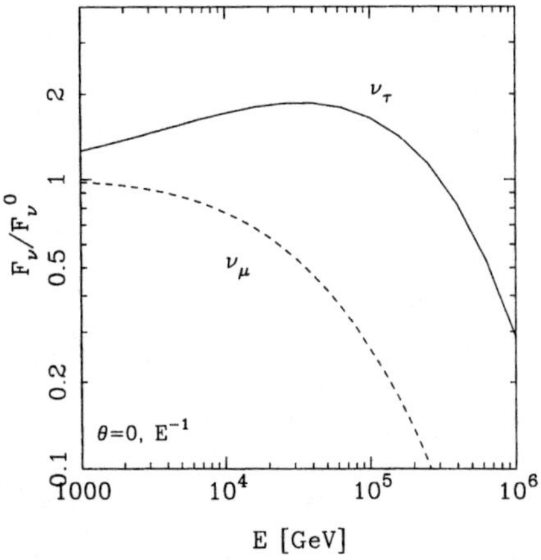

FIGURE 1. Ratio of attenuated flux to incident flux for $F_\nu^0 = 1/E$ at nadir angle of 0° for muon neutrinos (dashed) and tau neutrinos (solid).

$$
\begin{aligned}
&+\int_E^\infty dE_y G^{\tau\to\nu}(E,E_y,X) \\
\frac{\partial F_\tau}{\partial X} &= -\frac{F_\tau}{\mathcal{L}_\tau^{\text{int}}} - \frac{F_\tau}{\mathcal{L}_\tau^{\text{decay}}} + \int_E^\infty dE_y G^{\nu_\tau\to\tau}(E,E_y,X) \\
&+\int_E^\infty dE_y G^{\tau\to\tau}(E,E_y,X)
\end{aligned}
$$

where F_ν is the number of neutrinos/(cm^{2}s sr GeV), $\mathcal{L}$ is the interaction $(1/N_A\sigma_{\nu N})$ or decay length $(\gamma c\tau\rho)$ and X is the depth (usually in g/cm^2). The quantities $G^{i\to j}$ depend on the flux at higher energy $E_y = E/(1-y)$ and the differential cross section or decay width, for example,

$$
G^{\nu\to\nu}(E,E_y,X) = \left[\frac{F_\nu(E_y,X)}{\mathcal{L}_\nu^{\text{int}}}\right]\frac{dn^{NC}}{dE}(E_y,E) \tag{5}
$$

is the neutral current return of neutrinos at a lower energy E, initially of energy E_y.

The tau neutrino and charged tau lepton equations can be difficult to solve. We have made the simplifying approximation that we can ignore τ interactions. We keep only decays. This is valid for energies $E_\tau < 10^8 - 10^9$ GeV [7]. For muon neutrinos, one ignores the source of $\mu \to \nu$ since the muon neutrinos are returned at such low energies, so the two equations are decoupled.

To obtain a solution, we extend the procedure of Naumov and Perrone [8] to include not just neutral current return of neutrinos but also the contribution from tau decays. The starting point is considering a solution of the form:

$$
F_\nu(E,X) \quad = F_\nu^0(E)\exp\left[-\frac{X}{\Lambda_\nu(E,X)}\right] \tag{6}
$$

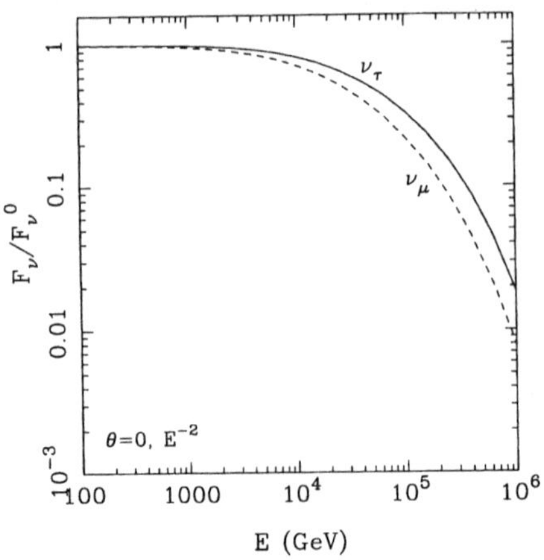

FIGURE 2. Ratio of attenuated flux to incident flux for $F_\nu^0 = 1/E^2$ at nadir angle of 0° for muon neutrinos (dashed) and tau neutrinos (solid).

$$\Lambda_\nu = \frac{\mathscr{L}_\nu^{\rm int}(E)}{1 - Z(E,X)} \,. \tag{7}$$

An iterative solution to $Z(E,X)$ is developed. Details of the solution appear in Ref. [9], including the Earth density model [10] and neutrino interaction lengths at high energy [11]. Here we show two examples of the results: tau neutrino attenuation for $1/E$ and $1/E^2$ fluxes with nadir angles equal to 0°. To aid with the numerical stability, we have cut off the $1/E$ flux at high energies: $F_\nu^0 = 1/E/(1+E/10^8\ \text{GeV})^2$ in all of the results labeled $1/E$. In Figs. 1 and 2, we show the ratio of the attenuated flux to incident flux for the two input spectra. The solid lines show the results for ν_τ. The dashed lines indicate the attenuated flux for ν_μ. The attenuation of tau neutrino and muon neutrino fluxes from a variety of flux models appears in Refs. [9] and [12].

DETECTION STRATEGIES

Detection of tau neutrinos is best achieved by looking for the production of the τ and its decay. For energies above a few PeV, the production vertex and the decay vertex will be well separated, so the signature of tau production and decay should be unmistakable [13]. At the energies considered here, namely from 1 TeV-1 PeV, the production of taus and their decay can only be inferred from the data. The strategy in this energy regime is to compare the upward hadronic/electromagnetic shower rate of contained events with the rate for upward muons. This is similar to the lower energy test of oscillation by comparing the neutral current event rate to the charged current event rate for incident neutrinos.

Both the muon rates and the shower rates have contributions from muon neutrinos and tau neutrinos. The upward muon event rates come from

$$\begin{aligned} \nu_\mu N &\to \mu X \\ \nu_\tau N &\to \tau X \to \nu_\tau \mu \nu_\mu X \,. \end{aligned} \tag{8}$$

The upward muon event rate, as usual, is enhanced by the long range of the muon, so the effective target volume is the detector area (here taken as 1 km^2) multiplied by the average muon range for an incident muon with energy E_μ to arrive at the detector with energy $E_\mu^{\min}$, $\langle R(E_\mu, E_\mu^{\min})\rangle$. Since the muon gets a fraction of the decaying tau's energy, the effect of the pileup in the tau neutrino flux is somewhat muted in the upward muon event rate. Depending on the nadir angle and the energy dependence of the incident flux, the muon rate assuming $\nu_\mu \to \nu_\tau$ oscillations with bimaximal mixing is between $\sim 0.6 - 0.9$ of the rate without oscillations. Given the uncertainties in the astrophysical neutrino fluxes, determining whether or not oscillations occur with underground upward muon rates is very difficult.

For upward shower event rates, there are a number of processes that contribute. These processes are

$$\begin{aligned} \nu_i N &\to \nu_i X, \qquad i = e,\ \mu,\ \tau \\ \nu_e N &\to eX \\ \nu_\tau N &\to \tau X,\ \tau \to \nu_\tau e \nu_e,\ \tau \to \nu_\tau + \text{hadrons} \,. \end{aligned} \tag{9}$$

Event rates discussed below are evaluated using a detection volume of 1 km^3. All neutrinos contribute neutral current events, but the shower induced by the neutral current interaction only has on average $E_{shr} = \langle y \rangle E_\nu \simeq 0.3 E_\nu$. More important are the charged current events with no muons (we assume that a high energy muon veto will be used). All of the incident ν_e energy goes into the hadronic/electromagnetic shower in charged current events, and most of the initial ν_τ energy goes into the showers when the tau decays hadronically. Since the attenuated ν_τ flux is larger than the attenuated ν_e flux due to the pileup effect discussed above, the net effect is a more enhanced shower rate. In particular, one can compare the ratio of upward shower rates to upward muon rates as a function of nadir angle and shower (or muon) energy threshold and see a distinct difference between the oscillation scenario and the no-oscillation scenario. The distinction does require some knowledge of the incident flux which, with large enough fluxes, will be ascertainable from the energy and angular dependence of the upward muons.

To be specific about event rates, we have used a number of flux predictions from different source categories. These fluxes are chosen in part to demonstrate the dependence of event rates on the energy dependence of the flux. The atmospheric flux, indicated in Fig. 3 by the hatched region, depends on angle, but the other fluxes are summed over sources and are isotropic. We include representative models of neutrinos from active galactic nuclei (AGN) from Stecker and Salamon (AGN_SS) [14] with a derating factor of 0.3 and Mannheim (Model A) (AGN_M95) [15]. A flux prediction from gamma ray bursters (GRB) from Waxman and Bahcall is shown (GRB_WB) [16], as are two models of neutrinos from the decays of super-heavy particles, from Wichoski, MacGibbon and

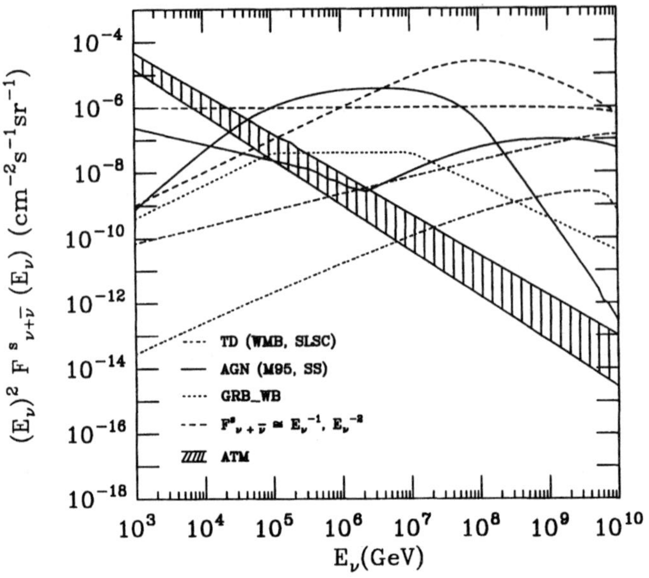

FIGURE 3. Isotropic muon neutrino plus antineutrino fluxes for a variety of sources (AGN_M95 upper solid line, AGN_SS lower solid line, GRB_WB dotted line, TD_WMB upper dot-dashed line, TD_SLSC lower dot-dashed line), scaled by the neutrino energy squared. No oscillations are assumed.

Brandenberger (TD_WMB) [17] and from Sigl, Lee, Schramm and Coppi (TD_SLSC) [18]. The E^{-1} and E^{-2} fluxes are normalized by $F_{\nu_\mu} = 10^{-12}\ (E/\mathrm{GeV})^{-1}/(\mathrm{GeV\,cm^2sr\,s})$ and $F_{\nu_\mu} = 10^{-6}\ (E/\mathrm{GeV})^{-2}/(\mathrm{GeV\,cm^2sr\,s})$, consistent with AMANDA limits [19]. The E^{-1} flux is smoothly cut off at high energies as indicated in the figure.

Without regard to statistical errors, we first plot in Fig. 4 the ratio of event rates (showers to muons) for contained upward showers in 1 km^3 and upward muons with an effective area of 1 km^2 for a variety of flux models and three energy thresholds. The separation of the fluxes is between steeply falling ($\sim E^{-2}$) and less-steeply falling ($\sim E^{-1}$). Experimentally, a separation between flux categories can be done by looking at the energy and nadir angle dependence of the upward muon rates.

In terms of detailed predictions of the ratios, statistical errors must be taken into account. Three flux examples are shown in Fig. 5, where the error on the ratio of events is statistical assuming a Poisson distribution and include the statistical error on the number of background events from the atmospheric flux. In the figure, *n* indicates the number of years of data taking. We have also evaluated the ratio, including statistical errors, for the case of four neutrino mixing assuming the fourth neutrino is a sterile neutrino[20].

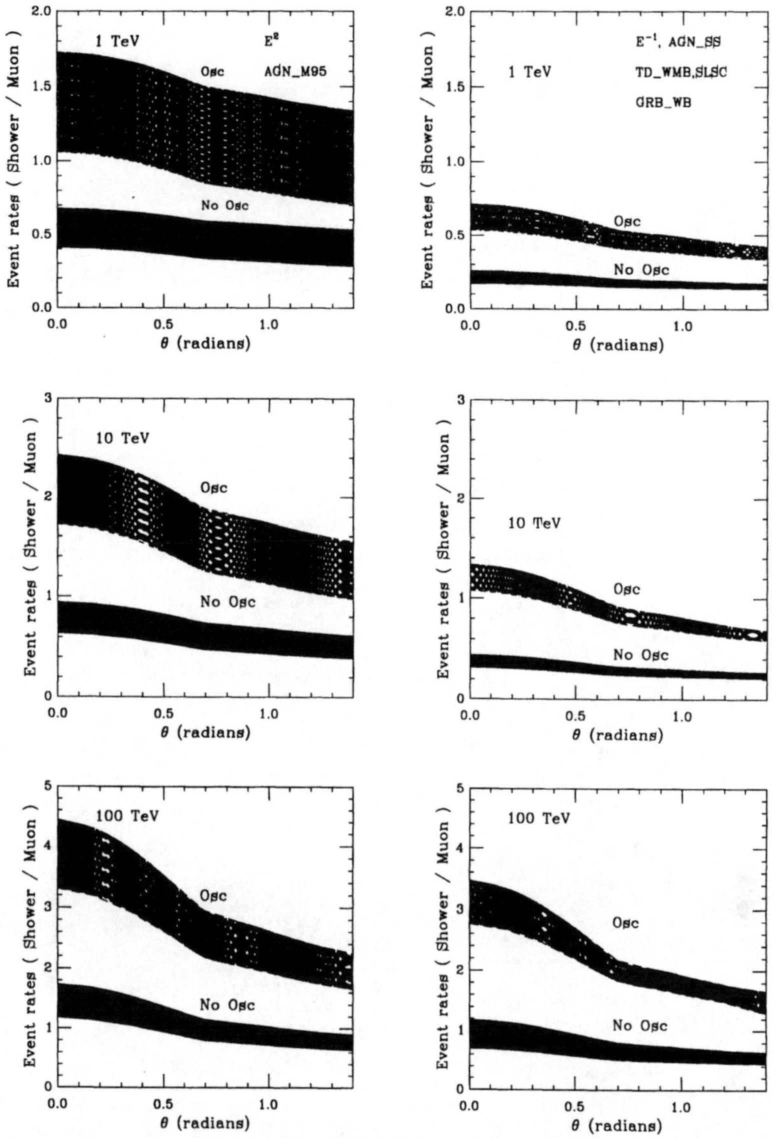

FIGURE 4. Ratio of the upward hadronic/electromagnetic shower event rate to upward muon event rate for oscillation (upper shaded area) and no-oscillation (lower shaded areas) scenarios as a function of nadir angle and energy threshold (1, 10, 100 TeV) for the fluxes indicated.

CONCLUSIONS

Event rates from astrophysical sources range from a handful to thousands of events, depending on the flux model and energy threshold. Kilometer-sized underground detectors

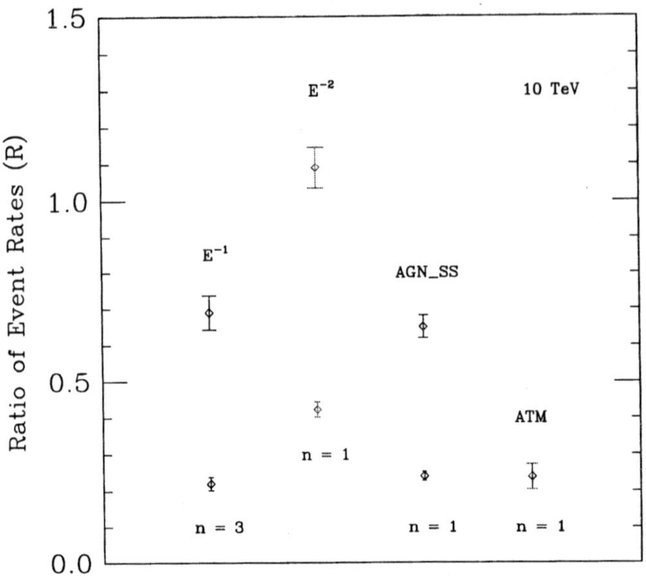

FIGURE 5. Ratio of the upward hadronic/electromagnetic shower event rate to upward muon event rate for oscillation (upper points) and no-oscillation (lower points) scenarios for a 10 TeV muon or shower threshold, assuming n years of data taking.

have the potential to detect tau neutrinos from $\nu_\mu \to \nu_\tau$ oscillations over astronomical distances using combined measurements of upward showers and upward muons in the 1 TeV-1 PeV energy range, opening up a new energy frontier for studying neutrino oscillations.

ACKNOWLEDGMENTS

The work of S.I.D. and I.S. was supported in part by the DOE under Contract DE-FG02-95ER40906. The work of M.H.R. was supported in part by NSF Grant No. PHY-9802403.

REFERENCES

1. Fukuda, Y. *et al.* [Super-Kamiokande Collaboration], Phys. Rev. Lett. **81**, 1562 (1998) [hep-ex/9807003]. Kajita, T. [Super-Kamiokande Collaboration], Nucl. Phys. Proc. Suppl. **100**, 107 (2001).
2. For a review, see for example, Learned, J. G. [SuperKamioKande Collaboration], "The atmospheric neutrino anomaly: Muon neutrino disappearance," [arXiv:hep-ex/0007056], in Caldwell, D. O. (ed.), Current aspects of neutrino physics, pp 89-130 (Springer-Verlag, 2001).
3. Pasquali, L., and Reno, M. H., Phys. Rev. D **59**, 093003 (1999) [arXiv:hep-ph/9811268].
4. Ahluwalia, D. V., Ortiz, C. A., and Adunas, G. Z., hep-ph/0006092.

5. Athar, H., Jezabek, M., and Yasuda, O., Phys. Rev. D **62**, 103007 (2000) [hep-ph/0005104]. Athar, H., Astropart. Phys. **14**, 217 (2000) [hep-ph/0004191]. Athar, H., Nucl. Phys. Proc. Suppl. **87**, 442 (2000) [hep-ph/9912417].
6. Halzen, F., and Saltzberg, D., Phys. Rev. Lett. **81**, 4305 (1998) [hep-ph/9804354];
7. Iyer Dutta, S., Reno, M. H., Sarcevic, I., and Seckel, D., Phys. Rev. D **63**, 094020 (2001) [hep-ph/0012350].
8. Naumov, V. A., and Perrone, L., Astropart. Phys. **10**, 239 (1999) [hep-ph/9804301].
9. Iyer, S., Reno, M. H., and Sarcevic, I., Phys. Rev. D **61**, 053003 (2000) [hep-ph/9909393]; see also, Becattini, F. and Bottai, S. astro-ph/0003179.
10. Dziewonski, A., Earth Structure, Global, in *Encyclopedia of Solid Earth Geophysics,* ed. David E. James (Van Nostrand Reinhold, New York)
11. See, for example, Gandhi, R., Quigg, C., Reno, M. H., and Sarcevic, I., Phys. Rev. D **58**, 093009 (1998) [hep-ph/9807264]; Gandhi, R., Quigg, C., Reno, M. H., and Sarcevic, I., Astropart. Phys. **5**, 81 (1996) [hep-ph/9512364].
12. Iyer Dutta, S., Reno, M. H., and Sarcevic, I., Phys. Rev. D **62**, 123001 (2000) [hep-ph/0005310].
13. Learned, J. G., and Pakvasa, S., Astropart. Phys. **3**, 267 (1995) [arXiv:hep-ph/9405296].
14. Stecker, F. W., and Salamon, M. H., Space Sci. Rev. **75**, 341 (1996) [astro-ph/9501064].
15. Mannheim, K., Astropart. Phys. **3**, 295 (1995).
16. Waxman, E., and Bahcall, J., Phys. Rev. D **59**, 023002 (1999) [hep-ph/9807282].
17. Wichoski, U. F., MacGibbon, J. H., and Brandenberger, R. H., hep-ph/9805419.
18. Sigl, G., Lee, S., Schramm, D. N., and Coppi, P., Phys. Lett. B **392**, 129 (1997) [astro-ph/9610221].
19. Andres, E., *et al.* [AMANDA Collaboration], Nucl. Phys. Proc. Suppl. **91**, 423 (2000) [astro-ph/0009242].
20. Iyer Dutta, S., Reno, M. H., and Sarcevic, I., Phys. Rev. D **64**, 113015 (2001) [arXiv:hep-ph/0104275].

Neutrino Reactions in Nuclei and a Connection to the Neutron Number

Stephan L. Mintz

Physics Department,Florida International University,Miami,Florida ,33157

Abstract. We compare the inclusive electron neutrino cross sections, $\nu_e + N_i \longrightarrow X + e^-$, averaged over the Michel spectrum for $^{12}C, ^{13}C, ^{56}Fe$ and ^{127}I. We use ^{56}Fe as a model and present a number of results for this nucleus. We choose these nuclei because terrestrial measurements at LSND and KARMEN have been undertaken for them and these measurements are in reasonable agreement with calculated values. We show that the cross sections increase linearly with neutron number for these nuclei and discuss what this might imply.

INTRODUCTION

This has been an interesting year for neutrino reaction studies in nucleons and nuclei. The Sudbury Neutrino Observatory (SNO) has finally begun taking data and has published its first results confirming the solar neutrino deficit of earlier experiments. SNO uses a heavy water target. On the other hand a terrible accident at Super-Kamiokande which resulted in the destruction of about three quarters of their photo-tube detectors has for the time being shut them down. Super-K uses a water target of very large size.

However in this paper we will be mostly concerned with neutrino reactions in more complicated nuclei. Over the past two decades many calculations for inclusive electron neutrino reactions in nuclei have been undertaken. We have undertaken calculations for neutrino reactions[1,2,3,4,5] in $^{12}C, ^{13}C, ^{56}Fe$ and ^{127}I because either these reactions were expected to be observed experimentally or in fact had actually been observed. Over this period of time these reactions have indeed been observed either by the LAMPF group or by the KARMEN collaboration. The experimental error bars for these reactions are quite large. However both calculated values for the cross sections averaged over the Michel spectrum and experimental values for these cross sections increase with increasing neutron number.

In this paper we examine the form of this increase and find that over a large range of neutron numbers the increase in this averaged cross section appears to be linear. We discuss our method of calculation making use of the example of iron. This will give us the opportunity to present our most recent results for both the electron neutrino and the electron antineutrino reactions on this nucleus. We will then list the available experimental results and make some comments on the linear relationship that we observe. Finally we shall discuss the limits of this linear relationship and what it implies.

CP624, *Cosmology and Elementary Particle Physics*, edited by B. N. Kursunoglu et al.

MATRIX ELEMENTS

Recently[5] we have completed a calculation for the process , $\nu_e + ^{56}Fe \longrightarrow e^- + X$. We made use of a phenomenological model which we have used for a number of other processes. The results for these calculations have appeared in the literature[1,2,3,4,6,7] and so we do not discuss the details of this model. However some of our assumptions will play a role in our comments so we give a brief presentation of the method used. It is adapted for iron but is used for all of the nuclei discussed here. We write the transition matrix element as:

$$M_{ki} = \frac{G}{\sqrt{2}} cos\theta_C \bar{u}_e \gamma^\lambda (1-\gamma_5) u_\nu < k|J_\lambda(0)|^{56}Fe > \quad (1)$$

where k is a particular final state and:

$$J_\mu(0) = V_\mu(0) - A_\mu(0) \quad (2)$$

We then write the cross section in a completely standard way as:

$$\sigma_c = \sum_k \frac{m_\nu}{2ME_\nu} \int d^3P_e \left|M_{ki}\right|^2 \frac{m_e}{E_e(2\pi)^3} \frac{d^3P_k}{2E_k(2\pi)^3} (2\pi)^4 \delta^4(P_k + p_e - p_\nu - P_i) \quad (3)$$

Here k stands for the possible final hadronic states,P_i^μ is the initial four momentum of the ^{56}Fe nucleus ,P_k^μ is the four momentum of the k state, and p_e^μ and p_ν^μ are the electron and neutrino four momenta respectively. The quantity $|M_{ki}|^2$ which appears in Eq.(3), is given by:

$$|M_{ki}|^2 = \frac{G^2 \cos^2\theta_C}{2m_\nu m_e} L^{\sigma\lambda} < k|J_\sigma(0)|^{56}Fe >< k|J_\lambda(0)|^{56}Fe >^* \quad (4)$$

The quantity, $L^{\sigma\lambda}$, is the lepton tensor appropriate to this process and is given by:

$$L^{\sigma\lambda} = p_e^\sigma p_\nu^\lambda - p_e \cdot p_\nu g^{\sigma\lambda} + p_\nu^\sigma p_e^\lambda - \varepsilon^{\alpha\sigma\beta\lambda} p_{e\alpha} p_{\nu\beta}. \quad (5)$$

We assume an average nuclear excitation of δ given by:

$$M_x - M_i = \delta \quad (6)$$

where δ is a function of the incoming neutrino energy. In general the value of δ will certainly increase with increasing incident neutrino energy. However we expect that above the giant dipole that this increase should be slow. For ^{56}Fe, the giant dipole resonance occurs at approximately[8] 18 MeV. From differential cross section data for electron scattering from ^{56}Fe available[8,9] for incident electrons from 150 MeV to 250 MeV at a scattering angle of 35 degrees,we obtained in the region immediately above the giant dipole region:

$$\delta(E_\nu) = .0215(E_\nu - m_e) + 17.58 \quad (7)$$

where of course E_ν is the incident neutrino energy and m_e is the electron mass. Below the giant dipole resonance we used a δ linear in E_ν which joins smoothly with Eq.(7) at 35 MeV. The result is:

$$\delta(E_\nu) = .6015(E_\nu - m_e) - 2.747 \tag{8}$$

Using these quantities we obtain the cross section as:

$$\begin{aligned} \sigma_c &= \frac{G^2 cos^2 \theta_C}{2ME_\nu} \int d\Omega_e \sum_k < k|J_\sigma(0)|^{56}Fe >< k|J_\lambda(0)|^{56}Fe >^* L^{\sigma\lambda} \\ &\times \frac{< |\vec{p_e}| >}{2M - 2E_e + 2E_e \cos\theta_e \frac{<E_e>}{<|\vec{p_e}|>}}. \end{aligned} \tag{9}$$

The hadronic part of Eq.(9) may be replaced by a tensor as:

$$< k|J_\sigma(0)|^{56}Fe >< k|J_\lambda(0)|^{56}Fe >^* \equiv Q_{\lambda\sigma}(P_i, < q >). \tag{10}$$

We have previously shown[1,2,3,4,6,7] that this tensor may be reduced to the simpler form:

$$Q^{\mu\nu} = \alpha g^{\mu\nu} + \frac{\beta}{M^2} P_i^\mu P_i^\nu . \tag{11}$$

The cross section then becomes:

$$\sigma_c = \frac{G^2 \cos^2(\theta_C)}{4\pi} \frac{< |\vec{p}_e| >< E_e > D}{M(M + E_\nu)} \tag{12}$$

where:

$$D = \beta - 2\alpha \tag{13}$$

and an impulse approximation based calculation[10,11] gives $D(q^2)$ as:

$$D = a_o - b_o q^2 . \tag{14}$$

We assume this simple q^2 dependence for D. Because total muon capture rates, $\mu^- + N_i \longrightarrow \nu_\mu + X$, are known for many nuclei and depend on the same D as does the cross section given by Eq.(12), it is possible to use muon capture data along with Eq.(14) to determine D either directly[1,2,6] or indirectly[3,4] and hence the cross sections completely. This brief description of the calculation is sufficient for our purposes here.

RESULTS AND DISCUSSION

Using the above outlined results we have obtained values for the cross section for the reaction $\nu_e + ^{56}Fe \longrightarrow e^- + X$. We give these results in figure 1 for neutrino energies from threshold to 240 MeV. We have also averaged our cross section over the Michel spectrum. The result is $< \sigma >= 214 \pm 25\% \times 10^{-42}\ cm^2$ in good agreement with

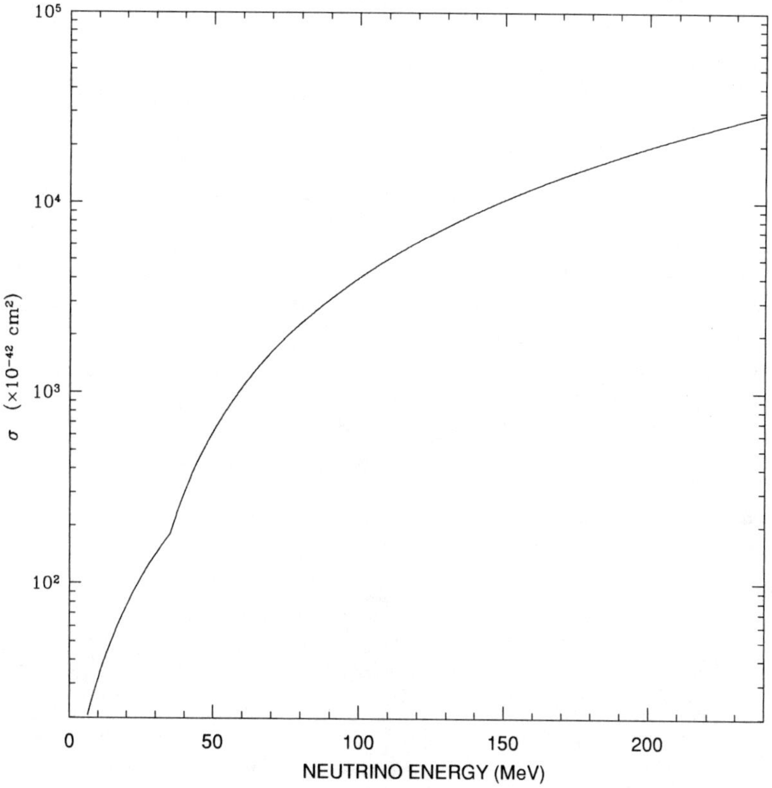

FIGURE 1. Plot of the differential cross section for the reaction, $\nu_e + ^{56}Fe \longrightarrow e^- + X$ as a function of incident laboratory neutrino energy

measured[12] values at KARMEN of $<\sigma>= 251 \pm 83\ (stat) \pm 42(sys)\ \times 10^{-42}\ cm^2$ and $<\sigma>= 256 \pm 108(stat) \pm 43(sys)\ \times 10^{-42}\ cm^2$ as well as with other calculations.

Experimentalists are also interested in the antineutrino reaction, $\bar{\nu}_e + ^{56}Fe \longrightarrow e^+ + X$. In figure 2 we present the cross section for this reaction. We also obtain the Michel spectrum averaged result for this case, namely $<\sigma>= 89.5 \pm 25\%\ x\ 10^{-42}\ cm^2$. These results help the experimentalists determine their errors.

Because it is known that inclusive neutrino reactions in nuclei increase with the increasing number of neutrons in the nucleus we were led to examine the form of this increase. To study how the inclusive neutrino reaction increases with the number of neutrons we looked at nuclei for which there were terrestrial measurements and theoretical calculations both in reasonable agreement. The nuclei for which this is true are $^{12}C, ^{13}C, ^{56}Fe$, and ^{127}I. Below we list the results in tabular form, where the theoretical and experimental references for the above nuclei are 2 and 12,13 for ^{12}C, 4 and 12 for ^{13}C, 5 and 12 for ^{56}Fe and 3 and 14 for ^{127}I respectively. In figure 3 we plot

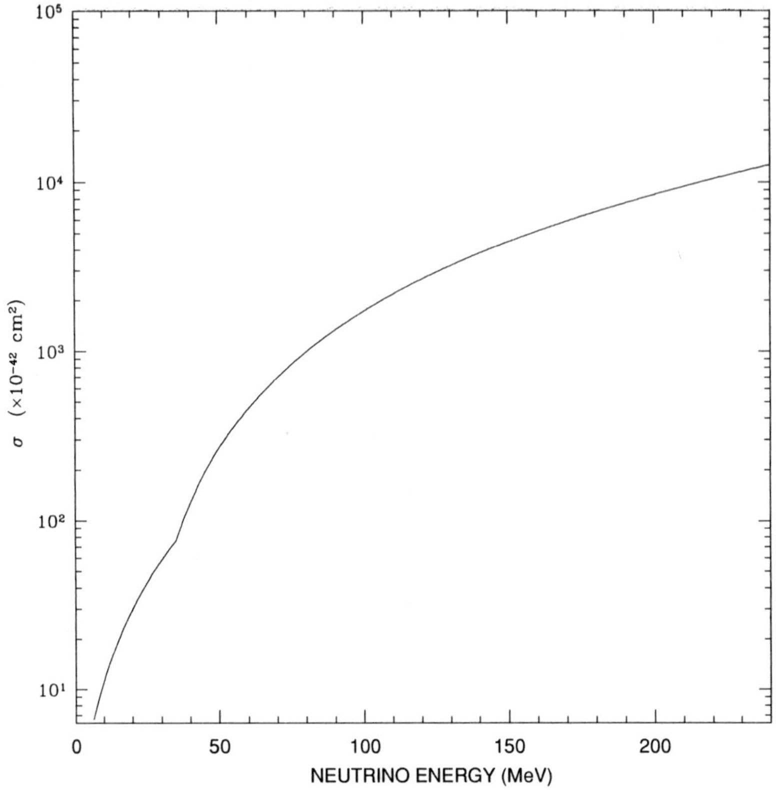

FIGURE 2. Plot of the differential cross section for the reaction, $\bar{\nu}_e + ^{56}Fe \longrightarrow e^+ + X$ as a function of incident laboratory neutrino energy

TABLE 1. Neutrino reaction cross sections averaged over the Michel spectrum for four nuclei. Both theoretical and experimental results are given as is the neutron number $A - Z$. All cross sections are $\times 10^{-42}$ cm^2. For the experimental results the first error is statistical and the second error is systematic.

Nucleus	$A-Z$	σ(theoretical)	σ(experimental)
^{12}C	6	16.85±25%	14.8 ± 1.0 ± 1.1
^{13}C	7	28.9± 25%	50± 37 ± 10
^{56}Fe	30	229± 27%	251 ± 83 ± 42
^{127}I	74	638 ± 30%	500 *to* 700

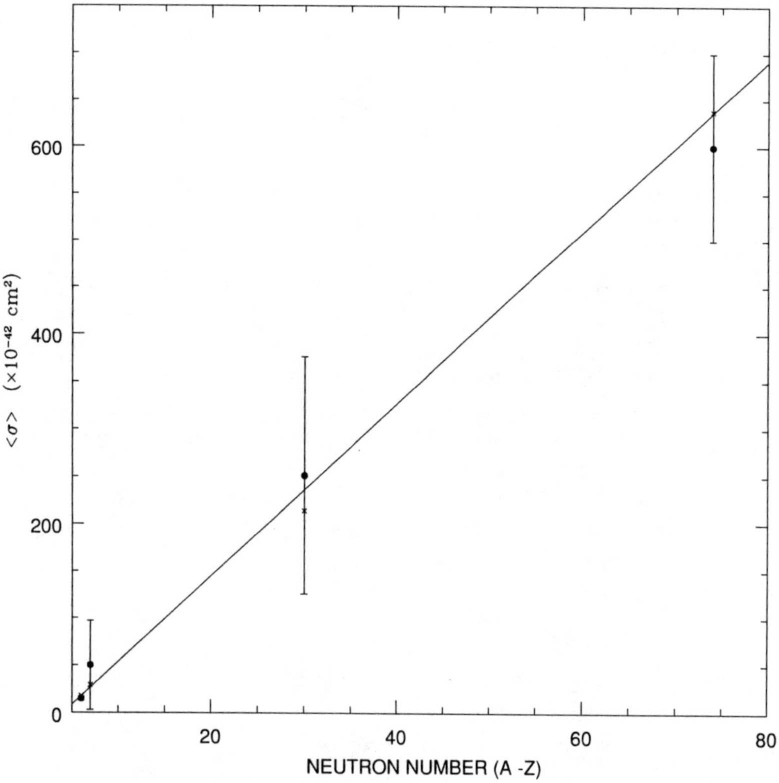

FIGURE 3. Plot of the differential cross section averaged over the Michel spectrum for the reaction, $\nu_e + N_i \longrightarrow e^- + X$ for the initial nuclei, N_i, ^{12}C, ^{13}C, ^{56}Fe and ^{127}I. The x's refer to the theoretical points and the error bars refer to the experimental results

the averaged cross sections given in the above table versus the neutron number, $A-Z$. We use our calculated values in above tables in hope that systematic errors might be the same. We find that the data is well fit by the straight line:

$$< \sigma >= [9.11(A-Z) - 37.23] \times 10^{-42}\ cm^2. \tag{15}$$

We do not want to read too much into this relation as the error bars are very large (in the 40 to 50 percent range for many of the experimental results) but it is nonetheless interesting. We should first remark that the cross sections have different electromagnetic final state interactions which might spoil the linearity apparent here. However for the electron case being considered here these corrections are small except at the lowest neutrino energies where the cross sections are very small and there is not much contribution to the averaged cross sections. For the ^{127}I case at 20 MeV, the final state interaction is approximately a 3 percent effect and for the case of lead, the contribution is of the order of 5 percent, where we have used the correction[15]:

$$C = \gamma/(1 - \exp(\gamma)) \qquad (16)$$

with $\gamma = 2\pi Z m_e / p_e$.

Another assumption which plays an important role in the calculations given here is the assumption that the states associated with the giant dipole resonances dominate the cross sections for the neutrino energy range of the Michel spectrum thus leading to Eqs. (7) and (8). This is obviously not true for the smallest nuclei which simply undergo break up and so we do not expect Eq. (15) to hold for them. The same would be true for very heavy nuclei for which break up states would again be very important and possibly dominant. Thus Eq.(15) might seriously underestimate these cases.

Nevertheless this relationship is an interesting one which suggests that the individual nucleons provide little shielding and that therefore the cross sections essentially scale in the number of neutrons over a fairly broad range. Hopefully the number of nuclei for which data becomes available will increase over time. This should be particularly true if the ORLaND project[16] goes forward as a substantial number of nuclear neutrino reactions are contemplated for it. Clearly additional work including more experimental results with reduced errors would be desirable.

REFERENCES

1. Mintz,S.L., and King,D.L., Phys. Rev. **C30**,1585(1984).
2. Mintz,S.L., and Pourkaviani,M., Nucl. Phys. **A573**501(1994).
3. Mintz,S.L., and Pourkaviani,M., Nucl. Phys. **A584**,665(1995).
4. Mintz,S.L.,Nucl. Phys. **A672**,503(2000).
5. Mintz,S.L.,Int. J. Mod. Phys. **E 10**,387(2001).
6. Mintz,S.L., and Pourkaviani,M., Nucl. Phys. **A589**,724(1995).
7. Mintz,S.L., and Pourkaviani,M.,Nucl. Phys. **A594**,346(1996).
8. Torizuka,Y., et al.,Proceedings of the International Conference on Nuclear Structure Studies Using Electron Scattering and Photoreactions, Sendai, Japan,Editors:K. Shoda and H. Ui,Tohoku University,Tomizawa,Sendai, Japan,171(1972).
9. Torizuka,Y., et al.,Proceedings of the International Conference on Photonuclear Reactions and Applications,Editor: Barry L. Berman,Ernest O. Lawrence Livermore Laboratory,University of California, Vol. 1,Livermore California,675(1973).
10. Goulard,B.,and Primakoff,H.,Phys. Rev.**135**,**B**1139(1964).
11. Goulard,B., and Primakoff,H.,Phys. Rev. **C10**,2034(1974).
12. Maschuw,R.,Prog. Part. Nucl. Phys. **40**,183(1998)
13. Athanassopoulos,C., et al.,Phys. Rev. **C54**,2717(1996).
14. Lande,K.,private communication.
15. Tzara,C.,Nucl. Phys. **B 18**,246(1970).
16. Report on the Workshop on Neutrino Nucleus Physics Using a Stopped Pion Neutrino Facility,the ORLaND Collaboration, Oak Ridge Tennessee,May 22-26,2000.

K-M matrix elements and Decays of the B meson to J/Psi

Richard Wilson

Harvard University
Cambridge, MA 02138
Work of the CLEO collaboration

Abstract. This talk discusses some of the last work on B meson decays of the CLEO collaboration, which work is, in fact, improvements in precision of much earlier work of the same collaboration. New theoretical developments have enabled us to present much improved numbers on the matrix elements V_{cb}, and V_{ub}. Also some recent work on the decay of B mesons to J/Psi plus other particles will be briefly presented.

INTRODUCTION

This talk was first prepared for a conference in Tashkent in September 2001. But the terrorist act on September 11th made me cancel the trip. My heart goes out to the 3000 victims of that attack as it does to the million or so starving Afghans.

I first became interested in the idea of colliding electron beams in 1956 when we informally proposed to study build an electron-electron ring at Harvard, (which proposal was delayed till our accelerator was completed when it became moot) and we proposed colliding electron-positron beams in 1959. Alas, our proposal was rejected in favor of another, and I was lucky somewhat later to be invited to participate in the Cornell Electron Storage Ring (CESR) starting in 1977 (figure 1).

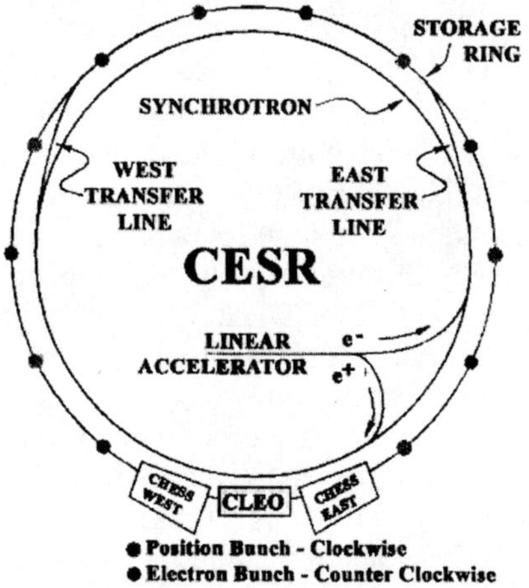

figure1. The CESR storage ring layout

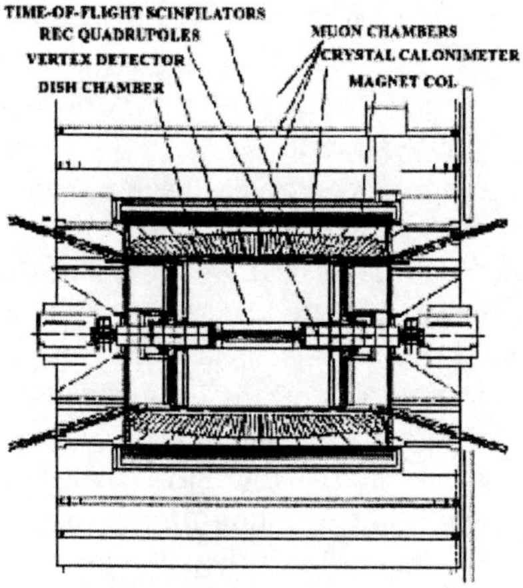

figure 2. The CLEO Mark II detector

CP624, *Cosmology and Elementary Particle Physics*, edited by B. N. Kursunoglu et al.

Interestingly the linear accelerator injector was the one we had ordered for the Cambridge Electron Accelerator in 1961 and which had been declared government surplus in 1972. Romantically the name Ceasar has often been accompanied by the name Cleopatra so the detector that I helped to build was called CLEO. (Figure 2) The important features for this experiment are: good particle momentum measurement, and absence of spurious high energy particles, good particle discrimination for electrons (by a CsI detector) and muons (by range), and gamma ray measurement (with the CsI crystals) with good resolution. All with high angular coverage and high efficiency. CLEO began operation in the fall of 1979 and in February 1980 we had our first new result - the production of the predicted upsilon resonance ϒ(4S). (Figure 3)

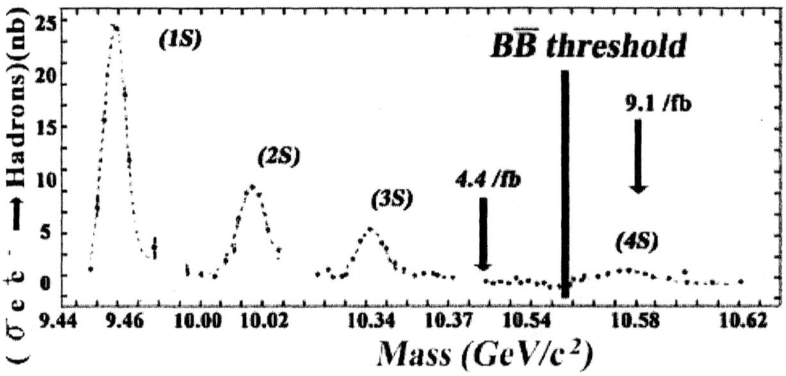

Figure 3. Mass spectrum of upsilon resources.

I had the honor and pleasure of being the first to describe this result, together with my other work (on muon scattering), on a visit to Beijing, China, in March of the same year (1980). The ϒ(4S) was just above the threshold for decay into particles and although the width of the upsilon ϒ(1S), ϒ(2S) and ϒ(3S) resonances was equal to the CESR resolution, the width of the ϒ(4S) is greater than the resolution indicating that it decays rapidly (into B particles). Historically it is interesting that the first extraction of the ϒ(4S) resonance used the Fox-Wolfram R2 shape variable to suppress continuum background. We still use this parameter for this purpose but combine it with several others, using a neutral net, to get a single parameter to reduce the continuum background. Since 1960 the study of these B particles has been the primary task of CESR and CLEO. This task was aided by the fact that over 20 years the skilled physicists at CESR produced the highest luminosity of any colliding beam facility in spite of the beams being injected from a synchrotron which (in the 1960s) many doubters had thought was too difficult. However, the colliding beams at Stanford and KEK have overtaken CESR, and have an asymmetric collision arrangement enabling identification of the B decay vertex. As a result the CLEO collaboration has officially stopped running electron-positron colliding beams at the energy of the Υ(4S). I will present the latest, and almost the last, of these results in what is the latest, and perhaps my last, of my own talks on high energy physics.

Right from the start of CESR and CLEO operation a major focus has been the study of the elements of the Cabibbo-Kobayashi-Miskawa (C-K-M) matrix (figure 4) and its convenient approximation describing the couplings between the various quarks.

$$V \equiv \begin{pmatrix} V_{ud} & V_{us} & V_{ub} \\ V_{cd} & V_{cs} & V_{tb} \\ V_{td} & V_{ts} & V_{tb} \end{pmatrix} \simeq \begin{pmatrix} 1-\frac{1}{2}\lambda^2 & \lambda & A\lambda^3(\rho - i\eta) \\ -\lambda & 1-\frac{1}{2}\lambda^2 & A\lambda^2 \\ A\lambda^3(1-\rho-i\eta) & -A\lambda^2 & 1 \end{pmatrix}$$

Figure 4. C-K-M matrix

Of these couplings V_{ud} is known from β decay of nucleons and nuclei, particularly of I = 0 to I = 0 nuclear transitions, and V_{us} is known from strange particle decays. V_{cb}, and V_{ub} were, until recently when BABAR and BELLE got going, accessible only to CLEO and PETRA. I will also present some more recent results on the process b → s which proceeds by a penguin diagram, and gave us some early information about the top quark. Finally I will conclude with a set of results over the last 2 years from my latest and perhaps the last, of my graduate students in this field. Alexei Ershov, from Serpukhov in Russia, and Daniel Kim, an American of Korean ancestry. Ours was truly a fine international collaboration.

The easiest way to measure V_{cb} is obviously to study the decays of B mesons into leptons. Indeed a study of the inclusive electron and muon spectra was first performed in 1980 (and presented in the High Energy Physics conference of that year at Madison, WI, USA by my former student Professor Edward Thorndike) by looking at the inclusive lepton spectrum assumed to derive mostly from decay into charmed particles ($B \rightarrow X_c\, l\, \nu$) where X_c represents any group of particles containing a charmed particle, either a μ meson or an electron. We first measured the branching ratio into leptons, and assuming that charged and neutral leptons were produced approximately equally found the decay width Γ. The measurements were compared to a theory which assumed that the quarks in the decaying B meson were free (a spectator model). The decay with for this channel is given by:
$\Gamma = (G_f^2|V_{cb}|^2 M_B^5/192\pi^3)0.3689$. This first result had an accuracy of about 25%. Although the theoretical derivation had an error of 25% or so, the experimental error at the time was also large. Now I present the latest result for V_{cb} which we believe to be accurate to 2 ½%. The improvements come from experimental improvements and form using the Heavy Quark Effective Theory (HQET). This leads to a double expansion of $(1/M_B)$ and α_s and terms in $(1/M_B)$, α_s , $(1/M_B)^2$ and α_s^2, and the first term in this expansion is the "simple" equation above. M_B is known to be 5.313 Gev

and α_s is also known. ρ_2 and τ_1 to τ_4 are set to (0 ±0.05 Gev), ρ_1 to [(0.5 Gev)^3 ± (0.5 Gev)^3] and λ_2 to 0.128 ± 0.01 but we need to measure the coefficients λ_1 and Λ.

The study of V_{ub} came (and still comes) from examining the electrons beyond the kinematical end point of the spectrum from decay into charmed mesons. This was first seen reliably by CLEO about 1978. Again the improvement comes from looking at higher terms in the expansion in $(1/M_b)$.

The measurement of $b \to s + \gamma$.

The improved measurement comes primarily from understanding the corrections and measurement of λ_1 and Λ but also a larger sample of B decays. We start by a look at the kinematically simpler reaction $b \to s + \gamma$ where similar HQET corrections appear and use these to measure λ_1 and $\overline{\Lambda}$ (Chen et al. 2001). Since the B mesons are produced almost at rest we expect a "peak" in the γ ray spectrum at about half the B mass, spread by the momentum distribution of the b quarks.

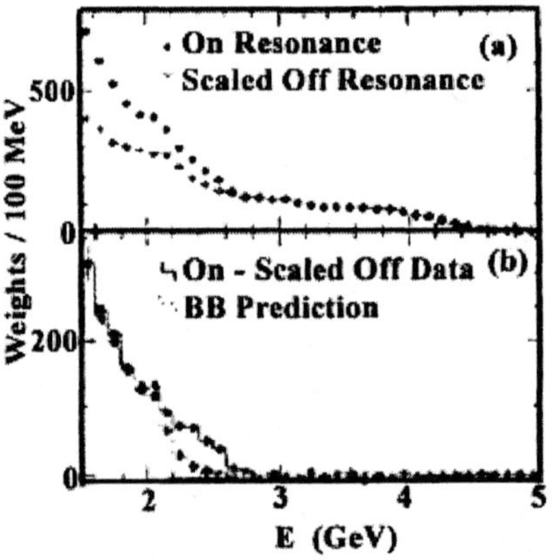

Figure 5. Photon energy spectra (weights per 100 MeV)The upper plot (a) shows the on $\Upsilon(4S)$ and scaled off- resonance spectra. The lower plot (b) show their difference and the spectrum estimated for all other B decay processes.

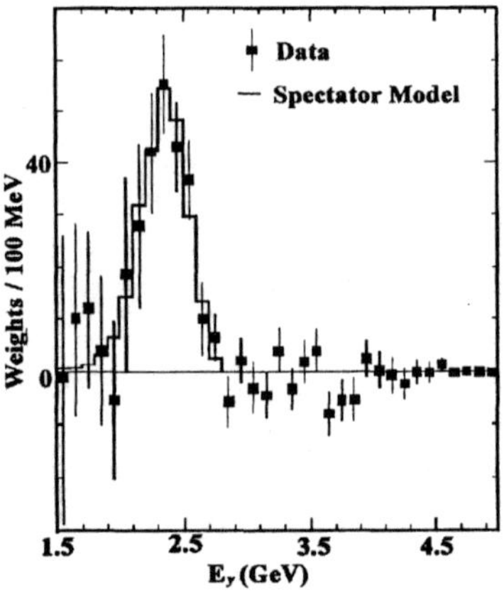

Figure 6. Fully subtracted measurement of γ rays in B production.

In the CLEO detector which shown in figure 2 the gamma rays are detected by a set of Cesium iodide (CsI) crystals in a barrel shaped arrangement. The resolution in energy is good. Most of the gamma rays came from the reaction $e^+ + e^-$ to a continuum rather than the reaction $e^+ + e^-$ to $\Upsilon(4S)$ with subsequent decay of the B particles. Since, the continuum also produces γ rays, the experiment relies on the detail of excluding these γ rays. The continuum was reduced by using an array of "shape cuts". These cuts rely upon the fact the continuum is isotropic, whereas the B particles from the the $\Upsilon(4S)$ are back to back. Several cuts were used and combined into a single variable using a neural net. The (reduced) background was determined and subtracted by runs at an energy just below the $\Upsilon(4S)$ resonance. The full spectrum and the continuance background after application of all cuts are shown in figure 5. We also exclude γ rays that could come from other known decays of the B such as from π^0 and η^0 mesons. These are shown in figure 5. The fully subtracted spectrum of gamma rays is shown in figure 6.

The spectator model (Ali and Greub 1991) suggests that we should find a spectrum of gamma rays near 2.4 Gev and we examine carefully include all gamma rays between 2.0 and 2.8 Gev. We indeed find a peak just below 2.4 Gev. The detail of the spectrum, and the calibration of our detector, are sufficient that, after correction for the excellent resolution, we can measure (for the first time) the first and second moments of the gamma ray spectrum:

$$\langle E\rangle = 2.346 \pm 0.034 \text{ GeV}$$
$$\langle E^2\rangle - \langle E\rangle^2 = 0.0226 \pm 0.0069 \text{ GeV}^2$$

The Heavy Quark Equivalent Theory (HQET), assuming parton-hadron duality, (Bauer, 1998; Ligeti et al., 1999; Falk and Ligeti, 2001) gives the following results for the first moment.

$$\langle E_y\rangle = \frac{M_B}{2}\left[1-.385\frac{\alpha_s}{\pi}-.620\beta_0\left(\frac{\alpha_s}{\pi}\right)^2-\frac{\overline{\Lambda}}{M_B}\left(1-.954\frac{\alpha_s}{\pi}-1.175\beta_0\left(\frac{\alpha_s}{\pi}\right)^2\right)\right.$$
$$\left.-\frac{13\rho_1-33\rho_2}{24M_B^2}-\frac{\tau_1+3\tau_2+\tau_3+3\tau_4}{8M_B^2}-\frac{\rho_2 C_2}{18M_D^2 C_7}+O\left(1/M_B^3\right)\right]$$

The expression for the second moment converges sufficiently slowly in $(1/M_B)^3$ that we do not try to extract parameters from it. Since most of the numbers are known, this gives the OPE parameter:

$$\overline{\Lambda} = (0.35 \pm 0.08 \ \pm 0.10) \text{ GeV}$$

An interesting result aside from the measurement of V_{cb}, is the measurement of Br $(b \to s + \gamma)$ also gives us a new value for the branching ratio:

$$\text{Br}\ (b \to s + \gamma) = (3.21 \pm 0.43 \pm 0.27^{+0.18}_{-0.10}) \times 10^{-4}$$

$$\text{or } (3.21 \pm 0.53) \times 10^{-4}$$

This may be compared to the theoretical result using the "standard model" for which we have two alternates

Br $(b \to s + \gamma) = (3.28 \pm 0.33) \times 10^{-4}$ and $(3.73 \pm 0.30) \times 10^{-4}$ depending upon which quark mass is used.

This result confirms the previous result that we find no room for "new physics" beyond the standard model but, of course, unless we know exactly what "new physics" we are looking for we cannot place a precise limit.

The next step in the argument comes from a careful reexamination of the lepton spectrum in $B \rightarrow X_c \, l \, \nu$ (Cronin-Hennesy et al. (2001). The use of the shape variables to reduce continuum background and the subtraction using data at an energy just below the $\Upsilon(4S)$ resonance is similar to the use in the study for $b \rightarrow s + \gamma$. From this we can determine the moments of the hadronic mass $(M_X)^2$ of X_c. This is aided by our extensive knowledge of some specific components $B \rightarrow D \, l \, \nu$ and $B \rightarrow D^* \, l \, \nu$ decays which are a large component thereof. The resultant spectrum of $M_X{}^2$ is shown in figure 7.

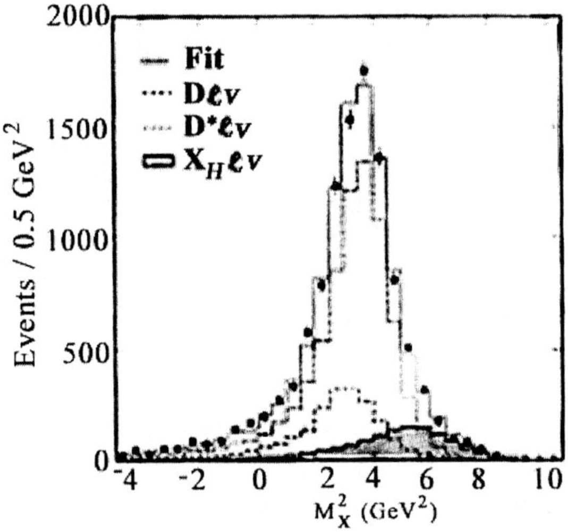

Figure 7. $B \rightarrow X\ell\nu$ spectrum shows contributions from different processes.

The moments become.

$<M_X{}^2 - M_D{}^2> \quad = \quad 0.251 \pm 0.023 \pm 0.062 \;\; GeV^2.$

$<(M_X{}^2 - M_{D^*}{}^2)^2> \quad = \quad 0.639 \pm 0.056 \pm 0.178 \;\; GeV^2$

$<(M_X{}^2 - <M_X{}^2>)^2> \quad = \quad 0.576 \pm 0.048 \pm 0.163 \;\; GeV^2$

The errors on these moments are dominated by systematic errors and we cannot hope for rapid improvements using better statistics. Following advice from theorists we only use the first moment in what follows.

Using HQET, the first moment can then be expressed in terms of several parameters of which $\overline{\Lambda}$ and λ_1 were the important unknowns:

$$\frac{\langle M_X^2 - \overline{M}_D^2 \rangle}{M_B^2} = \left[0.0272\frac{\alpha_s}{\pi} + 0.483\frac{\alpha_s^2}{\pi^2} + 0.207\frac{\overline{\Lambda}}{M_B}\left(1 + 0.43\frac{\alpha_s}{\pi}\right) + 0.193\frac{\overline{\Lambda}^2}{M_B^2} \right.$$

$$+ 1.38\frac{\lambda_1}{M_B^2} + 0.203\frac{\lambda_2}{M_B^2} + 0.19\frac{\overline{\Lambda}^3}{M_B^3} + 3.2\frac{\overline{\Lambda}\lambda_1}{M_B^3} + 1.4\frac{\overline{\Lambda}\lambda_2}{M_B^3} + 4.3\frac{\rho_1}{M_B^3} - 0.56\frac{\rho_2}{M_B^3}$$

$$\left. + 2.0\frac{\tau_1}{M_B^3} + 1.8\frac{\tau_2}{M_B^3} + \frac{\tau_3}{M_B^3} + \frac{\tau_4}{M_B^3} + O\left(1/M_B^4\right) \right]$$

Since we have already derived Λ we can insert the derived value into the formula and derive

$$\lambda_1 = -0.236 \pm 0.071 \pm 0.078 \text{ Gev}^2$$

Then we use:

(a) the measured branching ratio

$$\text{Br}(B \rightarrow X_c\, l\, \nu)^* = 10.39 \pm 0.46\%, \text{ from CLEO},$$

(b) the lifetimes for charged and neutral B mesons

$$1.58 \pm 0.032 \text{ ps and } 1.653 \pm 0.028 \text{ ps respectively},$$

(c) and the ratio of the charged and neutral production in $\Upsilon(4S)$ decays:

$$f_{\text{charged}} / f_{\text{neutral}} = 1.04 \pm 0.08,$$

to derive the semileptonic decay width:

$$\Gamma_{s1} = (0.427 \pm 0.020) \times 10^{-10} \text{ MeV},$$

Combining this with the theoretical value:

$$\Gamma_{sl} = \frac{G_F^2 |V_{cb}|^2 M_b^5}{192\pi^3} 0.3689(B \rightarrow \mu\nu X)$$

$$\left[1 - 1.5\frac{\alpha_s}{\pi} - 1.648\frac{\overline{\Lambda}}{M_B}\left(1 - 0.87\frac{\alpha_s}{\pi}\right) - 0.946\frac{\overline{\Lambda}^2}{M_B^2} \right.$$

$$- 3.185\frac{\lambda_1}{M_B^2} - 7.474\frac{\lambda_2}{M_B^2} - 0.298\frac{\overline{\Lambda}^3}{M_B^3} 3.28\frac{\overline{\Lambda}\lambda_1}{M_B^3} - 6.153\frac{\rho_1}{M_B^3}$$

$$\left. + 7.482\frac{\rho_2}{M_B^3} - 7.4\frac{\tau_1}{M_B^3} + 1.491\frac{\tau_2}{M_B^3} - 10.41\frac{\tau_s}{M_B^3} - 7.482\frac{\tau_4}{M_B^3} + O\left(1/M_B^4\right) \right]$$

we find at last: $|V_{cb}| = (4.04 \pm 0.09 \pm 0.05 \pm 0.08) \times 10^{-2}$
where the errors are (in order) from measuring Γ_{sl}, from measurement of $\overline{\Lambda}$ and λ_1, and from scale uncertainty in α.

The new measurement of V_{ub}
(Bornheim et al. 2002)

The measurement of $|V_{ub}|$ has consisted in looking for leptons of momentum above the cutoff for leptons coming from charm decays - between 2.2 and 2.6 GeV. The number of leptons in this region is quite small and there are two major experimental problems. We have to be sure that the tails of the resolution function of the detector are adequately understood, and that these are not a spill over from the decays due to charm. And we have to be sure that they are not one of the many hadrons masquerading as leptons. It was not until 1988 that we were sure of this excess above 2.3 Gev. Since also we only measure a small fraction of the leptons coming from the charmless decays, (the rest having momenta below 2.2 Gev) we also very dependent on the theory of the semileptonic decay.

The lepton spectrum from B decay after the continuum has been subtracted using the same complex neural net procedure used for determination of $|V_{cb}|$ (Figure 8).

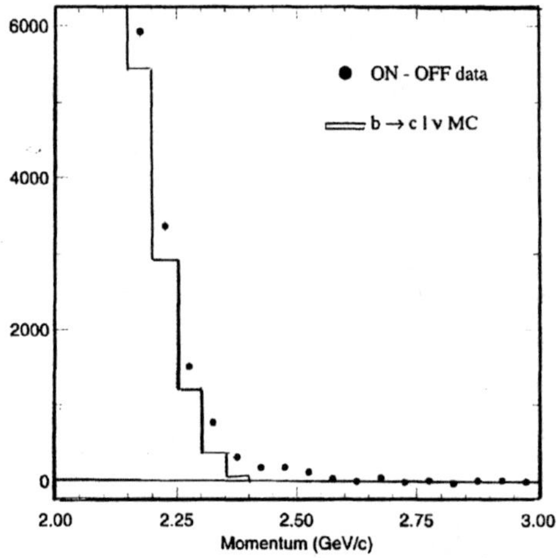

Figure 8. Lepton spectrum from B decays showing the subtraction of b→c events calculated by Monte Carlo from the total spectrum.

As before, the most important part of the measurement is the accurate subtraction of the b → c events. Also important is the correction for the b → u events missed by the cuts. Although ten years ago the measurement was limited by statistics, it is no longer and it is necessary to make the same careful examination as in measurement of $|V_{cb}|$. Again moments are calculated using the results of $b \to s + \gamma$. The result becomes:

$$|V_{ub}| = (4.08 \pm 0.34 \pm 0.44 \pm 0.16 \pm 0.24) \times 10^{-3}$$

where the first two errors are experimental (statistical and systematic) and the last two are theoretical. Combining them we get:

$$|V_{ub}| = (4.08 \pm 0.63) \times 10^{-3}.$$

These two measurements together show, as expected, that the components of the C-K-M matrix get smaller as one proceeds down the matrix. In addition, the ratio of the K-M matrix elements is an interesting quantity in discussing the unitarity triangle.

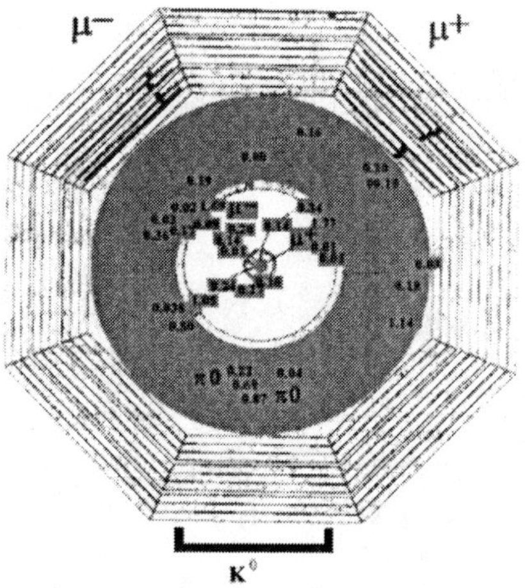

Figure 9. Beam eyeview of CLEO dectector. One of the events, $B \rightarrow (J/\psi)K^0$, and $(J/\psi) \rightarrow \mu^+\mu^-; K^0 \rightarrow \pi^0\pi^0$ was reconstructed from data. The tracks of the two muons are evident in the muon chambers. The four showers at 6 o'clock come from the K^0. The K^0 direction of flight is almost perpendicular to the beam direction.

Beauty decays to charm.

At age 75 the decay of Beauty is especially evident. It is encouraging that the principal decay is to Charm. Over the last few years it has been realized, primarily by Alexei Ershov, that the decays of B mesons to the channel $(J/\psi) + X$ is a powerful channel for examining the decay of B to Charmonium states. This arises because of the narrowness of the (J/ψ). This leads to a very easy identification by the decay into two leptons, and the consequent small background. Of course in this decay there are also γ rays produced from radiative correction effects and radiation of electrons in the beam pipe and other material. These are detected in the CsI crystals and if their directions are along the electron direction, their energy is added to that of the electrons. An event for $[B \rightarrow (J/\psi)K_s^0]$ is shown in figure 9. I will merely summarize the results here because they have mostly been published already.

A) Firstly we measured the decays:
$B \rightarrow \Xi_{c1}$ and $B \rightarrow \Xi_{c2}$ by looking for the decay Ξ_{c1} and Ξ_{c2} into $J/\psi + \gamma$ and the subsequent decay of J/ψ into two leptons. The results were:

$$Br[B \rightarrow \Xi_{c1} + X] = (3.83 \pm 0.31 \pm 0.4) \times 10^{-3}$$
$$Br[B \rightarrow \Xi_{c2} + X] < 1.7 \times 10^{-3}$$

Non relativistic QCD (NRQCD) suggests that the ratio of these two branching ratios should be 5 to 3 if a color octet model dominates, but 0 (as observed) if a color singlet model dominates.

B) We have found several exclusive decay modes.

$$Br[B \rightarrow (J/\psi)K_s^0] = (9.5 \pm 0.8 \pm 0.6) \times 10^{-4}$$
$$Br[B \rightarrow \Xi_{c1} K_s^0] = (3.9 \pm 0.5 \pm 0.4) \times 10^{-4}$$
$$Br[B \rightarrow (J/\psi) K\pi^0] = (2.5 \pm 1.0) \times 10^{-4}$$

These decay modes may be useful in measuring the parameter $\sin 2\beta$ in the K-M matrix but that must be left to our colleagues at BABAR and BELLE.

C. Dr Kim has just finished his study of the polarization and the spectrum of the (J/ψ) in the decays $B \rightarrow (J/\psi) + X$.

It is a long process, but the result for the average value of α, which parametrizes the polarization is:

$$\alpha(J/\psi) = -0.32 \pm 0.07 \pm 0.05$$
$$\alpha(\psi(2S)) = -0.45 \pm 0.21 \pm 0.04$$

Although these are preliminary until the CLEO collaboration agrees, it has been publicly presented in Dr. Kim's PhD thesis. Previous measurements were:

$$\alpha(J/\psi) = -0.92 \pm 0.16 \pm 0.09 \text{ (BELLE)}$$
$$\alpha(J/\psi) = -0.424 \pm 0.023 \text{ (BABAR)}$$

ACKNOWLEDGEMENTS

I am grateful for the good fellowship of the CLEO collaboration over nearly 25 years and acknowledge in particular the most recent Harvard collaborators, Hitoshi Yamamoto, George Brandenburg, Roy Briere, Yong-Sheng Gao, Alexei Ershov and Daniel Kim. Edward Thorndike and collaborators were responsible for the recent work on $b \rightarrow s + \gamma$ and Professor Poling and collaborators on $b \rightarrow u$ decays. The CESR designers and staff, were skillful, responsive and friendly. Of course it was critical that the excellent scientific administrators in DOE and NSF were responsive to our incessant requests for funds.

REFERENCES

Ali A. and Greub C. *Phys Lett. B* 259:182 (1991)
Chen S et al. (CLEO) (2001) "Branching Fraction and Photon Energy Spectrum for $b \rightarrow s\gamma$" *Phys. Rev. Letts*. 87(25); 251807
Ligeti Z., Luke, M., Manohar A.V. and Wise M.B. *Phys Rev D* 60:03409 (1999)\
Bauer C., *Phys Rev D* 57:5611 (1998)
Falk A. and Ligeti Z. Private Communication (2001)
Cronin-Hennesy et al. (CLEO) (2001) "Hadronic Mass Moments in Inclusive Semi-leptonic B Meson Decays" *Phys.Rev. Letts.* 87(25); 25108
Bornheim A. et al. (CLEO) "Improved measurement of $|V_{ub}|$ with Semi-leptonic B Decays" *Phys. Rev. Letts.* 88 in press (2002)
Ershov A.V. (2001) "Beauty Decays to Charmonium" Ph.D. thesis Harvard University.
Kim D. (2001) "Properties of Inclusive $B \rightarrow \psi$ production" Ph.D. thesis Harvard University.

CP Violation in Hyperon and Charged Kaon Decays

Presented by Michael Longo, University of Michigan, for the Hyper*CP* Collaboration

A. Chan[1], Y.C. Chen[1], C. Ho[1], P.K. Teng[1]; W.S. Choong[2], Y. Fu[2], G. Gidal[2], P. Gu[2], T.Jones[2], K.B. Luk[2], B. Turko[2], P. Zyla[2]; C. James[3], J. Volk[3]; J. Felix[4], G. Moreno[4], M. Sosa[4]; R. Burnstein[5], A. Chakravorty[5], D. Kaplan[5], W. Luebke[5], L. Lederman[5], H. Rubin[5], D. Rajaram[5], N. Solomey[5], Y. Torun[5], C. White[5], S. White[5]; N. Leros[6], J. P. Perroud[6]; H.R. Gustafson[7], M. J. Longo[7], F. Lopez[7], H. K. Park[7]; K. Clark[8], M. Jenkins[8]; C. Dukes[9], C. Durandet[9], R. Godang[9], T. Holmstrom[9], M. Huang[9], L.C. Lu[9], K. Nelson[9]

[1]*Institute of Physics, Academia Sinica, Taipei 11529, Taiwan, Republic of China*

[2]*Lawrence Berkeley National Laboratory and University of California, Berkeley, CA 94720*

[3]*Fermi National Accelerator Laboratory, Batavia, IL 60510*

[4]*University of Guanajuato, 37000 Leon, Mexico*

[5]*Illinois Institute of Technology, Chicago, IL 60616*

[6]*Université de Lausanne, CH-1015 Lausanne, Switzerland*

[7]*University of Michigan, Ann Arbor, MI 48109*

[8]*University of South Alabama, Mobile, AL 36688*

[9]*University of Virginia, Charlottesville, VA 22904*

Abstract. The primary purpose of the Hyper*CP* experiment at Fermilab is to test *CP* in hyperon decays by comparing the decay distributions for Ξ^- ("cascade") decays in the decay sequence: $\Xi^- \rightarrow \pi^- + \Lambda^0$, $\Lambda^0 \rightarrow \pi^- + p$, with those for the antiparticle $\overline{\Xi}^+$. In addition, we can test *CP* in charged kaon decays by comparing the slopes of the Dalitz plot for K^+ and K^- decays. We are also looking at rare decay modes of charged kaons and hyperons, particularly those involving muons. In two runs in 1997 and 1999, we collected approx. 500 million charged kaon decays, 2.5 billion Ξ^- and $\overline{\Xi}^+$ decays, and 19 million Ω^- and $\overline{\Omega}^+$ decays. This is the largest sample of fully reconstructed particle decays ever collected.

INTRODUCTION

In a general field theory *CP* violation is due to the presence of a complex phase between different fundamental fields. Within the Standard Model the only way to accommodate *CP* violation in the kaon system is by allowing some elements of the Cabibbo-Kobayashi-Maskawa (CKM) matrix that describes the charged-current weak interactions to be complex. With three generations of quarks, only one complex phase is allowed, which turns out to have a value of order unity.

Any extension of the Standard Model that introduces additional particles allows additional physically observable phases. General classes of such models that have been con-

CP624, *Cosmology and Elementary Particle Physics*, edited by B. N. Kursunoglu et al.

sidered are supersymmetric models, left-right symmetric models, and multi-Higgs models.

The role of *CP* symmetry in the theory of strong interactions is not well understood. The Lagrangian in Quantum Chromodynamics naturally contains a term that violates *CP*, parameterized by $\bar{\theta}$. The limits on neutron and mercury electric dipole moments put a limit $\bar{\theta} \lesssim 10^{-10}$, while the natural value is of order unity. Several mechanisms have been proposed to explain the smallness of $\bar{\theta}$. One of the most popular, Peccei-Quinn symmetry, predicts the existence of an additional pseudoscalar particle, the "axion". Many searches for the axion have been conducted with negative results. (See Sikivie's talk [1].)

Some evidence for *CP* violation beyond the Standard Model comes from cosmology. In Big Bang cosmology, matter/antimatter asymmetry has to be generated dynamically during cooling of the universe ("baryogenesis"). Baryogenesis requires:

- *CP* violation,
- Departure from thermal equilibrium,
- Baryon number violation.

Thus we are only here because of *CP* violation! [Or taking an anthropomorphic point of view, one might say *CP* violation is here because we are here....]

One of the most attractive scenarios of baryogenesis involves the electroweak phase transition (temperatures ~ few hundred GeV). Because the interactions at this energy scale are well known, one can make relatively reliable estimates of the baryon asymmetry. These estimates indicate that if the only source of *CP* violation is in the CKM matrix, the baryon asymmetry is smaller than the observed value by many orders of magnitude. However, extensions of the Standard Model, such as supersymmetry or multi-Higgs theories, which involve additional sources of *CP* violation, can naturally produce a baryon asymmetry of the correct magnitude. Generally, these theories predict possibly observable *CP* violation in systems other than K^0.

The possibility that direct *CP*ß violation is essentially zero and that *CP* violation occurs only in the mixing matrix was referred to as the "superweak" theory. I won't talk about *CP* violation in the K^0 system; there is a good, concise summary in the Particle Data Book [2].

Summarizing the present situation then, information on *CP* violation comes from–

- Neutral kaons: $\varepsilon'/\varepsilon = (1.72 \pm 0.18) \times 10^{-3} \Rightarrow$ Direct *CP* violation
- Electric Dipole Moment(EDM) of neutron and atoms $\Rightarrow$ QCD $\bar{\theta} \lesssim 10^{10}$
- B^0 mesons: World average $sin 2\beta = 0.79 \pm 0.10$(stat. + sys.) $\Rightarrow$ Not all *CP*-violating phases are small. (See Mattison's talk [3].)
- Cosmology: $n_{baryon}/n_\gamma = (5.5 \pm 0.5) \times 10^{-10} \Rightarrow$ Direct *CP* violation
- Other: (*e.g.*, hyperon and charged kaon decays)

THE HYPER*CP* EXPERIMENT

In the rest of this paper, I will only talk about the Hyper*CP* experiment at Fermilab. The primary purpose of this experiment is to test *CP* in hyperon decays by comparing the alpha parameter for Ξ^- ("cascade") decays in the decay sequence: $\Xi^- \rightarrow \pi + \Lambda^0$, $\Lambda^0 \rightarrow \pi^- + p$ with those for the antiparticle $\overline{\Xi}^+$. In addition, we can test *CP* in charged kaon decays by comparing the slopes of the Dalitz plot for K^+ and K^- decays. We are also looking at rare decay modes of charged kaons and hyperons, particularly those involving muons.

A plan view of the Hyper*CP* detector is shown in Fig. 1. Note that the transverse dimensions are exaggerated by a factor of 10. The charged beams are produced by directing an extracted 800 GeV/c proton beam onto a copper target. The channel through the Hyperon Magnet is designed to select a central momentum of 170 GeV/c. To change beam polarity, both the Hyperon and Analyzing magnet fields are reversed, so that the spectrometer presents the same geometry for both positive and negative beams. The beam polarity was changed every few hours. Decays occur in the 13-m long vacuum decay region.

The spectrometer consists of conventional fast wire chambers and scintillation counters used for triggering. The detectors and data acquisition were designed to allow very high data collection rates. Reference [4] gives more information on the detector.

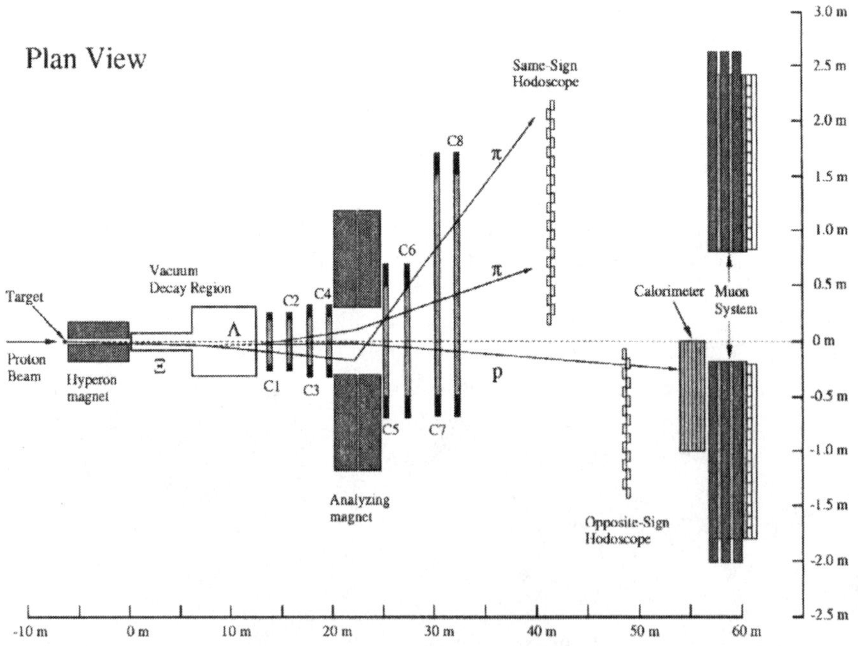

FIGURE 1. The Hyper*CP* detector

TABLE 1. Data Sample from the 1997 and 1999 Runs

	1997 Run	1999 Run
Number of Tapes	8980	20421
Data Volume	38 TB	82 TB

Projected number of reconstructed events:

Ξ^-	2 x 10^9	K^-	0.16 x 10^9	Ω^-	14 x 10^6
$\overline{\Xi}^+$	0.5 x 10^9	K^+	0.39 x 10^9	$\overline{\Omega}^+$	4.9 x 10^6

The data sample taken in two runs in 1997 and 1999 is summarized in Table 1. The total data set comprises about 100 terabytes of information. To put this in perspective, it is about an order of magnitude more information than that contained in the Library of Congress. This is the largest sample of fully reconstructed particle decays ever collected. With 2.5 x 10^9 Ξ and $\overline{\Xi}^+$ decays we anticipate an eventual precision of $\delta A_{\Xi\Lambda} \approx 2 \times 10^{-4}$ in the Ξ decay *CP* test.

CP TEST IN Ξ HYPERON DECAYS

In nonleptonic ΔS=1 decays of a spin-$\frac{1}{2}$ strange baryon (*e.g.*, Λ^0 or Ξ) into another spin-$\frac{1}{2}$ baryon and pion, angular momentum conservation allows only two final-state amplitudes. These are a parity violating *S*-wave and a parity conserving *P*-wave. The decay parameters α, β. and γ can be written in terms of the *S*- and *P*-wave amplitudes. The angular distribution of the final-state baryon in the rest frame of a parent baryon with polarization *P* depends on α [2] ,

$$\frac{dn}{d\Omega} = \frac{1}{4\pi}(1 + \alpha P \cos\theta) \qquad (1)$$

If *CP* holds, $\alpha = -\overline{\alpha}$ where $\overline{\alpha}$ is that for the antiparticle. We can define a parameter $A \equiv (\alpha + \overline{\alpha})/(\alpha - \overline{\alpha})$ as a measure of *CP* violation. From Eq, 1, to test *CP* it is necessary to know the polarization of the particle and antiparticle samples. The Hyper*CP* technique is to use polarized Λ ($\overline{\Lambda}$) that are produced from decays of unpolarized Ξ ($\overline{\Xi}$), so that the Λ's ($\overline{\Lambda}$'s) have a known polarization. Thus we compare the decay distribution of protons in the decay sequence $\Xi^- \to \pi^- + \Lambda^0$, $\Lambda^0 \to \pi^- + p$, with that for antiprotons from $\overline{\Xi}^+$. In effect, we measure

$$A_{\Lambda\Xi} = \frac{\alpha_\Lambda \alpha_\Xi - \alpha_{\overline{\Lambda}} \alpha_{\overline{\Xi}}}{\alpha_\Lambda \alpha_\Xi + \alpha_{\overline{\Lambda}} \alpha_{\overline{\Xi}}} \approx A_\Lambda + A_\Xi \qquad (2)$$

Theoretical predictions for A_Λ and A_Ξ vary over a wide range, as shown in Table 2.

TABLE 2. Predictions for A_Λ and A_Ξ in various models [5],[6]

	A_Λ	A_Ξ
Superweak	0	0
CKM ("Standard model")	$(-0.6 \text{ to } 6.8) \times 10^{-5}$	$(-0.1 \text{ to } 1) \times 10^{-5}$
2-Higgs	$\approx -2 \times 10^{-5}$	$\approx -3 \times 10^{-4}$
Left-Right	$\lesssim 5 \times 10^{-4}$	$\lesssim 10^{-4}$
Supersymmetric	$\lesssim 1.9 \times 10^{-3}$	$\lesssim 10^{-4}$

Fortunately most models predict that A_Λ and A_Ξ will have the same sign. Typically $|A_\Lambda|$ is predicted to be considerably larger than $|A_\Xi|$. The expected precision from Hyper*CP* when all the data are analyzed is $\approx 2 \times 10^{-4}$.

In Hyper*CP* the switch between Ξ^- and $\overline{\Xi}^+$ was made merely by reversing the polarity of the Hyperon and Analyzing magnets (Fig. 1). Thus no changes in the geometry were required. This greatly reduced potential systematic effects in the *CP* test. Furthermore, geometric biases are greatly reduced because the events are analyzed in the Λ helicity frame, which changes for each event. Thus, for example, the effect of a dead wire in a chamber would be washed out in the angular distributions. A thinner production target was used for producing the $\overline{\Xi}^+$ so that the total beam intensity through the detector was approximately the same for positive and negative beams, ≈ 13 MHz.

The main systematics come from the fact that the experiment itself is, of necessity, not charge symmetric. The chambers are made of protons and neutrons, and the beam is produced by protons on a copper target. The difference between the absorption of the decay products in the chambers is very small and can be corrected for. The largest potential source of systematic error is due to the slightly different momentum spectra for the Ξ^- and $\overline{\Xi}^+$. This causes small differences in the acceptance *vs.* proton(antiproton) angle which can cause non-negligible differences in the measured α's if corrections are not made. In the preliminary results discussed below, an equalization procedure was used to remove the systematic effects due to the difference between the Ξ^- and $\overline{\Xi}^+$ spectra and related differences between the beam distributions in the decay region [7].

Also, the backgrounds for Ξ^- and $\overline{\Xi}^+$, Λ^0 and $\overline{\Lambda}^0$ may differ somewhat, so these have to be handled carefully. Data taken with polarized Ξ's provide important checks for systematic biases. These also serve as a verification that we are sensitive to small asymmetries.

Figure 2 shows preliminary results based on a sample with $\approx 1 \times 10^6$ $\overline{\Xi}^+$ and 11×10^6 Ξ^- decays, which is $\approx 0.5\%$ of the total data sample. This gives

$$A_{\Xi\Lambda} = (-1.6 \pm 1.3 \pm 1.6) \times 10^{-3}.$$

Even with this small sample, this is already about a factor of 5 improvement over the best previous result for $A_{\Xi\Lambda}$ from Fermilab E756 [8].

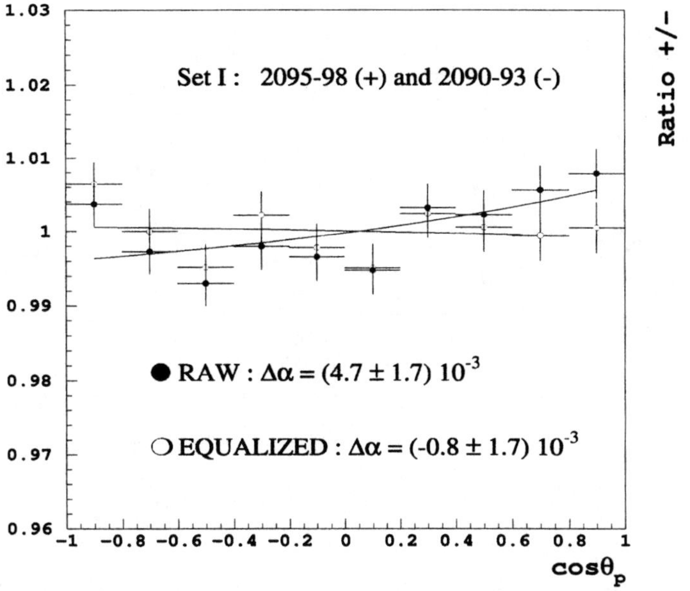

FIGURE 2. Ratio of *cos* θ_p distributions for $\overline{\Xi}^+$ and Ξ^- decays based on about 0.5% of the total data sample. The solid points are before the equalization procedure described above, and the open points after equalization.

CP VIOLATION IN $K^{\pm} \rightarrow \pi^{\pm}\pi^{+}\pi^{-}$

The decay amplitudes for $K^{\pm} \rightarrow \pi^{\pm}\pi^{+}\pi^{-}$ depend only on two independent kinematical variables, *X* and *Y*, which are related to the momenta of the pions in the rest frame of the kaon. The invariant matrix element can be parametrized as $|M^2| \propto 1 + gY + \ldots$. If $g_{K^-} \neq g_{K^+}$, then *CP* symmetry is broken. We can define a parameter to measure possible *CP* violation as $\delta g \equiv (g - \overline{g})/(g + \overline{g})$. Theoretical predictions for δg range from $\sim 10^{-6}$ [9] to $\sim 10^{-3}$ [10].

Systematics in this *CP* test are more of a problem than in the hyperon decay test because any difference in acceptance for K^+ and K^- can directly affect the *Y*-distributions. This requires very careful simulations using the observed K^+ and K^- spectra and beam distributions. This is challenging because $\sim 10^9$ kaon decays must be simulated. The *X* dependence of the distributions serve as a check since they are even under *CP*.

Figure 3 shows a preliminary study of the *CP* violation test in $K^{\pm} \rightarrow \pi^{\pm}\pi^{+}\pi^{-}$ decays [11]. In this sample, which includes 41.8×10^6 K^+ and 12.4×10^6 K^- or $\approx$10% of the total, we compare the slope *g* for runs with positive and negative kaons. The lower figure shows a histogram of the values for the runs in the upper figure. The few runs that show a large scatter are short runs with large statistical errors. Overall for this sample,

$$\Delta g/2g = (2.2 \pm 1.5) \times 10^{-3},$$

in good agreement with *CP*.

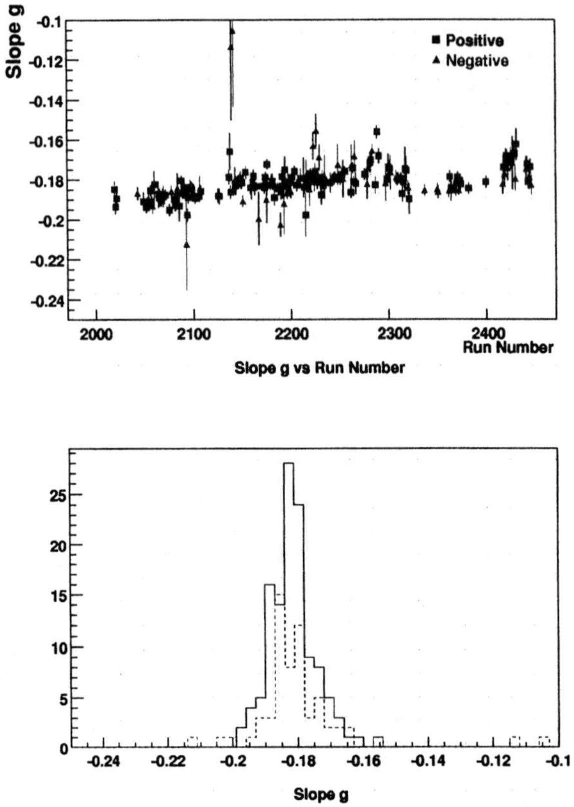

FIGURE 3. Preliminary study of $K^{\pm}$ *CP* test based on $\approx 54 \times 10^6$ $K^{\pm}$ events (10% of total sample).

SUMMARY

• *CP* violation is one of the most important topics in particle physics and cosmology. It is now clear that direct *CP* violation (*i.e.*, "new physics") occurs. Its origin is still unknown.

• Hyper*CP* is providing a rich trove of data on hyperon and charged kaon decays including such timely topics as *CP* violation and lepton flavor violating decays. It is the only dedicated experiment to look for *CP* violation in hyperon decays. Based on preliminary results from small samples of the total data, we see no evidence for *CP* violation in either hyperon or charged kaon decays.

ACKNOWLEDGMENTS

The authors are indebted to the staffs of Fermilab and the participating institutions for their vital contributions. This work was supported by the U.S. Department of Energy and the National Science Council of Taiwan, R.O.C.

REFERENCES

1. Sikivie, P, "Status of the Axion", this conference.
2. Groom, D.E. *et al.*, European Physical Journal **C15**, 1 (2000).
3. Mattison, T., "Recent Results from BABAR for *CP* Violation in the *B* System", this conference.
4. Park, H.K. *et al.*, to be published in Physical Review Letters.
5. X.-G. He, "*CP* Violation in Hyperon Decays", KAON 2001 conference, hep-ph/0108164.
6. Pakvasa, S, 3rd International Conference on *B* Physics and *CP* Violation (BCONF99), *Taipei 1999, B physics and CP violation*, pp. 423-433.
7. Leros, N., "Recherche de la violation *CP* dans les désintégrations d'hypérons $\Xi^-/\overline{\Xi}^+$ et $\Lambda/\overline{\Lambda}$", doctoral thesis, Université de Lausanne, 2001.
8. K.B. Luk *et al.*, Phys. Rev. Lett. 85, 4860-4863 (2000).
9. Isidori, G., Maiani,L., and Pugliese,A., Nucl. Phys. **B**, 522 (1992).
10. Belkov, A. *et al.*, Phys. Lett., **B300**, 283 (1993).
11. Choong, W. S., "A Search for Direct *CP* Violation in $K^{\pm} \rightarrow \pi^{\pm}\pi^{\pm}\pi^{\mp}$ Decays", doctoral thesis, University of California, 2000.

Results from SNO

Peter Skensved

Queen's University, Kingston Ontario, Canada
For the SNO Collaboration[1]

Abstract. Neutrinos from the decay of ^{8}B in the core of the Sun have been observed in the SNO detector via the Charged Current (CC) and Elastic Scattering (ES) reactions. Comparison with the ES result from Super-Kamiokande (SK) provides very strong evidence that the flux detected on Earth has a non-electron type component.

INTRODUCTION

The Sudbury Neutrino Observatory (SNO) is a heavy water Čerenkov detector located at a depth of 6800 feet in a 22 m diameter by 34 m high cavity in the INCO Ltd. Creighton mine [2]. The cavity holds approximately 7000 tonnes of ultra pure H_2O and in it is hung a 17 m diameter geodesic structure which holds 9456 20 cm diameter Hamamatsu photomultiplier tubes (PMTs) equipped with light concentrators. In the center is a 5 cm thick, 12 m diameter acrylic sphere which contains 1000 tonnes of ultra-pure D_2O. The effective photocathode coverage is approximately 55%.

The primary motivation for the experiment is the long standing Solar Neutrino Problem [3, 4, 5, 6, 7, 8] - the fact that over the last 30 years other Solar neutrino experiments have consistently detected fewer electron neutrinos than predicted by all standard models of the Sun. The basic fusion reactions in the core of the Sun convert Hydrogen into Helium and the only byproduct to escape is electron neutrinos. The bulk of these are very low in energy and their flux is constrained in the models by other observables such as the total luminosity. A less well determined, much weaker but more energetic flux comes from the decay of ^{8}B into two α's. This is the flux to which SNO is sensitive.

SIGNALS IN SNO

Many explanations for the observed deficit of ν_e's have been proposed over the years. Among the commonly accepted ones are that either the Standard Solar Model (SSM) [9, 10] is wrong or that the ν_e's change flavour between the core of the Sun and the detectors. Depending on the mixing parameters these changes could manifest themselves in the form of spectral distortions, changes in the overall ν_e flux or even in different fluxes during day and night. SNO is in a unique position to distinguish between many of these scenarios in that we can measure not only the total flux and the shape and of the ν_e spectrum but also the total flux of *all* active neutrinos thus avoiding any reliance on models. This is due to the unique properties of our detection medium, the D_2O . In

CP624, *Cosmology and Elementary Particle Physics,* edited by B. N. Kursunoglu et al.

addition to the ordinary neutrino elastic scattering observed in other water Čerenkov detectors[4] :

$$\nu + e^- \rightarrow \nu + e^- \quad (ES)$$

SNO also sees charged and Neutral Current reactions on the deuteron :

$$\nu_e + d \rightarrow p + p + e^- \quad (CC)$$

$$\nu_x + d \rightarrow p + n + \nu_x' \quad (NC)$$

The CC reaction has weak directional sensitivity (1 - 1/3 $cos\theta$), is sensitive to ν_e only and has a threshold of 1.44 MeV. Elastic scattering on the other hand has a strong directional sensitivity (forward peaked), has some sensitivity to ν_μ and ν_τ (about 14%) but is a very poor measure of the neutrino energy. Finally, the NC reaction has equal sensitivity to *all* active flavours, no directional sensitivity and a threshold of 2.22 MeV. It is analogous to photodisintegration of the deuteron which in fact is the major source of background to our NC measurement. In SNO we have three different methods of detecting NC corresponding to 3 different phases of the experiment : pure D_2O, D_2O with 0.2% NaCl and ^{3}He counters. In the first two we detect neutron by detecting γ-rays :

$$n + d \rightarrow t + \gamma(6.25MeV) \qquad \varepsilon \simeq 25\%$$

$$n +^{35} Cl \rightarrow^{36} Cl + \Sigma E_\gamma = 8.6MeV \quad \varepsilon \simeq 86\%$$

and in the third we use discrete ^3He proportional counters deployed on a 1 meter grid throughout the D_2O to give us a unique 'neutron' signal with approximately 45% efficiency.

As mentioned earlier SNO is a Čerenkov detector. It is sensitive to light emitted from any relativistic electron (or positron) above 260 keV kinetic energy whether they are created from the neutrino interactions above, from radioactivity or by Compton interactions from γ-rays. The number of hit phototubes is a good measure of the electron energy, the individual photon arrival times are used to reconstruct the event vertex and the direction of the Čerenkov cone is strongly correlated with the initial electron direction which for the ES reaction points back toward the Sun.

Figure 1 shows a Monte Carlo simulation of the signals and backgrounds. The energy scale is approximately 9 hits per MeV.

CALIBRATION

SNO is calibrated optically with a pulsed Nitrogen laser. The fundamental wavelength of the light is 337 nm and dyecells enable us to produce light at 365, 386, 421, 500 and 620 nm which covers from the UV cutoff in the acrylic to near the upper end of the sensitivity of the PMTs. The laserlight is injected into a fiber and diffused in a 10 cm diameter ball which can be moved in two perpendicular planes within the acrylic vessel using a remotely controlled manipulator system. By using many source positions at all wavelengths and different laserball orientations it is possible to extract

Signals in SNO

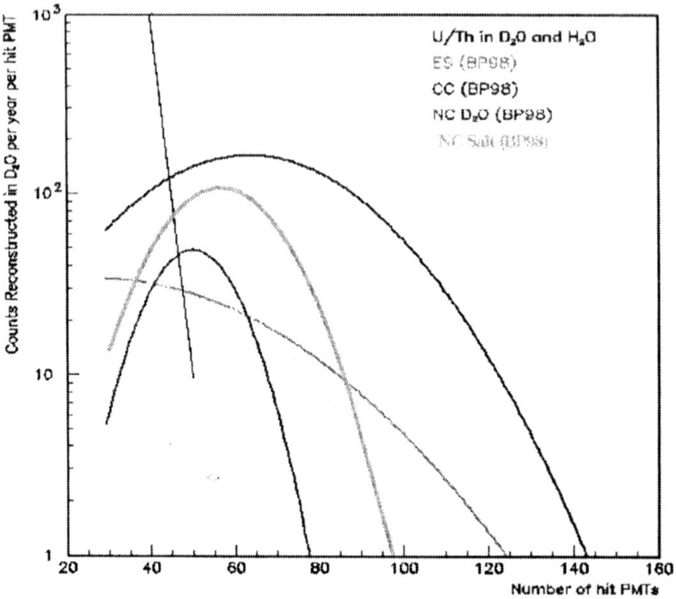

FIGURE 1. Monte Carlo simulation of signals and backgrounds in SNO

the fundamental optical parameters such as the total attenuation length in the D_2O, the combined attenuation length in the acrylic and the H_2O and the PMT/Concentrator angular distribution without apriori knowledge of the PMT efficiencies or the laserball angular distribution. This allows us to calculate the energy response of SNO to within a single absolute scale factor which we determine from the ^{16}N source. Fast neutrons from a D-T generator are used to make a radioactive ^{16}N gas which is piped through a capillary into a small decay volume hung from the source manipulator system. A scintillator coupled to the decay volume (triggering on the β's) is used for tagging. Other sources such as a ^{8}Li β source, a source of 19.8 MeV γ's from ^{3}H(p,γ), Th and U sources and γ's from captured neutrons (from ^{252}Cf) have been used to verify that our modeling of the detector is correct. A composite picture showing the agreement between the various sources and Monte Carlo calculations is shown in figure 2.

The energy resolution is well described by a Gaussian with a standard deviation σ_E which varies as $-0.4620 + 0.5470 \times \sqrt{E} + 0.008722 \times \mathrm{E}$. For reference this corresponds to 13.5 % at 10 MeV. The position resolution is extracted from ^{16}N data using Monte Carlo to correct for the e^--γ differences. It is approximately 16 cm at 5.5 MeV e^- and varies rather slowly with energy. The angular resolution is determined by looking at γ-rays which convert far from the ^{16} source to be 26.7°. Finally, the PMT timing resolution

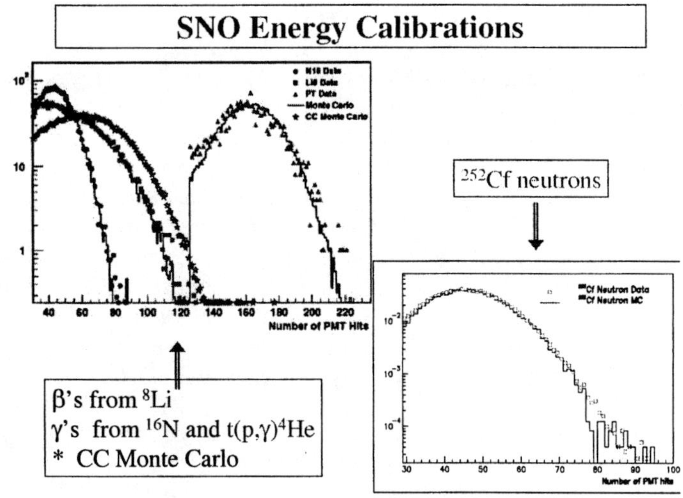

FIGURE 2. Source data and Monte Carlo simulations

averaged over the entire detector is 1.53 ns again determined with the ^{16}N source. This includes (very) small components due to reconstruction and walk corrections. The timing measured with the laserball is somewhat worse (1.8 ns) due to dispersion in the 30 meter long fiber.

BACKGROUNDS

The success of SNO depends critically on being able to identify and remove any backgrounds from the neutrino signals. One category of backgrounds is instrumental effects such as electrical pickup in neighbouring channels and electrical micro discharges inside the PMTs. These discharges cause a large pulse in the originating PMT and a characteristic light pattern in the PMTs on the other side of the detector. Such events (referred to as 'flashers') are easily distinguished from real Čerenkov events based on their unique charge, time and angular distribution. They occur at a rate of the order of one per day per PMT. Blasting in the mine can induce these.

Another source of backgrounds is due to β's and γ's from the decay chains of ^{238}U and ^{232}Th. These isotopes occur at levels from several times 10^{-6} g/g in the rock to around 10^{-12} g/g in the acrylic vessel to a few times 10^{-15} g/g n the D_2O. In the D_2O and H_2O the U chain is also fed by the 3.8 day ^{222}Rn from the cover gas. Only the decays at the bottom of the chains (decays from ^{208}Tl and ^{214}Bi) are energetic enough to be of any concern. A finite hardware threshold and the Čerenkov threshold helps us here. Tails from β-γ's originating in the acrylic vessel and the H_2O extend in to the outer regions of the D_2O. Since they are concentrated near the edge of the D_2O radial cuts can be used to remove or reduce them. β-γ's from the phototube support structure (PSUP) are attenuated sufficiently well by the H_2O so as not to be a problem. High energy γ's

($E_\gamma > 4$ MeV) from the cavity and PSUP are effectively shielded by the H_2O as well and a limit on the leakage has been measured with the ^{16}N source to be less than 0.8%.

The most serious background in SNO arises from the fact alluded to above that any γ-ray with an energy above 2.2 MeV can photo-disintegrate the deuteron thus making a free neutron which when captured leads to events in the signal region. They are indistinguishable from NC events. There are three ways to deal with these- reduce the Th/U levels, impose an energy threshold or determine the number independently and subtract them. We use all three. The water is recirculated on a regular basis and assayed by passing it through HTiO and MnO_x filters which picks up Ra, Th and Pb. These filters are then removed and analyzed by doing either extraction and elution or Rn emanation followed by counting. Figure 3 shows the results of the assays. The D_2O is also monitored continuously by looking at the rate of low energy events in the center. This rate is totally dominated by radioactivity from the within D_2O.

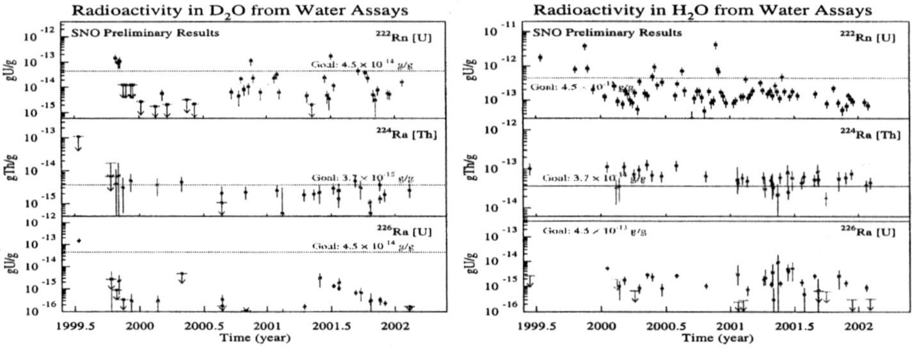

FIGURE 3. Water Assays

DATA REDUCTION

Data taken for a total of 241 live days is reported on here. This is the same dataset reported on in [12]. Table 1 shows the reduction in number of events as the different cuts are applied. The total number of triggers includes a 5 Hz component of induced triggers used for livetime and PMT dark rate checks. It also includes triggers where the summed PMT charge exceeds a preset value. Such triggers are mainly due to flashers. The rest are mostly triggers where more than 16 PMTs fire (N_{hit} triggers). In line 4 we apply the instrumental cuts referred to in the section above (flasher cut etc.) and in line 5 we apply a time cut after each muon in order to remove (mostly) neutrons from spallation products. The 'high level cuts' include cuts such as the ratio of in time to out of time light and mean angle between hit PMTs etc. Finally we apply a radial cut at 550 cm and an energy cut of 6.75 MeV kinetic energy. The effect of the high level cuts were tested by applying them to all the different source calibration data. The various cuts lead

TABLE 1. Data reduction steps.

Analysis step	Number of events
Total event triggers	355 320 964
N_{hit} triggers	143 756 178
$N_{hit} \geq 30$ triggers	6 372 899
Instrumental background cuts	1 842 491
Muon followers	1 809 979
High level cuts	923 717
Fiducial volume cut	17 884
Threshold cut	1 169
Total events	1 169

to an estimated loss of $1.4^{+0.7}_{-0.6}\%$ in neutrino signal with a residual contamination from instrumental backgrounds of less than 0.2 % .

Monte Carlo calculations were used to confirm that the contributions from external γ-rays and neutrons are negligible with the energy and radial cuts imposed above.

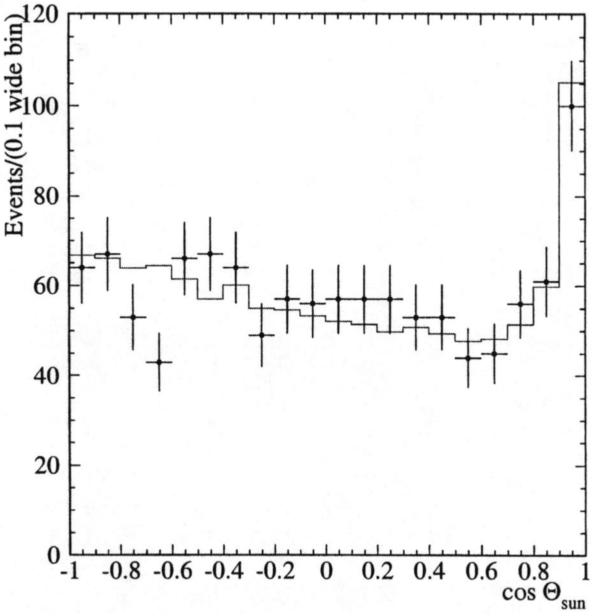

FIGURE 4. $\cos\theta_{\odot}$

As a first step one might ask whether SNO indeed observes neutrinos from the Sun. The answer is in figure 4 which shows a histogram of the cosine of the angle of the events with respect to the Sun. The solid line is a Monte Carlo calculation for an undistorted ^{8}B spectrum. The shape is as expected - a flat background, a $\simeq$ 1 - 1/3 $cos\theta$ term and a

TABLE 2. Extracted Fluxes.

Component	Number of events	Flux Φ (10^{-6} cm^{-2} sec^{-1}	Stat. (%)	Syst. (%)
CC	975.4	1.75	± 4.1	$^{+7.0}_{6.2}$
ES	106.1	2.39	± 14	$^{+6.8}_{-5.7}$
Neutrons	87.5		± 28	

forward-peaked ES signal. With a set of fairly simple and not very stringent cuts on the data SNO indeed does see the Sun. SNO is a very low background experiment.

The extraction of the different signal components was done using probability density functions (PDFs) in T_e, $cos\theta_{\odot}$, and $(R/R_{AV})^3$ generated by Monte Carlo assuming a standard undistorted ^{8}B neutrino spectrum (without *hep* neutrinos) and by employing the extended maximum likelihood technique. There are three free parameters : the CC flux, the ES flux and the number of neutrons. The extended ML technique ensures that the total number of events is constrained. The radial projection of the decomposition is shown in figure 5 and the extracted fluxes together with percentage statistical and systematic uncertainties are summarized in table2.

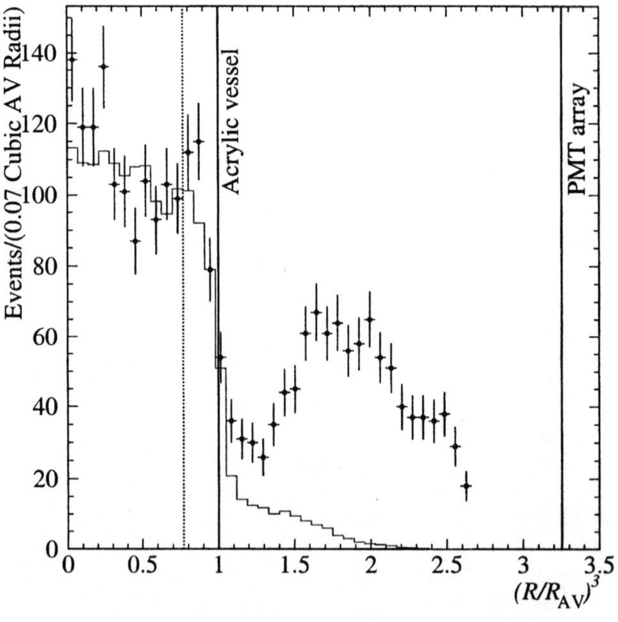

FIGURE 5. Radial decomposition of PDFs

The systematic uncertainties are largely dominated by the contribution from the determination of the energy scale ($^{+6.1\%}_{-5.2\%}$ for CC and $^{5.4\%}_{-3.5\%}$ for ES) with the next largest from the limit set on the external high energy γ background ($^{+0.0\%}_{-1.8\%}$ and $^{+0.0\%}_{-1.9\%}$

respectively). Not included is the cross section uncertainty ($\pm 3\%$ for CC and $\pm 0.5\%$ for ES) and the uncertainty in the Solar model ($^{+20\%}_{-16\%}$). No correction was made for radiative effects.

The ES reaction has a 14% sensitivity to non-electron type neutrinos whereas the CC reaction is sensitive exclusively to ν_e's. We can use this difference with our measured fluxes to perform the null hypothesis test that neutrinos from the Sun do not change flavour between the Sun and our detector A non-zero value of Φ^{ES} - Φ^{CC} would indicate a non-electron type component in the flux. For SNO this difference is at the 1.6σ level which on its own is not particularly strong evidence. However, given that our determination of $\Phi^{ES}(\nu_e)$ is in good agreement with the one measured by Super-Kamiokande of $2.32 \pm 0.03(\text{stat})^{+0.06}_{-0.07}(\text{sys}) \times 10^{-6}$ cm^{-2} sec^{-1}[4] we can combine our data with the SK data and calculate Φ^{ES}_{SK} - Φ^{CC}_{SNO}. This difference is $0.57 \pm 0.17 \times 10^{-6}$ cm^{-2} sec^{-1} or 3.3σ. The probability for this to be a statistical fluctuation is only 0.04% . This is very strong evidence that the Solar neutrino flux as observed on Earth contains an active non-electron type component and it is the first such direct evidence. It should be noted that this conclusion is totally independent of Solar models. The only assumption is that the ^{8}B spectrum is undistorted.

Oscillation exclusively to sterile neutrinos is also strongly disfavoured since the CC flux above 6.75 MeV should be consistent with the ES flux above 8.5 MeV in Super-Kamiokande [11]. The difference is $0.53 \pm 0.17 \times 10^{-6}$ cm^{-2} sec^{-1} or 3.1σ and again the probability for this to be a statistical fluctuation is very small at 0.13%.

One can summarize the SNO and Super-Kamiokande data in a $\Phi(\nu_\mu \tau)$ versus $\Phi(\nu_e)$ plot where the shaded regions are the combined statistical and systematic uncertainties (figure 6). The ellipses are the joint probability contours for 1, 2 and 3σ respectively. The best fit is for $\Phi(\nu_{\mu\tau}) = 3.69 \pm 1.13 \times 10^{-6}$ cm^{-2} sec^{-1} which leads to a total flux of $5.44 \pm 0.99 \times 10^{-6}$ cm^{-2} sec^{-1} in very good agreement with the SSM prediction of $5.05 \times 10-6$ cm^{-2} sec^{-1}[9]. The diagonal lines show the total active $\pm 1\sigma$ flux bands based on either SNO or the combined SNO plus SK data.

CONCLUSIONS

SNO has detected neutrinos from the decay of ^{8}B in the core of the Sun. We have measured the flux of ν_e via both the CC and the ES reactions. Comparison with the ES result from Super-Kamiokande provides strong evidence (at the 3.3σ level) for an active non-electron type component in the observed flux. Oscillation to sterile neutrinos are disfavoured at the 3.1σ level and the total flux of active ^{8}B neutrinos is in good agreement with the SSM.

ACKNOWLEDGMENTS

This research was supported by the Natural Sciences and Engineering Council of Canada, Industry Canada, National Research Council of Canada, Northern Ontario Heritage Fund Corporation and the Province of Ontario, the United States Department of

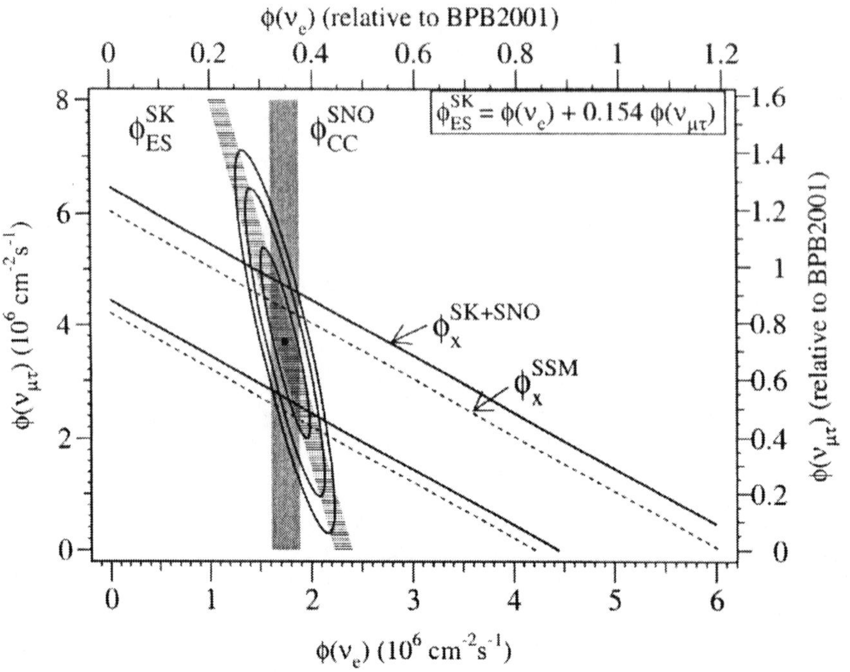

FIGURE 6. SNO / Super-Kamiokande combined data

Energy, and in the United Kingdom by the Science and Engineering Council and the Particle Physics and Astronomy Research Council. Further support was provided by INCO, Ltd, Atomic Energy of Canada Limited (AECL), Agra-Monenco, Canatom, Canadian Microelectronics Corporation, AT&T Microelectronics, Northern Telecom and British Nuclear Fuels,Ltd. The heavy water was loaned by AECL with the cooperation of Ontario Power Corporation.

REFERENCES

1. The SNO collaboration includes :
B. Sur, **Atomic Energy of Canada Limited, Chalk River, ON, Canada**;
J. Heise, R.L. Helmer, R.J. Komar, T. Kutter, C.W. Nally, C.E. Waltham, **University of British Columbia, Vancouver, BC, Canada**;
J. Boger, R.L. Hahn, J.K. Rowley, M. Yeh, **Brookhaven National Laboratory, Upton, NY**;
F. Dalnoki-Veress, D.R. Grant, C.K. Hargrove, I. Levine, K. McFarlane, A.J. Noble, D. Sinclair,N. Starinsky, **Carleton University, Ottawa, ON, Canada**;
T. Andersen, M.C. Chon, P. Jagam, J. Law, I.T. Lawson, J.J. Simpson, **University of Guelph, Guelph, ON, Canada**;
J. Farine, E.D. Hallman, S. Luoma, M.H. Schwendener, R. Tafirout, C.J. Virtue, **Laurentian University, Sudbury, ON, Canada**;

Y.D. Chan, X. Chen, M.C.P. Isaac, K.T. Lesko, A.D. Marino, E.B. Norman, C.E. Okada, A.W.P. Poon A.R. Smith, R.G. Stokstad, **Lawrence Berkeley National Laboratory, Berkeley, CA**;
M.G. Boulay, T.J. Bowles, S.J. Brice, M. Dragowsky, M.M. Fowler, A. Hamer, A. Hime, J.B. Wilhelmy, J.M. Wouters, **Los Alamos National Laboratory, Los Alamos, NM**;
S. Biller, M.G. Bowler, J. Cameron, B.T. Cleveland, X. Dai, G. Doucas, J. Dunmore, H. Fergami, A.P. Ferraris, K. Frame, H. Heron, N.A. Jelley, S. Majerus, N. McCauley, G. MacGregor, N. Tagg, D.L. Wark, N. West, J. Wilson, **Oxford University, Oxford, United Kingdom**;
E.W. Beier, D.F. Cowen, W. Frati, W.J. Heintzelman, P.T. Keener, J.R. Klein, C.C.M. Kyba, D.S. McDonald, M.S. Neubauer, S.M. Oser, V.L. Rusu, R. Van Berg, R.G. Van de Water, P. Wittich, **University of Pennsylvania, Philadelphia, PA**;
E. Bonvin, M. Chen, F.A. Duncan, E.D. Earle, H.C. Evans, G.T. Ewan, R.J. Ford, A.L. Hallin, P.J. Harvey, J.D. Hepburn, C. Jillings, H.W. Lee, J.R. Leslie, H.B. Mak, A.B. McDonald, B.A. Moffat, T.J. Radcliffe, B.C. Robertson, P. Skensved, **Queen's University, Kingston, ON, Canada**;
Q.R. Ahmad, T.V. Bullard, T.H. Burritt, G.A. Cox, P.J. Doe, C.A. Duba, S.R. Elliott, J.V. Germani, A.A. Hamian, R. Hazama, K.M. Heeger, M. Howe, J. Orrell, R.G.H. Robertson, K.K. Schaffer, M.W.E. Smith, J.F. Wilkerson, **University of Washington, Seattle, WA**;

2. The SNO Collaboration, Nucl. Instr. and Meth. **A449**, 172 (2000).
3. B.T. Cleveland *et al.*, Astrophys. J. **496**, 505 (1998).
4. K.S. Hirata *et al.*, Phys. Rev. Lett. **65**, 1297 (1990); K.S. Hirata *et al.*, Phys. Rev. D **44**, 2241 (1991), **45** 2170E (1992); Y. Fukuda *et al.*, Phys. Rev. Lett. **77**, 1683 (1996).
5. J.N. Abdurashitov *et al.*, Phys. Rev. C **60**, 055801, (1999).
6. W. Hampel *et al.*, Phys. Lett. B **447**, 127 (1999).
7. S. Fukuda *et al.*, Phys. Rev. Lett. **86**, 5651 (2001).
8. M. Altmann *et al.*, Phys. Lett. B **490**, 16 (2000).
9. J.N. Bahcall, M. H. Pinsonneault, and S. Basu, astro-ph/0010346 v2. The reference ^{8}B neutrino flux is 5.05×10^6 cm^{-2}s^{-1}.
10. A.S. Brun, S. Turck-Chièze, and J.P. Zahn, Astrophys. J. **525**, 1032 (1999); S. Turck-Chièze *et al.*, Ap. J. Lett., v. **555** July 1, 2001.
11. G. L. Fogli, E. Lisi, A. Palazzo, and F.L. Villante Phys. Rev. D **63**, 113016 (2001); F.L. Villante, G. Fiorentini and E. Lisi Phys. Rev. D **59** 01300x6 (1999).
12. Q.R. Ahmad *et al.* Phys. Rev. Lett. **87**, 071301 (2001)

A Detector for Neutrino Oscillations and Other Neutrino Interactions

Ali R. Fazely

Department of Physics, Southern University, Baton Rouge, Louisiana 70813

Abstract. A multipurpose neutrino detector is described to measure neutrino oscillation with high sensitivity and other neutrino cross sections, of interest to astrophysics. This detector could use the high intensity, fast-spill beam produced by the Spallation Neutron Source under construction at the Oak Ridge National Laboratory.

INTRODUCTION

Searches for neutrino oscillations and other neutrino induced reactions, as a comprehensive test of the Standard Model (SM) of particle physics can be performed at the Spallation Neutron Source (SNS) under construction at the Oak Ridge National Laboratory. The principal motivation of these measurements would be to perform a class of experiments to rigorously test the validity of the SM of particle physics using neutrinos as probes. Measurements include the appearance of $\bar{\nu}_e$ from $\bar{\nu}_\mu$ produced in the beam-stop from muons decaying at rest. One could also search for $\nu_\mu \to \nu_e$ oscillations with super high sensitivity. The detector will also be sensitive to the disappearance of ν_e's when the second, identical proton storage ring is complete. Recent reported neutrino oscillation signal observed by the Liquid Scintillator Neutrino Detector [1][2][3] (LSND) can also be scrutinized with an improved sensitivity of more than an order of magnitude. These measurements would be performed using the newly approved high-intensity short beam-spill spallation source. The experiment would yield data with high statistical precision and small systematic errors due to the precise measurement of the known $\nu_e + ^{12}C \to ^{12}N_{g.s.} + e^-$ cross section. Measurements could also be made of other neutrino-nucleus cross sections with super high precision which are of interest to fundamental physics as well as nuclear physics and astrophysics.
Observation of the decays $\pi^0 \to \nu\bar{\nu}$ and $\eta \to \nu\bar{\nu}$ would imply new physics. Momentum and angular-momentum conservation require that the decay ν and $\bar{\nu}$ possess the same helicity. Therefore, in the SM of particle physics where the neutrinos are assumed to be massless and purely left handed (antineutrinos purely right handed) the above-mentioned processes are strictly forbidden. These measurements could be done with an order of magnitude more sensitivity than LSND.

CP624, *Cosmology and Elementary Particle Physics,* edited by B. N. Kursunoglu et al.

NEUTRINO OSCILLATIONS

Neutrino oscillations phenomenon is a class of lepton-number violating process which can be utilized to test the validity of the SM of the electroweak sector in particle physics. Neutrino oscillations can only occur if the physical neutrinos are not pure quantum mechanical systems, but rather a superposition of different mass eigenstates. This immediately implies that at least one of the neutrino flavor eigenstates has to be massive. This is obviously in contradiction with the SM where the neutrinos are assumed to be massless. In the past two decades, questions regarding the masses of neutrinos and the extent of mixing of different flavor eigenstates have received a great deal of attention. Evidence for neutrino oscillations from the LSND [1][2] and flux deficit in solar [4][5]neutrino experiments have intensified the debate. The atmospheric neutrino experiments with recent results from Super K [6] has also reported deficit in the ν_μ flux and angular distribution of muon neutrinos consistent with neutrino oscillations. Although LSND, solar neutrino and atmospheric neutrino results do not point to the same conclusion regarding neutrino masses and mixing within the traditional two-neutrino mixing analysis approach, one could reconcile these two measurements with more sophisticated analysis of three-neutrino mixing with an extra "sterile" neutrino. It is, therefore, essential to check the validity of these measurements. Amongst these measurements, LSND by far is the most intriguing. The LSND collaboration has reported 22 events consistent with $\bar{\nu}_e$ appearance resulting from $\bar{\nu}_\mu \to \bar{\nu}_e$ oscillations. The most attractive aspect of the LSND experiment is the full understanding of the neutrino flux and systematics of the experiment. Fig. 1 shows the energy spectrum for oscillations and Fig. 2 shows the areas of probable neutrino oscillations in the customary δm^2 vs. $sin^2 2\theta$.[7]

Decay-at-Rest Neutrino Oscillations

At SNS pions are copiously produced in the Hg target at a rate of approximately 0.1 π^+ per incident proton. The majority of these pions (99%) come to rest inside the target and decay. The decay chain produces the well-known neutrino spectrum shown in Fig. 3. We, thus, have:

$$\pi^+(\tau = 26ns) \to \mu^+ + \nu_\mu \tag{1}$$

$$\mu^+(\tau = 2.2\mu s) \to e^+ + \bar{\nu}_\mu + \nu_e \tag{2}$$

The search is done for the neutrino flavor $\bar{\nu}_e$. Note that this flavor is missing in the above decay scheme. Only the negative pions that are not captured and decay can produce this flavor of neutrinos. We estimate this number to be suppressed by 10^4. Therefore, if $\bar{\nu}_e$ is observed in a detector, then the most likely conclusion is that $\bar{\nu}_\mu$'s have oscillated into $\bar{\nu}_e$'s.

The detection method for this mode of neutrino oscillation is the detection of a positron followed by a neutron. The reaction inside a scintillator-based detector with H as the

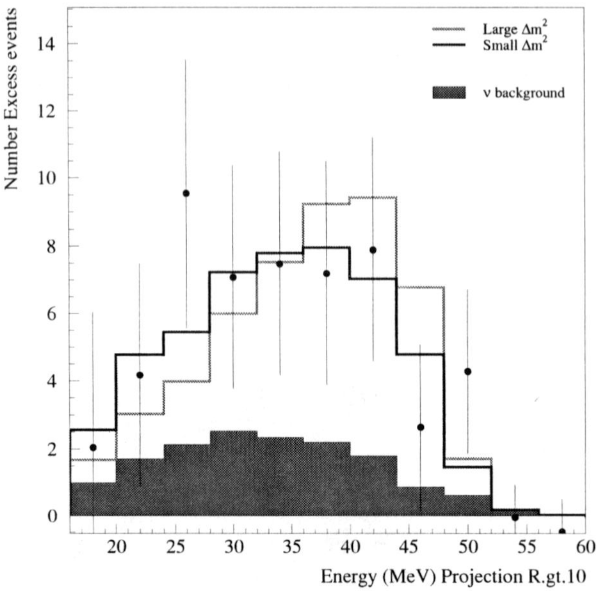

FIGURE 1. Electron energy spectrum for the LSND signal. There are two calculations indicating low and high mass solutions.

target for $\bar{\nu}_e$ is the usual inverse β-decay process;

$$\bar{\nu}_e + p \rightarrow e^+ + n \tag{3}$$

The positron has a Michel energy spectrum and provides the primary trigger. The detection of the neutron is usually done in a so-called slow coincidence technique such as LSND. In this method a gate of a few hundreds μ-seconds is opened to allow for the neutron to thermalize and a capture process on H or an additive such as Gd would produce low energy γ's constituting the signature for the slow coincidence. In a fast coincidence method one looks for recoil protons which can only be done in a fine modular detector such as the one described here. As the neutron slows down, it collides with protons in the liquid scintillator and produces light. This method is background free and requires fast electronics. Note that these two methods are not mutually exclusive and a gold-plated event would, ideally, have both signatures in coincidence with the primary positron.

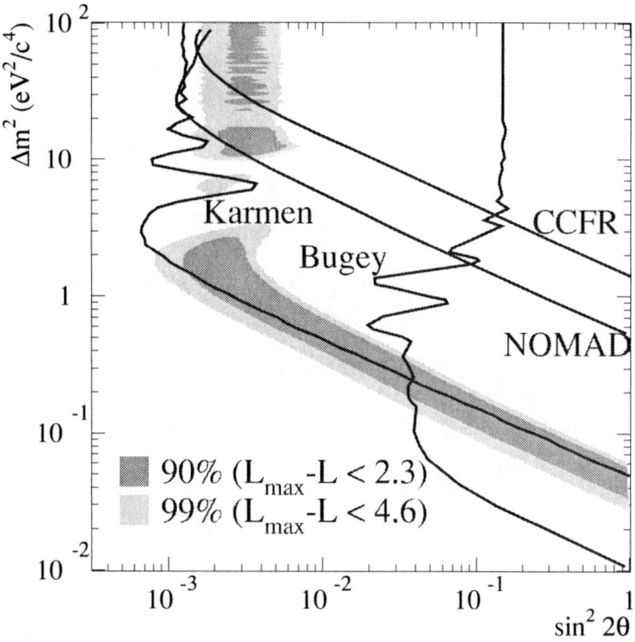

FIGURE 2. Probable region of $\bar{\nu}_\mu \to \bar{\nu}_e$ oscillation for 2.3 σ and 4.6 σ. Other neutrino oscillations experimental limits are also displayed.

Decay-in-flight Neutrino Oscillations

We could use the accumulator ring beam-dumps at SNS for π^+ decay-in-flight (DIF) neutrino oscillation studies. These beam-dumps are not used for material science studies. We could build a neutrino horn to direct the desired neutrinos to the detector.

$\nu_\mu \to \nu_e$ Oscillations Analysis Using ^{12}N, inclusive states. In this mode of oscillation search, electron-like events, above 60 MeV of total energy are considered. In this mode of data analysis, the identified electrons are free from the Michel electrons from the cosmic ray stopped muon decay, however, π^o production by high-energy neutrons can produce substantial background. In a finely segmented detector, like the one described here, the π^o events can be rejected because they have double tracks. In LSND, these events were rejected by the fact that neutron induced electromagnetic events had more scintillation light than Cerenkov light. The LSND reports 10.5 and 10.1 in two different analyses for the inclusive states in ^{12}N [8].

$\nu_\mu \to \nu_e$ Oscillations Analysis Using the Ground State of ^{12}N. The experiment in the pion decay-in-flight (DIF) region, is mainly sensitive to reactions of ν_e from $\nu_\mu \to \nu_e$ oscillations in the beam. The ν_e's are detected by the inverse beta decay reaction. The detector is also sensitive to detecting positrons from β^+-decay of the $^{12}N_{g.s.}$ with an end

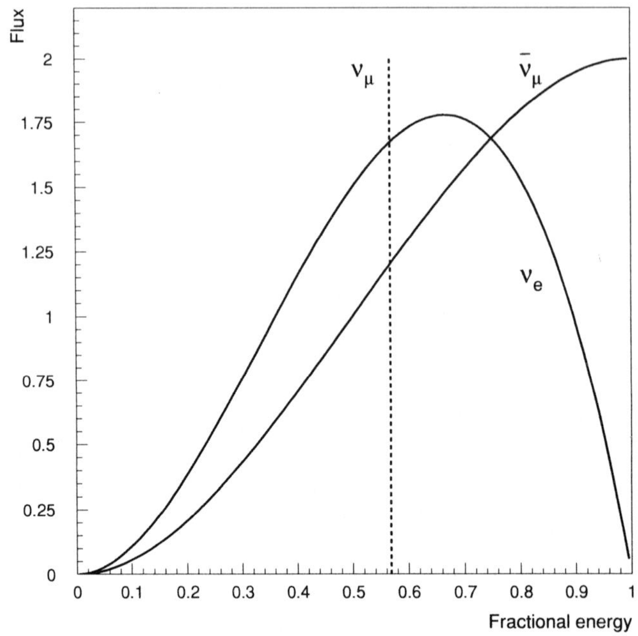

FIGURE 3. Neutrino energy spectra for pion and muon decay-at-rest

point energy of 16.826 MeV. This constitutes a unique signature for the DIF oscillations search. The LSND experiment observed a few such events above 55 MeV with virtually zero background [10][8].

OTHER TESTS OF THE SM

A class of measurements can be performed at the SNS to test the SM and to study spin physics as well as nuclear physics and its impact on astrophysics. These measurements include rare neutral meson decay search, ν-p elastic scattering and neutino-nucleus scattering cross sections.

Limits on $\pi^o \to \nu\bar{\nu}$, and $\eta \to \nu\bar{\nu}$, Decays

The observation of the decays $\pi^o \to \nu\bar{\nu}$ and $\eta \to \nu\bar{\nu}$ would imply physics beyond the SM. Momentum and angular-momentum conservation require that the decay ν and $\bar{\nu}$ possess the same helicity. Therefore, in the standard electroweak interaction model, where the neutrinos are assumed to be massless and purely left handed (antineutrinos purely right handed) the above-mentioned processes are strictly forbidden. One also can argue that since π^o is a pseudo-scalar meson (0^-), and vacuum is 0^+, this transition is strictly forbidden, within the V-A interaction. These decays, however, can proceed

if pseudo-scalar weak interaction exists, or both right and left-handed neutrinos exist and/or the total lepton number is not conserved. The information derived from $\pi^o \rightarrow \nu\bar{\nu}$ and $\eta \rightarrow \nu\bar{\nu}$ are complementary because the former is sensitive only to the isovector neutral-current interactions while the latter is sensitive to the isoscalar neutral-current interactions.[11] Furthermore, π^o decays involve only u- and d-quarks while η decays additionally involve the s-quark and perhaps other heavier quarks.[11]
Equally interesting are the limits on decays $\pi^o \rightarrow \nu_e\nu_\mu$ and $\eta \rightarrow \nu_e\nu_\mu$, which violate individual lepton number conservation in the neutral current sector.
An experimental upper limit, $\Gamma(\pi^o \rightarrow \nu_e\bar{\nu}_e)/\Gamma(\pi^o \rightarrow all) \leq 2.4 \times 10^{-5}$ (90% CL), was set by Herczeg and Hoffman [12] who used the $K^+ \rightarrow \pi^+\nu\bar{\nu}$ data and focused on the $K^+ \rightarrow \pi^+(108MeV)\nu\bar{\nu}$ region of the decay spectrum. Hoffman [13] has further set limits for the branching ratio for $\pi^o \rightarrow \nu\bar{\nu}$ by using the data from several beam-dump experiments. Similar limits were obtained by Dorenbosch et al.[14] These upper limits are shown in Table II. To date, no limits on $\eta \rightarrow \nu\bar{\nu}$ exist. Other limits from $K^+ \rightarrow \pi^+\nu\bar{\nu}$ data sets an inclusive tighter limit of $\Gamma(\pi^o \rightarrow \nu_e\bar{\nu}_e)/\Gamma(\pi^o \rightarrow all) \leq 8.3 \times 10^{-7}$ (90% CL).[15] These limits can be improved by two orders of magnitude at SNS.

Strange Quark Contribution to the Spin of the Nucleon, $\nu p \rightarrow \nu p$

The νp elastic scattering is a simple semileptonic reaction that provides a powerful tool to measure the strange quark content, Δs of the proton and its relative contribution to the proton spin. Cross sections for these reactions are well known, aside from small theoretical uncertainty, and small radiative corrections, characteristic of neutral-current neutrino-induced reactions.
The only two observables in $\nu p \rightarrow \nu p$ are the recoil proton kinetic energy and the angle of the recoil proton with respect to the incident neutrino. We assume that the recoil neutrino and the recoil proton polarization are not measurable in a practical way. The proton kinetic energy is given by $T_p = Q^2/2m_p$, where Q_2 is the momentum transfer squared of the reaction and m_p is the proton mass. The angle of the recoil proton in the lab frame can be expressed as $cos(\theta) = (1 + m_p/E_\nu)/(1 + 2m_p/T_p)^{1/2}$, where E_ν in the neutrino energy. At SNS with a fine segmented detector, we will be able to measure the proton angle at high momentum transfers and this together with the proton kinetic energy will provide the signature for νp elastic scattering.
All the form factors relevant to $\nu p \rightarrow \nu p$ scattering are well understood, and the third, G_1 is the axial vector. All these form factors are functions of Q^2, therefore,

$$F_1(Q^2 = 0) = 0.034 - F_1{}^s/2 \tag{4}$$

$$F_2(Q^2 = 0) = 1.017 - F_2{}^s/2 \tag{5}$$

Note that $F_1{}^s = 0$ at $Q^2 = 0$ and $F_2{}^s$, the weak strange magnetism has been measured by HAPPEX at JLAB [16] in the forward direction and the SAMPLE experiment at BATES [17] in the backward direction. The measured value

is $G_M^Z = 0.34 \pm 0.09 \pm 0.04 \pm 0.05$ n.m.(nuclear magnetons) at $Q^2 = 0.1(\text{Gev/c})^2$. $G_1(0) = -g_A/2 + G_1{}^s/2$, where $g_A = 1.26$ from neutron beta decay and $G_1{}^s = \Delta s$ is the strange quark contribution to the spin of the proton. We assume only first class current and therefore, F_3 and G_2 are zero and $G_3 = 0$, because the mass of the neutrino is very light compared to the electron mass.
The cross section at low Q^2's is:

$$\frac{d\sigma}{dQ^2} = \frac{G_F^2}{2\pi}[(-0.63 + \frac{G_1^s}{2})^2(1 + \frac{Q^2}{4E_\nu^2}) + 0.001156(1 - \frac{Q^2}{4E_\nu^2})] \quad (6)$$

In the above expression, the only unknown is G_1^s. A measurement of the above cross section will determine this strange quark contribution to the proton spin.
Fig. 4 shows a calculation for νp elastics for bound and free protons. It also shows the νn results. Note that in νn, the trigger requirement and the associated energy of neutron collision throughout the detector provide important tools to identify neutrons and protons in a fine segmented detector.

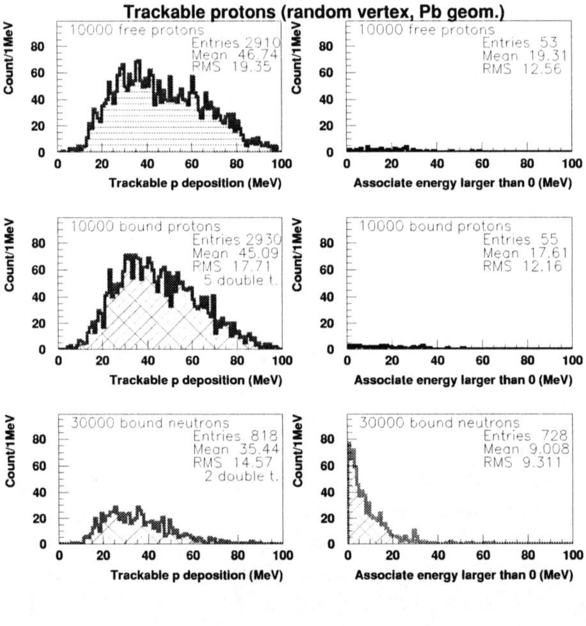

FIGURE 4. Calculated proton and neutron energy spectra for bound and free protons and neutrons.

Nuclear Physics Program

A segmented detector provides the ideal laboratory to measure neutrino-nucleus cross sections for nuclei of interest in astrophysics or nuclear physics. Measurement of such cross sections are important in order to get a better understanding of, for example, the dynamics of super novae. Some nuclei may have larger or smaller cross sections than what calculations predict and may provide tools by which we can design more efficient neutrino detectors.

Fig. 5 shows the electron energy spectra for Al, Si, iron and lead. Our calculations show that given sufficient mass of a given nuclear target, a better than a 10% cross section measurement is possible. As mentioned above these measurements will have important astrophysical consequences.

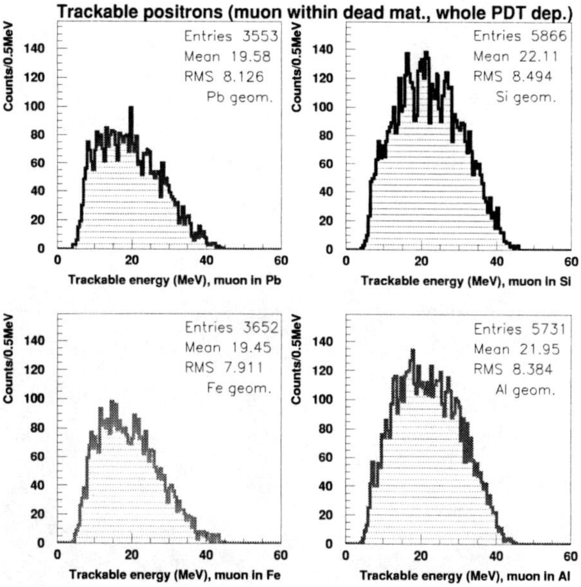

FIGURE 5. Calculated electron spectra for neutrino charged-current reactions on Al, Si, Fe and Pb.

THE DETECTOR

With the experience from LSND and the LAMPF beam, we believe the best and the most suitable place to extend the sensitivity of LSND and KARMEN measurements, is the SNS being built at ORNL. The 1 GeV beam energy at SNS would be perfect for copious pion production in an Hg target. This energy, being higher than LANSCE by about 20%, provides higher pion flux and is also well below kaon production threshold, thus keeping the oscillation signal free from neutrinos resulting from K-decay. The pion decay at rest neutrino spectrum is well known (Fig. 2) and pion production at these energies have been measured and/or can be calculated reliably. Furthermore, the high-Z

Hg target insures higher π^- and μ^- capture rate than Cu beam-stop at LANSCE, thus reducing the beam-related $\bar{\nu}_e$'s.
A segmented detector can be designed containing long-thin plastic modules 5 meters long by 1 cm by 20 cm cross section. These modules will be arranged in x layers to provide calorimetry followed by x-y planes of proportional drift tubes (PDT) for tracking. Each optically-isolated module will be read out with green wave-shifter fibers at both ends. The outside of each module will be painted with Gd_2O_3 paint for slow neutron detection.

Fig. 6 shows the schematic of a fine segmented detector. This detector is designed to have nuclear targets as inserts.

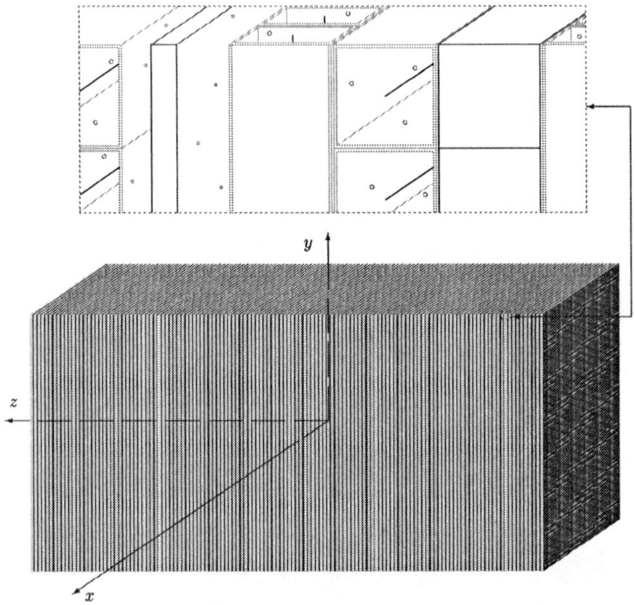

FIGURE 6. Schematic view of a fine segmented neutrino detector at the SNS

The modular scintillator design has also the capability of providing particle identification. This is due to observation of ionization of particles via dE/dx technique. At energies below 100 MeV the electrons and protons are well separated and an electron-proton rejection of 10^{-4} is achievable as it was done in E645 at LAMPF. A segmented detector also provides the capability of fast neutron detection for the DAR neutrino oscillations measurement. Note that this is not mutually exclusive with the delayed neutron capture scheme. In fact, a "gold plated" DAR neutrino oscillations event has both fast and slow neutron signal in coincidence with the primary positron trigger.

ACKNOWLEDGMENTS

This work was supported by an LDRD grant from the Oak Ridge National Laboratory.

REFERENCES

1. Athanassopoulos, C. A. et al., *Phys. Rev. Letters* **75**, 2650 (1995).
2. Athanassopoulos, C. A. et al., *Phys. Rev. Letters* **77**, 3082 (1996).
3. Aguilar, A. et al., *Phys. Rev. D* **64**, 112007(2001).
4. Anselmann, P. et al., *Phys. Letters* **327B**, 377 (1994).
5. GALLEX Collaboration, *Phys. Letters* **285B**, 376 (1992).
6. Fukuda, Y. et al., *Phys. Rev. Letters* **81**, 1562 (1998).
7. Louis, W. C., private communication.
8. Athanassopoulos, C. A. et al., *Phys. Rev. Letters* **81** 1774 (1998).
9. Athanassopoulos, C. A. et. al., *Phys. Rev. C* **58** 2489 (1998).
10. Fazely, A. R., *LSND Technote 108* (1998).
11. Herczeg, P., *Proceedings of the Workshop on Production and Decay of Light Mesons*, ed. P. Fleury, World Scientific Publishers, Singapore(1988).
12. Herczeg P. and Hoffman, C. M., *Phys. Lett. B* **100**,347 (1981).
13. Hoffman, C. M., *Phys. Lett. B* **208**,149 (1988).
14. Dorenbosch, J. et al., *Z. Phys. C* **40**, 497 (1988).
15. Atiya, M. S. et al., *Phys. Rev. Letters* **66** 2189 (1991).
16. Aniol, K. A. et al., *Phys. Rev. Letters* **82** 1096 (1999).
17. Spayde, D. T. et al., *Phys. Rev. Letters* **84** 1106 (2000).

V. STUDENT PAPERS

Rare K decays in a model of quark and lepton masses.

P.Q. Hung* and A. Soddu*

*Dept. of Physics, University of Virginia,
382 McCormick Road, P. O. Box 400714, Charlottesville, Virginia 22904-4714

Abstract. An extension of a model of neutrino masses to the quark sector provides an interesting link between these two sectors. A parameter which is important to describe neutrino oscillations and masses is found to be a crucial one appearing in various "penguin" operators, in particular the so-called Z penguin. This parameter is severely constrained by the rare decay process $K_L \to \mu^+\mu^-$. This in turn has interesting implications on the masses of the neutrinos and the masses of the vector-like particles which appear in our model.

(Presented by A. Soddu)

INTRODUCTION

In the last few years, we have witnessed a flurry of far-reaching experimental results, among which are the neutrino oscillation data [1] [2], data on direct CP-violation such as ε'/ε [3] [4] and upper bounds on Flavour-Changing-Neutral-Current (FCNC) rare decays of the kaons. On the one hand, the neutrino oscillation data clearly points to possible physics beyond the Standard Model (SM). On the other hand, it is still not clear if the new results on ε'/ε, which differ roughly by a factor of two from present calculations within the SM, imply any new physics since the aforementioned calculations are still plagued with non-perturbative uncertainties. For the kaon's FCNC rare decays, the experimental situation is still far from giving evidences of physics beyond the SM or to confirm the SM itself. Nevertheless, whatever new physics, which might be responsible for giving rise to neutrino masses, could, in principle, affect the quark sector, and hence, also quantities such as ε'/ε or the branching ratios of the kaon's FCNC rare decays. If this is the case, results from the quark sector could then be used to put constraints on the lepton sector itself since it is possible that both sectors have a common set of parameters.

NEUTRINO SECTOR

There are strong indications that neutrinos have a mass, albeit a very tiny one, and, as a result, oscillate. The exact nature of the masses, Majorana or Dirac, as well as the oscillation angles, is an important subject which is under intense investigation. Most

CP624, *Cosmology and Elementary Particle Physics*, edited by B. N. Kursunoglu et al.

efforts have been done on the problem of neutrino masses, at least on the model-building front. There is one common assumption present in many of such models, which is one in which light neutrino masses arise from a see-saw mechanism. The smallness of neutrino masses would come from an expression that goes like m_D^2/M, where m_D is a Dirac mass, and M is a Majorana mass, which typically is very much larger than m_D and could be some kind of Grand Unified scale. It is a wide spread opinion that with Majorana neutrinos and the see-saw mechanism, one could "easily" obtain small neutrino masses. Nevertheless, Majorana neutrinos will imply the existence of lepton number violating processes such as neutrinoless-double-beta-decays or $K^+ \to \pi^- e^+ e^+$. At the moment no evidences of lepton number violating processes have been found.

Now if the mass were to be of the Dirac type, one can straightforwardly write down a gauge-invariant Yukawa coupling in the SM itself (endowed with right ended neutrinos). But to obtain a small neutrino mass, one has to put "by hand" a Yukawa coupling which is incredibly small, of the order of 10^{-11}. Such a fine tuning is commonly believed to be highly unnatural. A possible solution using only Dirac neutrinos has been presented in Ref. [5]. The SM is extended introducing two new gauge symmetries, taking the gauge structure:

$$SU(3)_c \otimes SU(2)_L \otimes U(1)_Y \otimes SO(4) \otimes SU(2)_{\nu_R}. \tag{1}$$

In this model there are four families, the neutrino of the fourth generation is required to have mass bigger than $M_Z/2$, and the right-handed neutrinos transform as doublets under $SU(2)_{\nu_R}$, all the other particles being singlets under this symmetry. The symmetry $SU(2)_{\nu_R}$ forbids tree level mass terms for the three light families. The three light neutrinos acquire mass dynamically through one-loop level diagrams. In Fig. 1 we present one of the two one-loop diagrams which generates the general element m_{ij} of the three light neutrinos mass matrix block. The particles propagating in the loop are the vector-like fermions F^l and $\mathcal{M}_2^l$, whose quantum numbers are reported in Table 1, and the Nambu Goldstone, (NG) bosons $\tilde{\Omega}_k'$, with $k = 1,2,3$, being the longitudinal components of the family gauge bosons for the three light families sector.

The other of the two one-loop diagrams has the pseudo NG bosons $Re\tilde{\rho}_k$, with $k = 1,2,3$, propagating in the loop instead of the NG bosons $\tilde{\Omega}_k'$. See the appendix of Ref. [5] for a detailed description of how $SO(4) \otimes SU(2)_{\nu_R}$ is spontaneously broken.

An important feature of this model is that the matrix elements m_{ij} for the three light neutrino sector depend on ratios of the propagating particles masses. As it is shown in Fig. 2,3,4 and 5 of Ref. [5], it is possible to make the mass of the three light neutrinos very small, requiring that the propagating particle masses have ratios with values in certain ranges.

The NG bosons $\tilde{\Omega}_k'$ and the pseudo NG bosons $Re\tilde{\rho}_k$ can mix with different flavours, making the off diagonal mass matrix elements m_{ij} for the three light neutrinos different from zero. As a consequence the three light neutrinos are not degenerate, with differences for their squared masses given by:

$$\Delta m_{32}^2 = m_2 (m_N 2b \frac{\sin(2\beta)}{32\pi^2} \Delta I(G,P,-b)), \tag{2}$$

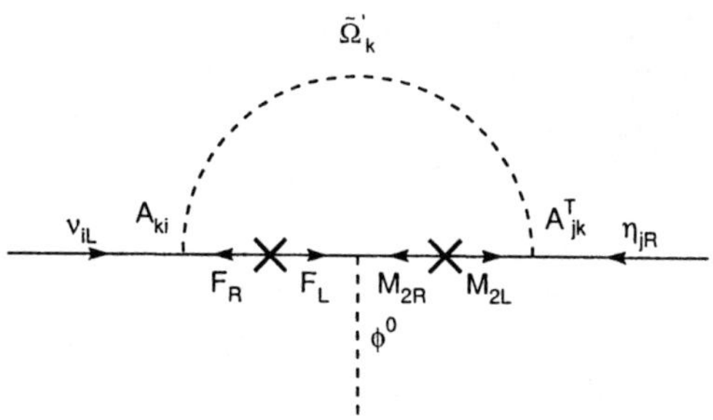

FIGURE 1. One-loop diagram for the three light neutrinos mass matrix elements m_{ij}

$$\Delta m^2_{21} = m_2(m_N 2b\frac{\sin(2\beta)}{32\pi^2}\Delta I(G,P,b)). \tag{3}$$

The mixing angle β and the function $\Delta I(G,P,-b))$ are introduced respectively in Eq. (10c), (10d) and Eq. (43) of Ref. [5]. m_N is the mass of the fourth neutrino and b is a parameter smaller than one which enters through the mass spectrum of the NG bosons $\tilde{\Omega}'_k$ and the pseudo NG bosons $Re\tilde{\rho}_k$, see Eq. (45) of Ref. [5]. When b goes to zero, the NG bosons $\tilde{\Omega}'_k$ have the same mass, and the same thing happens for the pseudo NG bosons $Re\tilde{\rho}_k$. The NG bosons $\tilde{\Omega}'_k$ and the pseudo NG bosons $Re\tilde{\rho}_k$ no longer mix with different flavours and the three light neutrinos reasult degenerate, as it can be seen explicitly making b go to zero in the expressions for Δm^2_{32} and Δm^2_{21} in Eq. (2) and (3). Looking at the expressions for Δm^2_{32} and Δm^2_{21} it is evident that in order to match the oscillation data, one needs the function $\Delta I(G,P,x))$ to depend strongly on the argument x. It comes instead that

$$\Delta I(G,P,b)) \simeq \Delta I(G,P,-b)), \tag{4}$$

making Δm^2_{32} and Δm^2_{21} almost degenerate. A possible solution to remove the degeneracy in $\Delta m^{2\prime}_{ij}s$ comes asking that the mass matrix elements m_{4j} and m_{j4} connecting the fourth generation with the three light families sector are non zero for at least one family j. This could be obtained for example by introducing an extra scalar, see Eq. (77) of Ref. [5]. In Table 2, for the particular choice $m_{34} = m_{43} = 0.8 \cdot 10^{-7}(100GeV)$ two different cases are presented corresponding to two different values of b. The masses of the propagating particles, M_P being the central value for the pseudo NG bosons $Re\tilde{\rho}_k$ masses, M_G the central value for the NG bosons $\tilde{\Omega}'_k$ masses, and M^l_2 the mass for the vector-like fermion $\mathscr{M}^l_2$, are expressed in units of $M_F = 200GeV$, the mass of the vector-like fermion F^l. They have been chosen properly, correspondingly to b, in order to match the oscillation data. Looking at Table 2 it is evident that the parameter b with the only limitation of

TABLE 1. Particle content and quantum numbers of $SU(3)_c \otimes SU(2)_L \otimes U(1)_Y \otimes SO(N_f) \otimes SU(2)_{\nu_R}$

Standard Fermions	$q_L = (3,2,1/6,N_f,1)$ $l_L = (1,2,-1/2,N_f,1)$ $u_R = (3,1,2/3,N_f,1)$ $d_R = (3,1,-1/3,N_f,1)$ $e_R = (1,1,-1,N_f,1)$
Right-handed ν's	Option 1: $\eta_R = (1,1,0,N_f,2)$ Option 2: $\eta_R = (1,1,0,N_f,2)$; $\eta'_R = (1,1,0,1,2)$
Vector-like Fermions for the lepton sector	$F^l_{L,R} = (1,2,-1/2,1,1)$ $\mathscr{M}^l_{1L,R} = (1,1,-1,1,1)$ $\mathscr{M}^l_{2L,R} = (1,1,0,1,1)$
Vector-like Fermions for the quark sector	$F^q_{L,R} = (3,2,1/6,1,1)$ $\mathscr{M}^q_{1L,R} = (3,1,-1/3,1,1)$ $\mathscr{M}^q_{2L,R} = (3,1,2/3,1,1)$
Scalars	$\Omega^\alpha = (1,1,0,N_f,1)$ $\rho^\alpha_i = (1,1,0,N_f,2)$ $\phi = (1,2,1/2,1,1)$

TABLE 2. Δm^2_{32} and Δm^2_{21} for two different values of b

Case I	Case II
$M_F = 1 \quad M_P = 5$	$M_F = 1 \quad M_P = 5$
$M_G = 10^6 \quad M_2 = 2.5 \cdot 10^9$	$M_G = 10^4 \quad M_2 = 1.2 \cdot 10^9$
$b = 0.035$	$b = 0.000095$
$\lvert m_1\rvert \sim \lvert m_2\rvert \sim \lvert m_3\rvert \sim 1.58eV$	$\lvert m_1\rvert \sim \lvert m_2\rvert \sim \lvert m_3\rvert \sim 1.38eV$
$\lvert m_4\rvert = 100GeV$	$\lvert m_4\rvert = 100GeV$
$\Delta m^2_{32} = 2.02 \cdot 10^{-3} eV^2$	$\Delta m^2_{32} = 1.77 \cdot 10^{-3} eV^2$
$\Delta m^2_{21} = 5.5 \cdot 10^{-6} eV^2$	$\Delta m^2_{21} = 5.4 \cdot 10^{-6} eV^2$

being smaller than unity, can take a big range of values, and from the oscillation data one cannot have any insight in order to reduce this range.

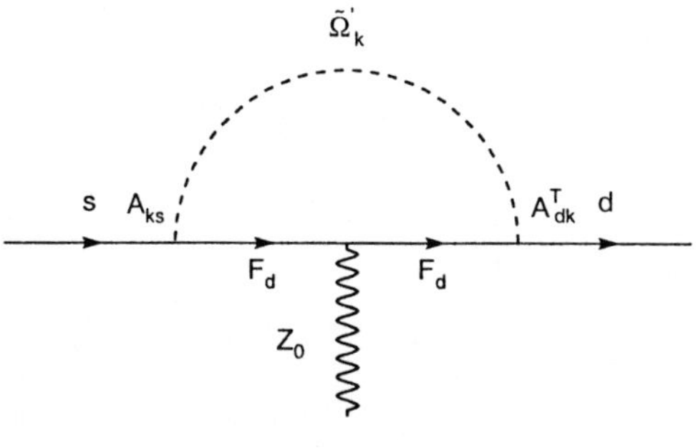

FIGURE 2. Z penguin diagram for the transition $s \to d$ in the model of Ref. [5]

QUARK SECTOR

The way the SM has been extended in Ref. [5], expecially through the symmetry $SO(4)$, can have important consequences other than explaining small neutrino masses. The invariance under rotation in the family space of the leptonic Lagrangian , see Eq. (3) of Ref. [5], is shared also by the Lagrangian of the quark sector, see Eq. (3) of Ref. [6]. The same NG bosons $\tilde{\Omega}'_k$ and the pseudo NG bosons $Re\tilde{\rho}_k$ introduced in the lepton sector can mediate processes involving quarks. In particular, because of the couplings of quarks of different flavours to the NG bosons $\tilde{\Omega}'_k$ and the pseudo NG bosons $Re\tilde{\rho}_k$, new contributions to FCNC processes are made possible. In Fig. 2 a FCNC process in the model of Ref. [5] involving s and d quarks through the exchange of a Z boson, the so called Z penguin diagram, is presented. The vector-like fermion F^q propagating in the loop has color charge. See Table 1 for a complete list of the particles for both lepton and quark sectors. FCNC processes involving s and d quarks can also happen through the exchange of a photon or a gluon, as well as through box diagrams. In Ref. [6] a detailed study of FCNC processes involving s and d quarks in the model of Ref. [5] has been presented.

The main result is that all the FCNC processes at the lowest order depend linearly on the same parameter b introduced in the lepton sector. In particular it has been shown that the Z penguin dominates over all the other new contributions, being the last ones $1/M_G^2$ suppresed. M_G is the central mass of the family gauge bosons, and it is required to be bigger than $\mathcal{O}(100TeV)$ in order for the tree level FCNC processes to be suppressed. Taking into account only the effect of the Z penguin operator and neglecting all the other new contributions, one can now write the modified effective Lagrangian for processes involving s and d quarks with the exchange of a Z boson.

$$\mathcal{L}_Z = \frac{g}{16\pi^2 \cos(\theta_W)} g^2 Z_{ds} \bar{s}\gamma^\mu (1-\gamma^5) d Z_\mu + h.c., \tag{5}$$

where

$$Z_{ds} = Z_{ds}^{SM} + \frac{g_F^2}{g^2} Z_{ds}^F \tag{6}$$

is the modified effective Z coupling for s, d quark transitions. g is the $SU(2)_L$ coupling and g_F is the family $SO(4)$ coupling.

$$Z_{ds}^{SM} = V_{ts}^* V_{td} C_0^{SM}(x_t) \tag{7}$$

is the effective Z coupling as derived in the SM, $C_0^{SM}(x_t)$ is defined in Eq. (30) of Ref. [6], $V_{ij}'s$ are the Cabibbo-Kobayashi-Maskawa (CKM) matrix elements, and

$$Z_{ds}^F = b C_0(x_F) \tag{8}$$

is the new contribution to Z_{ds} coming from our extension of the SM. The function $C_0(x_F)$ has been introduced in Eq. (20) of Ref. [6]. The nature of Z_{ds}^F, complex or real, is essential to determine which physical quantities will be affected by new physics. In particular Z_{ds}^F comes to be real, being both b and $C_0(x_F)$ real. This implies that our extension of the SM does not affect the quantity ε'/ε, which measures the direct over the indirect CP violation in the kaon sector. On the other hand it affects the branching ratio $BR(K_L \to \mu^+\mu^-)$ in his short-distance (SD) contribution. From the experimental value of $BR(K_L \to \mu^+\mu^-) = (7.2 \pm 0.5)10^{-9}$ [7], and using the analysis in Ref. [8] of the long-distance (LD) and SD contributions to $BR(K_L \to \mu^+\mu^-)$, the highest possible value for $BR(K_L \to \mu^+\mu^-)_{SD}$ is derived

$$BR(K_L \to \mu^+\mu^-)_{SD} < 2.8 \cdot 10^{-9}. \tag{9}$$

Now one can use the above upper limit to constrain the parameter b finding

$$b < 2.5 \cdot 10^{-4}. \tag{10}$$

It is important to notice here that using the constraint for b, one can now rule out a lot of configurations for the masses of the particles propagating in the loop diagram for the neutrinos mass matrix elements. At the same time one could have an insight on the possible values of the three light neutrino masses, also if this aspect is far away from being experimental tested.

ACKNOWLEDGMENTS

I would like to thank Prof. Behram N. Kursunoglu for giving me the possibility to partecipate in the Coral Gables Conference on Cosmology and Elementary Particle Physics. I would like also to thank my advisor Prof. P.Q. Hung for guiding my research, and the UVa theoretical group for sponsoring my participation at the conference.

REFERENCES

1. Super-Kamiokande Collaboration, *Phys. Rev. Lett.*, **81**, 1562 (1998).
2. SNO Collaboration, *Phys. Rev. Lett.*, **87**, 071301 (2001).
3. V. Fanti et al., *Phys Lett. B*, **465**, 335 (1999).
4. A. Alavi-Harati et al., *Phys. Rev. Lett.*, **83**, 22 (1999).
5. P. Q. Hung, *Phys. Rev. D*, **62**, 053015 (2000).
6. P. Q. Hung, A. Soddu, *hep-ph/0108119, to appear in Phys. Rev. D.* (2002).
7. A. P. Heinson et al., *Phys. Rev. D*, **51**, 985 (1995).
8. A. J. Buras, L. Silvestrini, *Nucl. Phys. B*, **546**, 299 (1999).

The Reaction $\mu^- + p \longrightarrow \Lambda + \nu_\mu$ and the Contributions of the Pseudoscalar Form Factor

Michael A. Barnett* and Stephan L. Mintz *

*Physics Department, Florida International University, Miami, Florida, 33199

Abstract. We calculate the differential cross section for the weak, strangeness changing, muon scattering process, $\mu^- + p \longrightarrow \nu_\mu + \Lambda$, for incoming muon energies of 2.0,0.275 and 0.265 GeV. We obtain as well contributions of the individual form factors to the differential cross sections but focus on the pseudocalar form factor which has not been well studied in strangeness changing processes. In particular we assess the possibility of observing pseudoscalar form factor contributions at low q^2 where the mass of the muon may accentuate these contributions. We find that the differential cross sections peak as the maximal scattering angle for the Λ is approached and that the peak height increases as the muon energy is increased. The behavior of the differential cross section near the maximal angle is discussed as is the possibility of observing second class current form factors.

INTRODUCTION

Recently there has been interest in establishing facilities for generating muon beams. The purpose of such facilities would be primarily to produce high intensity, high energy neutrino beams. However the existence of high energy muon beams might lead to the possibility of performing other muon physics experiments. It is with this in mind that we consider a preliminary study of the reaction $\mu^- + p \longrightarrow \Lambda + \nu_\mu$. We have already considered the related process[1] $e^- + p \longrightarrow \Lambda + \nu_e$ but this reaction is completely insensitive to the contributions to the pseudoscalar form factor. The use of muon beams would change the kinematics , especially low q^2. In particular terms involving F_p the pseudoscalar form factor, and F_S, the scalar form factor are suppressed in the electron reaction due to the small size of the lepton mass. These terms could conceivably be large enough to observe in the muon induced reaction. In addition the relative weights of the larger terms might be changed as well. Finally the results might tell us if it would be worthwhile to consider the use of polarization either in the target, or the final state Λ to isolate particular pieces of the interaction.

Partial conservation of the axial current (PCAC) has not been systematicallystudied in strangeness changing reactions, particularly in the energy ranges of interest here. If terms proportional to F_p can be observed, reactions of this sort would open up the possiblility of systematic study of this question. It is for these reasons that we undertake this preliminary work. Furthermore we shall make the calculation as phenomenological as possible to give some idea as to the practicality of such an experiment.

In this paper we calculate the differential cross section for the process, $\mu^- + p \longrightarrow \Lambda + \nu_\mu$ for energies of 2.0 GeV, 0.275 GeV, and 0.265 GeV. The former energy is for comparison purposes with the electron scattering case and the latter two energies are

CP624, *Cosmology and Elementary Particle Physics,* edited by B. N. Kursunoglu et al.

much closer to threshold. We shall where possible make use of the Cabibbo model to obtain the vector current form factors necessary for the calculation and of experimental results[2] to obtain the the axial current form factor at $q^2 = 0$. The Cabibbo model has worked well for Λ beta decay which takes place at $|q^2|$ up to approximately 0.1 GeV^2 , and it would be interesting to see at what energy range the Cabibbo model might break down.

There is also an interesting kinematical effect near the maximal angle of the outgoing Λ which leads to a mild singularity in the differential cross section and hence apparently unbounded results. This can be corrected by an appropriate wave packet for the outgoing Λ as will be discussed later.

Finally it might be asked why a weak process of this sort which is suppressed by the sine squared of the Cabibbo angle should be undertaken rather than a non-strangeness changing process. The answer to this is two-fold. In this reaction, the background is less difficult than in most non-strangeness changing electron induced weak processes, and comes principally from $\mu^- + p \longrightarrow e + \Lambda + K^+$. This background process can probably be separated via missing mass analysis from the reaction of interest here. Also much less is known about strangeness changing processes including how well the conserved vector current hypothesis works for the weak currents and how accurate the Cabibbo relations are. The process of interest here might prove to be very useful in this respect.

In the next section of this paper we shall obtain the matrix elements necessary to perform the calculation indicated here. In the final section of this paper we shall calculate the differential cross section including contributions for the energies indicated and present our preliminary results and briefly discuss them.

MATRIX ELEMENTS

The process we are considering can be well described as a first order weak interaction. Although values of q^2 as high as 15 GeV^2/c^2 are possible for the energies being considered here, this is still small compared to the mass squared of the intermediate vector boson and so we are justified in writing the interaction as:

$$< \nu\Lambda|H_w|\mu^- \; p >= \frac{G}{\sqrt{2}} \sin\theta_C \bar{u}_\nu \gamma^\lambda (1-\gamma_5) u_\mu < \Lambda|J^\dagger_\lambda(0)|p > . \tag{1}$$

If the quantity $< \Lambda|J^\dagger_\lambda(0)|p >$ were known it would be immediately possible to obtain a differential cross section. We note that here the hadronic current,$J_\mu(0)$ is written as:

$$J_\mu(0) = V_\mu(0) - A_\mu(0) \tag{2}$$

where V_μ and A_μ are the vector and axial vector parts of the weak strangeness changing hadronic current.

There are almost as many notations as papers concerning the $p \leftrightarrow \Lambda$ transition. We use here a notation similar to that used for $p \leftrightarrow n$ transitions which in any case is transparent.

In this notation, the weak current matrix elements may be written as follows:

$$< \Lambda|V_{\mu}^{\dagger}(0)|p >= \bar{u}_f[\gamma_{\mu}F_V(q^2) + i\frac{F_M(q^2)\sigma_{\mu\nu}q^{\nu}}{2m_p} - F_S(q^2)\frac{q_{\mu}}{2m_p}]u_i \quad (3)$$

and

$$< \Lambda|A_{\mu}^{\dagger}(0)|p >= \bar{u}_f[\gamma_{\mu}\gamma_5 F_A(q^2) + \frac{q_{\mu}\gamma_5 F_P(q^2)}{m_{\pi}} + \frac{iF_E(q^2)\sigma_{\mu\nu}q^{\nu}\gamma_5}{2m_p}]u_i \quad (4)$$

where i is the initial particle and f is the final particle, p and Λ respectively. The structure of the particles is contained, of course, in the six form factors,$F_V(q^2)$,$F_M(q^2)$,$F_S(q^2)$,$F_A(q^2)$, $F_P(q^2)$,and $F_E(q^2)$. We note that the form factors F_S and F_E would be due to contributions from second class currents. Thus if we are able to determine these six form factors we can write and evaluate a transition matrix element for the $p \leftrightarrow \Lambda$ transition.

Because this is a muon induced process, and the terms of the transition matrix element squared containing either F_P or F_S are proportional to the lepton mass squared, we may unlike the electron case be able to observe these contributions. Therefore all form factors need to be determined. Unlike the case of of the non-strangeness changing weak current, SU(3) relations rather than SU(2) relations must be used to obtain the unknown form factors. These results are well known and we may in general express the form factors as:

$$F_r^{ijk} = -if^{ijk}\tilde{F}_r + d^{ijk}\tilde{D}_r \quad (5)$$

where we use a tilde to distinguish the SU(3) functions from the form factors used in Eqs. (3) and (4). Here i refers to the current octet number,k and j refer to the initial and final baryon octet numbers and r stands for V,M,A, E, or 1,2,and 3 if an electromagnetic current is being described. For our process Eq.(5) reduces to:

$$F_r = \frac{-1}{\sqrt{6}}(3\tilde{F}_r + \tilde{D}_r) \quad (6)$$

where we have suppressed the q^2 behavior of the form factors in Eq.(5) and Eq.(6).

Making use of Eq.(6) for the electromagnetic current,$V_{\mu}^3 + \frac{1}{\sqrt{3}}V_{\mu}^8$, and for first the proton and then the neutron cases one obtains:

$$\tilde{D}_V = 0 \quad (7)$$

$$\tilde{F}_V = F_1^p \quad (8)$$

$$\tilde{D}_M = \frac{3}{2}F_2^n \quad (9)$$

$$\tilde{F}_M = -F_2^p - \frac{1}{2}F_2^n \quad (10)$$

where we have again suppressed the q^2 dependence of the form factors. Using these relations,and Eq.(6) we obtain the following expressions for the vector form factors:

$$F_V(q^2) = F_V(0)/(1 - q^2/M_V^2)^2 \quad (11)$$

with $F_V(0) = 1.2247$ and $M_V = .98 GeV/c^2$ and

$$F_M(q^2) = F_M(0)/(1 - q^2/M_M^2)^2 \tag{12}$$

with $F_M(0) = 1.793/2m_p$ and $M_M = .71 GeV/c^2$. This completely determines the vector current matrix element for the purposes of this calculation.

The axial current matrix element is more difficult to obtain. Although there exist relations given by Eq.(6), there is no corresponding electromagnetic current of course. However there is very useful experimental data[2] from Λ beta decay,$\Lambda \longrightarrow p + e^- + \bar{\nu}_e$, which gives in the notation used here:

$$\frac{F_A(0)}{F_V(0)} = 0.718 \pm 0.015. \tag{13}$$

Furthermore,these measurements[3,4,5,6] are consistent with a dipole fit given by:

$$F_A(q^2) = F_A(0)/(1 - q^2/M_A^2)^2 \tag{14}$$

with $M_A = 1.25 GeV/c^2$ and from Eqs.(11) and (13),$F_A(0) = .8793$. For F_P we use a standard form first given by Nambu, namely

$$F_P(q^2) = -F_A m_k (m_i + m_f)/(q^2 - m_k^2) \tag{15}$$

This completes the axial current form factors.

Finally we estimate a values for F_E and F_S. From a theoretical reference[7] we obtain an estimate for $F_E(0) = .705/2m_p$ in the notation used here. From the same reference $F_S(0) = .344\ F_E(0)$ Making use of our experience that[8] F_E and F_M have similar q^2 dependence and making a similar assumption for F_S, we write:

$$F_E(q^2) = F_E(0)/(1 - q^2/M_M^2)^2 \tag{16}$$

$$F_S(q^2) = F_S(0)/(1 - q^2/M_M^2)^2 \tag{17}$$

where M_M is given in Eq.(12) and $F_E(0)$ and $F_S(0)$ are given above.

Thus we have obtained all of the necessary form factors for evaluating the differential cross section which is given by:

$$\begin{aligned} |M|^2 &= \frac{1}{m_e m_\nu}\Big[\frac{4|F_V|^2}{m_i m_f}[p_f \cdot \nu p_i \cdot e + p_f \cdot e p_i \cdot \nu - e \cdot \nu m_i m_f] + \\ & \frac{4|F_A|^2}{m_f m_i}[p_f \cdot \nu p_i \cdot e + p_f \cdot e p_i \cdot \nu + e \cdot \nu m_i m_f] + \\ & \frac{2|F_M|^2}{m_p^2 m_i m_f}[e \cdot \nu (p_i \cdot e p_f \cdot e + p_i \cdot \nu p_f \cdot \nu + e \cdot \nu m_i m_f)] + \\ & \frac{8 F_V F_A}{m_i m_f}[p_f \cdot \nu p_i \cdot e - p_i \cdot \nu p_f \cdot e] + \end{aligned} \tag{18}$$

$$
\begin{aligned}
&\frac{4F_MF_V}{m_pm_f}[e\cdot v(p_f\cdot v-p_f\cdot e)]+\\
&\frac{4F_MF_V}{m_pm_i}[e\cdot v(p_i\cdot e-p_i\cdot v)]+\\
&\frac{4F_AF_M}{m_pm_f}[e\cdot v(p_f\cdot e+p_f\cdot v)]+\\
&\frac{4F_AF_M}{m_pm_i}[e\cdot v(p_i\cdot e+p_i\cdot v)]+\\
&\frac{|F_E|^2}{m_p^2m_im_f}[e\cdot v(2(p_f\cdot ep_i\cdot e+p_f\cdot vp_i\cdot v)+q^2p_f\cdot p_i]+\\
&\frac{4F_EF_V}{m_pm_i}[e\cdot v(p_i\cdot v+p_i\cdot e)]-\\
&\frac{4F_EF_V}{m_pm_f}[e\cdot v(p_f\cdot v+p_f\cdot e)]+\\
&\frac{4F_EF_A}{m_pm_f}[e\cdot v(p_f\cdot v-p_f\cdot e)]+\\
&\frac{4F_EF_A}{m_pm_i}[e\cdot v(p_i\cdot v-p_i\cdot e)]+\\
&\frac{4F_S^2}{(2m_p)^2}[m_\mu^2e\cdot v(p_f\cdot p_i+m_fm_i]+\\
&\frac{2F_P^2}{m_pm_fm_k^2}[m_\mu^2e\cdot v(p_f\cdot p_i-m_fM_i]+\\
&\frac{8F_AF_P}{m_im_pm_k}[m_\mu^2p_i\cdot vm_f-p_f\cdot vm_i]\Big].
\end{aligned}
\tag{19}
$$

In the above equation we use m_i to denote the mass of the initial hadron, here the proton and m_f to denote the final hadron which is the Λ. We use m_p to denote the proton mass which occurs in the definition of current matrix elements so that Eq.(19) may be more readily adapted to other processes. We note that every term in this matrix element squared is proportional to the neutrino and to the electron energy. The differential cross section can now be calculated by standard methods. The result is:

$$
\frac{d\sigma}{d\Omega}=\frac{m_\mu m_\nu G^2 m_f p_f|M|^2}{(2\pi)^2E8|m_i+E-\frac{EE_f\cos\theta}{p_f}|}
\tag{20}
$$

where p_f and E_f are here the magnitude of the three momentum and the energy of the the final state Λ respectively. We note here the presence of E in the denominator of Eq.(20) which cancels much of the direct dependence of the matrix element squared ,Eq.(19), on the incoming muon energy. We are now ready to calculate our results.

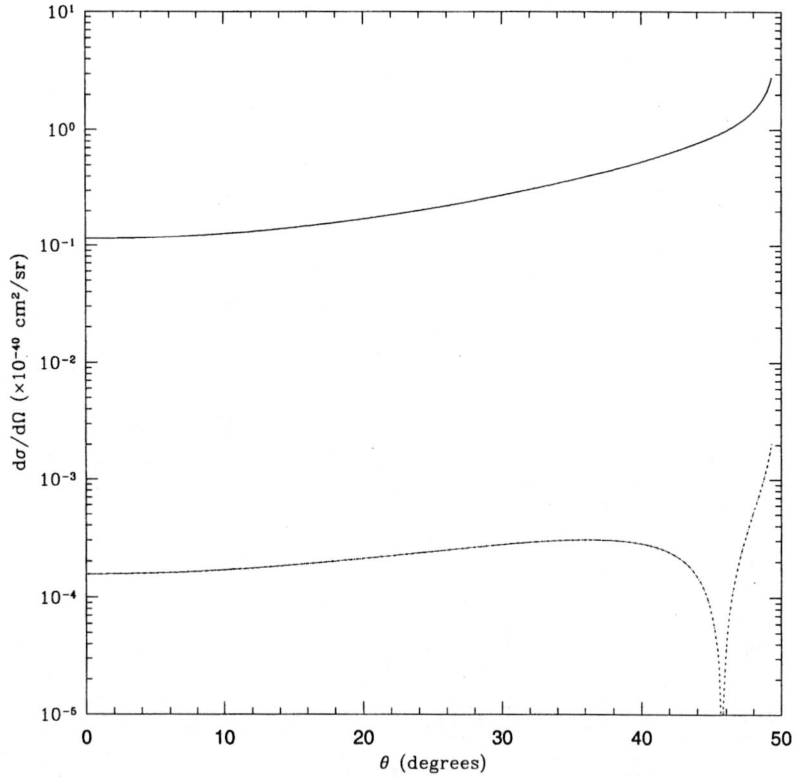

FIGURE 1. Plot of the differential cross section for the reaction, $\mu^- + p \longrightarrow \Lambda + \nu_\mu$ as a function of the laboratory angle of the outgoing Λ for an incident muon energy of 2.0 GeV. The solid curve indicates the whole differential cross section and the dashed curve indicates the absolute value of the contribution from the $F_A F_P$ interference term

RESULTS AND DISCUSSION

Using the results of the last section we are able to calculate the differential cross sections for incoming muons with energies of 2.0 Gev, 0.275 GeV and 0.265 GeV. We plot these results in figures 1, 2 and 3 respectively. We also plot the contributions of the absolute values of the $F_A F_P$ interference term for each of these energies. This interference term is the largest containing F_P.

We see from the above graphs that at higher muon energies the $F_A F_P$ intereference term is too small to seriously consider observing. Its contribution to the differential cross section for 2.0 GeV incident muons is of the order of 0.1 percent. However the situation is improved for both the 0.275 GeV and 0.265 GeV cases where the contribution is of the order of 2 to 3 percent. This is still a small number. Furthermore for the 0.275 and

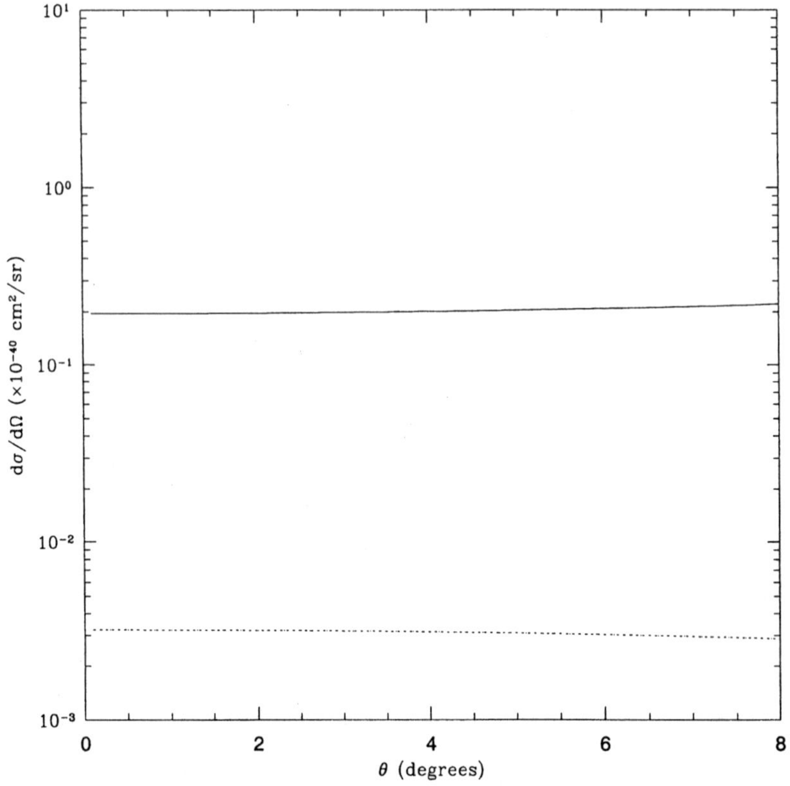

FIGURE 2. Plot of the differential cross section for the reaction, $\mu^- + p \longrightarrow \Lambda + \nu_\mu$ as a function of the laboratory angle of the outgoing Λ for an incident muon energy of 0.275 GeV. The solid curve indicates the whole differential cross section and the dashed curve indicates the absolute value of the contribution from the $F_A F_P$ interference term

the 0.265 GeV cases both the differential cross sections and the $F_A F_P$ interference terms are very flat so that there is no a priori prefered region for studying the interference term. However this does enable one to study this processes away from the maximal angle without sacrificing much in terms of magnitude. This is very different from the case for higher energy muons where the differential cross section grows rapidly near the maximal angle.

Finally we remark that the maximal angle changes for the muon case in comparison to the electron case because muon mass terms must be included. We will discuss this in a forthcoming paper. We again empasize that are results are very preliminary but do seem to be in line with expectations.

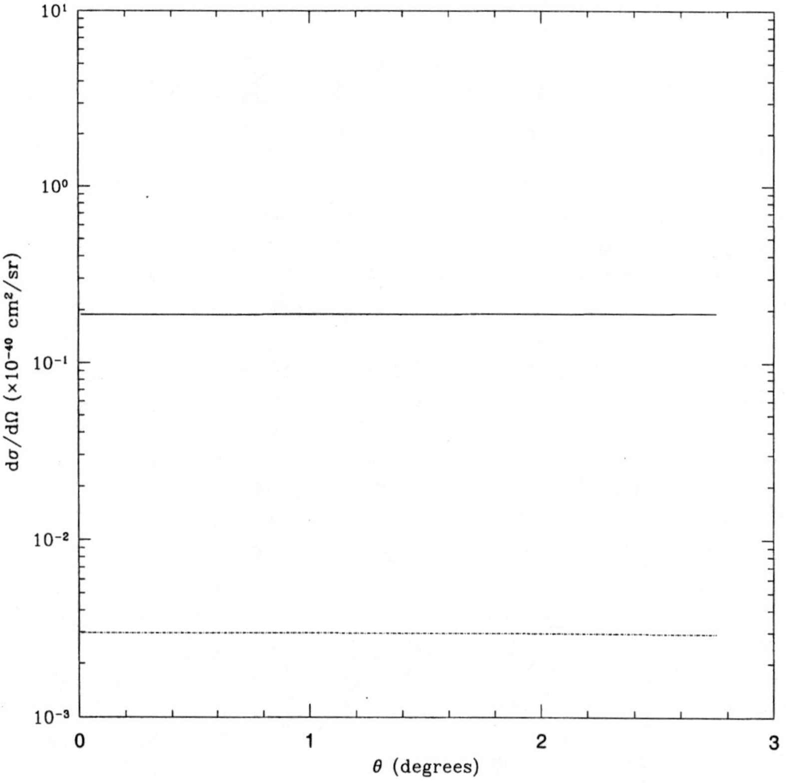

FIGURE 3. Plot of the differential cross section for the reaction, $\mu^- + p \longrightarrow \Lambda + \nu_\mu$ as a function of the laboratory angle of the outgoing Λ for an incident muon energy of 0.265 GeV. The solid curve indicates the whole differential cross section and the dashed curve indicates the absolute value of the contribution from the $F_A F_P$ interference term

REFERENCES

1. Mintz,S.L., Nucl. Phys. **A 657**,303(1999).
2. Finn,J. M., private communication.
3. Particle Data Group,Eur. Phys.J **C 3**,1(1998).
4. Dworkin,J., et al.,Phys. Rev. **D 41**,780(1990).
5. Wise,J., et al., Phys. Lett. **98 B**,123(1981).
6. Bourquin,M., et al.,Z. Phys. **C 12**,307(1982).
7. Carson,L.J.,Oakes,R. J., and Wilcox,C.R.,Phys. Rev. **D 37**,3197(1988).
8. Hwang,W-Y.P., and Henley,E. M.,Phys. Rev. **D 38**,798(1988).

Quantum Discontinuity for Massive Gravity with a Cosmological Term

Hisham Sati[1]

Michigan Center for Theoretical Physics,
Randall Laboratory, Department of Physics, University of Michigan,
Ann Arbor, MI 48109-1120

Abstract. We report on the recent work on the van Dam-Veltman-Zakharov discontinuity for massive and partially massless gravitons in A(dS) space at one-loop.

The question of whether the graviton has a small non-zero mass or exactly zero mass has been addressed by van Dam and Veltman, and Zakharov[1]. The difference is physically detectible, e.g. by effect on bending of light by the sun. The case for non-zero Λ has been studies by references [2] and [3] for de-Sitter and Anti-deSitter space respectively. In the latter it was found that the massive graviton has a smooth limit. An interesting phenomenon [4] that occurs in (A)dS space is partial masslessness with reduced degrees of freedom

Theory	D.O.F.	Unitary?
$M^2 < \frac{2}{3}\Lambda$	5	No
$M^2 = \frac{2}{3}\Lambda$	4	Yes
$M^2 = 0$	2	Yes

All the above analysis is classical. One might ask the question what happens when loop effects are taken into account. It was shown [5] that the discontinuity persists in the flat space case. In [6] it was shown that this is also true in the case $\Lambda \neq 0$.

The starting point of the analysis is our starting point will be the action

$$S[h_{\mu\nu}, T_{\mu\nu}] = S_L[h_{\mu\nu}] + S_M[h_{\mu\nu}] + S_T[h \cdot T]$$

where S_L the Einstein-Hilbert action $S_E = -\frac{1}{16\pi G}\int_M \mathrm{d}^4x\sqrt{\hat{g}}(\hat{R} - 2\Lambda)$, linearized about a background metric $g_{\mu\nu}$ according to $\hat{g}_{\mu\nu} = g_{\mu\nu} + \kappa h_{\mu\nu}$. If the background metric is Einstein, then the linearized action for $h_{\mu\nu}$ is

$$S_L = \int \mathrm{d}^4x\sqrt{g}\left[\tfrac{1}{2}\tilde{h}^{\mu\nu}\left(-g_{\mu\rho}g_{\nu\sigma}\Box - 2R_{\mu\rho\nu\sigma}\right)h^{\rho\sigma} - \nabla^\rho \tilde{h}_{\rho\mu}\nabla^\sigma \tilde{h}^\mu_\sigma\right] \tag{1}$$

where $\tilde{h}_{\mu\nu} = h_{\mu\nu} - \frac{1}{2}g_{\mu\nu}h^\sigma_\sigma$. The linearized Lagrangian S_L has a diffeomorphism symmetry which is broken when we add the Pauli-Fierz spin-2 mass term

$$S_M = \frac{M^2}{2}\int \mathrm{d}^4x\sqrt{g}\left[h^{\mu\nu}h_{\mu\nu} - (h^\mu_\mu)^2\right]$$

[1] hsati@umich.edu

CP624, *Cosmology and Elementary Particle Physics*, edited by B. N. Kursunoglu et al.

The source term for $T_{\mu\nu}$ (assumed conserved) is given by

$$S_T = \int d^4x \sqrt{g}\, h_{\mu\nu} T^{\mu\nu}$$

In order to neatly compare the massive (and partially massless) case with the massless one, we use the Stückelberg formalism to restore gauge symmetry. This is done by introducing an auxiliary vector field V_μ and then making a change of field integration variables $h_{\mu\nu} \to h_{\mu\nu} - \frac{2}{M}\nabla_{(\mu}V_{\nu)}$. The gauge invariances are then

massive	partially massless
$h_{\mu\nu} \to h_{\mu\nu} + 2\nabla_{(\mu}\xi_{\nu)}$	$h_{\mu\nu} \to h_{\mu\nu} + 2\nabla_{(\mu}\xi_{\nu)} + \frac{2}{3}\Lambda g_{\mu\nu}\alpha$
$V_\mu \to V_\mu + M\xi_\mu$	$V_\mu \to V_\mu + M[\xi_\mu - \nabla_\mu \alpha]$

Note that the parameters ξ_μ and α correspond to diffeomorphisms and Weyl rescalings respectively. For gauge-fixing we use a lagrangian quadratic in the constraint

$$MV_\mu - \nabla^\rho \tilde{h}_{\rho\mu} = 0\,.$$

with the additional constraint $h = 0$ for the partially massless case. We decompose the metric fluctuation into traceless and scalar parts given respectively by $\phi_{\mu\nu} \equiv h_{\mu\nu} - \frac{1}{4}g_{\mu\nu}h^\sigma_\sigma$, and $\phi \equiv h^\sigma_\sigma$ so that

$$\begin{aligned} h_{\mu\nu} &= \phi_{\mu\nu} + \tfrac{1}{4}g_{\mu\nu}\phi\,, \\ \tilde{h}_{\mu\nu} &= \phi_{\mu\nu} - \tfrac{1}{4}g_{\mu\nu}\phi\,, \\ h^\mu_\mu &= \phi \end{aligned}$$

The source will similarly be split into its irreducible components $T_{\mu\nu} = j_{\mu\nu} + \frac{1}{4}g_{\mu\nu}j$. The corresponding differential (Lichnerowicz)operators are given as

spin	dimension	operator
2	9	$\Delta(1,1)\phi_\mu\nu = -\Box\phi + R_{\mu\tau}\phi^\tau_\nu + R_{\nu\tau}\phi^\tau_\mu - 2R_{\mu\rho\nu\tau}\phi^{\rho\tau}$
1	4	$\Delta(\frac{1}{2},\frac{1}{2})\xi_\mu \equiv -\Box\xi_\mu + R_{\mu\nu}\xi^\nu$
0	1	$\Delta(0,0) \equiv -\Box$

At this stage one has to take into account the vector ghosts and the corresponding Faddeev-popov determinant. Now we restore yet another gauge invariance, namely that for the "massive" spin-1 field V^μ. As before, we impose gauge symmetry by multiplying by a path-integral $\int \mathcal{D}\chi$ over a new decoupled scalar field χ and then making the change of variables $V_\mu \to V_\mu - M^{-1}\nabla_\mu\chi$. By construction, the action resulting from the previous step is invariant under the gauge-transformation

$$\begin{aligned} V_\mu &\to V_\mu + \nabla_\mu\zeta \\ \chi &\to \chi + M\zeta \end{aligned}$$

A useful choice for the gauge-condition is to associate the longitudinal component of V with χ. We now have a Faddeev-Popov scalar ghost and the resulting gauge-fixed action is

$$\tilde{S} = \int d^4x\sqrt{g}\Big[\tfrac{1}{2}\phi^{\mu\nu}\left(\Delta(1,1) - 2\Lambda + M^2\right)\phi_{\mu\nu} + V^\mu\left(\Delta(\tfrac{1}{2},\tfrac{1}{2}) - 2\Lambda + M^2\right)V_\mu$$
$$-\tfrac{1}{8}\left(\tfrac{-2\Lambda+3M^2}{-2\Lambda+M^2}\right)\phi\left(\Delta(0,0) - 2\Lambda + M^2\right)\phi + \tfrac{-2\Lambda+M^2}{M^2}\chi\left(\Delta(0,0) - 2\Lambda + M^2\right)\chi$$
$$+\tfrac{1}{2}\phi_{\mu\nu}j^{\mu\nu} + \tfrac{1}{8}\phi j$$

for the massive case, and

$$\tilde{S} = \int d^4x\sqrt{g}\Big[\tfrac{1}{2}\phi^{\mu\nu}\left(\Delta(1,1) - \tfrac{4}{3}\Lambda\right)\phi_{\mu\nu} + V^\mu\left(\Delta(\tfrac{1}{2},\tfrac{1}{2}) - \tfrac{4}{3}\Lambda\right)V_\mu$$
$$-2\chi\left(\Delta(0,0) - \tfrac{4}{3}\Lambda\right)\chi + \tfrac{1}{2}\phi_{\mu\nu}j^{\mu\nu} + \tfrac{1}{8}\phi j$$

for the partially massless case. The resulting one loop contributions $\Gamma^{(1)}[g] = -\ln Z[g,0]$ are tabulated as

$M^2 \to 0$	$M^2 = 0$
$-\frac{1}{2}\ln\mathrm{Det}\left(\Delta(\frac{1}{2},\frac{1}{2}) - 2\Lambda + M^2\right)$ $+\frac{1}{2}\ln\mathrm{Det}\left(\Delta(1,1) - 2\Lambda + M^2\right)$	$-\ln\mathrm{Det}\left(\Delta(\frac{1}{2},\frac{1}{2}) - 2\Lambda\right)$ $+\frac{1}{2}\ln\mathrm{Det}\left(\Delta(1,1) - 2\Lambda\right)$ $+\frac{1}{2}\ln\mathrm{Det}\left(\Delta(0,0) - 2\Lambda\right)$
$M^2 \to \frac{2}{3}\Lambda$	$M^2 = \frac{2}{3}\Lambda$
$-\frac{1}{2}\ln\mathrm{Det}\left(\Delta(\frac{1}{2},\frac{1}{2}) - \frac{4}{3}\Lambda\right)$ $+\frac{1}{2}\ln\mathrm{Det}\left(\Delta(1,1) - \frac{4}{3}\Lambda\right)$ $-\frac{1}{2}\ln\mathrm{Det}\left(\Delta(0,0) - \frac{4}{3}\Lambda\right)$	$-\frac{1}{2}\ln\mathrm{Det}\left(\Delta(\frac{1}{2},\frac{1}{2}) - 2\Lambda + M^2\right)$ $+\frac{1}{2}\ln\mathrm{Det}\left(\Delta(1,1) - 2\Lambda + M^2\right)$

It remains to check that there is no conspiracy among the eigenvalues of these operators that would make these expressions coincide. To show this, it suffices to calculate the coefficients in expansion for the graviton propagator associated with $S_L + S_M$, and compare it with the massless and the partially massless cases as can be determined from [7]. The trace of the heat kernel

$$\mathrm{Tr}\, e^{-\Delta^{(\Lambda)}t} = \sum_{k=0}^{\infty} t^{(k-4)/2}\int d^4x\sqrt{g}\, b_k^{(\Lambda)}$$

gives $b_4^{(\Lambda,M)}$ which determines the number of zero modes and is given by

$M^2 \to 0$	$M^2 = 0$
$\left[b_4^{(\Lambda)}(1,1) - b_4^{(\Lambda)}(\frac{1}{2},\frac{1}{2})\right]$ $= 200R_{\mu\nu\rho\sigma}R^{\mu\nu\rho\sigma} - 1740\Lambda^2$	$\left[b_4^{(\Lambda)}(1,1) - 2b_4^{(\Lambda)}(1/2,1/2) + b_4^{(\Lambda)}(0,0)\right]$ $= 212R_{\mu\nu\rho\sigma}R^{\mu\nu\rho\sigma} - 2088\Lambda^2$
$M^2 \to \frac{2}{3}\Lambda$	$M^2 = \frac{2}{3}\Lambda$
$\left[b_4^{(\Lambda)}(1,1) - b_4^{(\Lambda)}(\frac{1}{2},\frac{1}{2})\right]$ $= 200R_{\mu\nu\rho\sigma}R^{\mu\nu\rho\sigma} - 740\Lambda^2$	$\left[b_4^{(\Lambda)}(1,1) - b_4^{(\Lambda)}(1/2,1/2) - b_4^{(\Lambda)}(0,0)\right]$ $= 199R_{\mu\nu\rho\sigma}R^{\mu\nu\rho\sigma} - 1096\Lambda^2$

We see that the full quantum theory for each of the massive and partially massless theories is discontinuous [6]. This is perhaps not surprising considering the different degrees of freedom in the three cases.

Recently the case of the gravitino has been studies classically for $\Lambda \neq 0$ [8] with conclusion similar to those in the case for the graviton. The quantum case is currently under investigation.

ACKNOWLEDGEMENTS

I would like to thank the organizers of the conference for invitation and for the interesting atmosphere. I also thank F. Dilkes, M.J. Duff and J.T. Liu for collaboration on the subject of this talk and for discussions and suggestions. I also thank UM Rackham School for travel support.

REFERENCES

1. van Dam,H. and Veltman, M., *Nucl. Phys.* **B22**, 397-411 (1970); Zakharov, V. , *JETP Lett.* **12** 312-314 (1970).
2. Higuchi, A., *Nucl. Phys.* **B325**, 745-765 (1989).
3. Porrati, M., *Phys. Lett.* **498B**, 92-96 (2001); Kogan, I. Mouslopoulos, S. and Papazoglou, A., *Phys. Lett.* **503B**, 173-180 (2001)
4. Deser, S. and Waldron, A., *Phys. Rev. Lett.* **87** 031601, (2001).
5. Christensen, S.M. and Duff, M.J., *Phys. Lett.* **76B**, 571-574 (1978).
6. Dilkes, F.A.,Duff, M.J., Liu, J.T. and Sati, H., *Phys. Rev. Lett.* **87** 041301 (2001); Duff, M.J., Liu, J.T. and Sati, H., *Phys. Lett.* **516B**,156-160 (2001).
7. Christensen, S.M. and Duff, M.J., *Nucl. Phys.* **B154**, 301-342 (1979); *ibid* **B170**, 480-506 (1980).
8. Grassi, P. and van Nieuwenhuizen, P., *Phys. Lett.* **499B**, 174-178 (2001); Deser, S. and Waldron, A., *Phys.Lett.* **501B**, 134-139 (2001).

Possible Conformal $\mathcal{N} = 0$ $d = 4$ Gauge Theories from AdS/CFT Superstring Duality?

William F. Shively

Department of Physics and Astronomy, University of North Carolina, Chapel Hill, NC 27599-3255

Abstract. In this article, I will discuss conditions for which non-supersymmetric $d = 4$ gauge theories, arising from superstring duality on a manifold $AdS_5 \times S_5/Z_p$, also have vanishing two-loop gauge $\beta-$functions. Such is necessary for finding possible non-supersymmetric ultraviolet-finite theories that may accommodate the standard model.

This talk is based on a previous paper[1] that I coauthored with Dr. Paul Frampton.

Over the past few years, much excitment has arisen from the possibility that information contained in superstring theory is encoded in a four dimensional gauge field theory, including the non-perturbative sector. Specifically, $\mathcal{N} = 4$ supersymmetric Yang-Mills theory, with a gauge group broken from $U(N)$ to $SU(N)$, corresponds to N coincident D3-branes with 4-dimensional world volume theories and superconformal symmetry. This theory is dual via strong-weak coupling to a type IIB superstring theory within a spacetime with geometry $AdS_5 \times S_5$[2, 3, 4, 5]. $\mathcal{N} = 4$ $SU(N)$ gauge theory has been known for quite some time to be ultraviolet finite, not only in the asymptotic limit of N, but also for finite N as well[6]. Hence, the RGE β-functions for $\mathcal{N} = 4$ gauge theory vanish to all orders.

The question remains if there exists a corresponding $\mathcal{N} = 0$ $SU(N)$ theory which is ultraviolet-finite as well. The possibility was explored[7] that an $\mathcal{N} = 0$ theory, arising from compactification [8, 9, 10, 11, 12, 13, 14, 15, 16, 17] on the orbifold $AdS_5 \times S_5/\Gamma$ could be conformal and even accommodate the standard model. In this work we systematically catalog the available $\mathcal{N} = 0$ theories for Γ an abelian discrete group $\Gamma = Z_p$. We also find the subset which has $\beta_g^{(2)} = 0$, a vanishing two-loop $\beta-$function for the gauge coupling, according to the criteria of [7]. Note also that the one-loop $\beta-$functions satisfy $\beta_Y^{(1)} = 0$ and $\beta_H^{(1)} = 0$ for the Yukawa and Higgs couplings respectively, because they are leading order in the planar expansion[18, 19, 20, 21]. All one-loop $\mathcal{N} = 0$ calculations coincide with those of the conformal $\mathcal{N} = 4$ theory, but beyond one-loop this coincidence ceases, in general.

To begin with, the S^5 of $AdS_5 \times S^5$ has an $SU(4)$ ($SO(6)$) isometry. We can break the supersymmetries of the theory by orbifolding, that is, by factoring out of S^5 some discrete group Γ that encapsulates the desired symmetry. For our purposes, we will consider only abelian groups, $\Gamma = Z_p$; for a look at nonabelian groups, see eg.[22].

CP624, *Cosmology and Elementary Particle Physics*, edited by B. N. Kursunoglu et al.

The embedding of Z_p in the real six-dimensional space, or equivalently, complex three-dimensional space $\mathscr{C}^3$, can be conveniently specified by three integers $a_i = (a_1, a_2, a_3)$. The action of Z_p on the three complex coordinates (X_1, X_2, X_3) is then:

$$(X_1, X_2, X_3) \xrightarrow{Z_p} (\alpha^{a_1} X_1, \alpha^{a_2} X_2, \alpha^{a_3} X_3) \tag{1}$$

where $\alpha = exp(2\pi i/p)$ and the elements of Z_p are α^r $(0 \leq r \leq (p-1))$. For $\Gamma \subset SU(2)$, there remains $\mathscr{N} = 2$ supersymmetry; $\Gamma \subset SU(3)$, $\mathscr{N} = 1$ supersymmetry; and for $\Gamma \not\subset SU(3)$, no supersymmetry ($\mathscr{N} = 0$).

Thus, to ensure that $\Gamma \not\subset SU(3)$, we therefore set

$$a_1 + a_2 + a_3 \neq 0 \ (mod \ p) \tag{2}$$

To ensure the correct behavior of spinors, we must also set

$$a_1 + a_2 + a_3 = 0 \ (mod \ 2) \tag{3}$$

This orbifolding procedure leads to an identification of p points in C^3, and the N coinciding D-branes converge on all p copies. Hence, the gauge group becomes $SU(N)^p$. Therefore, in the corresponding field theory, the scalars will have the representations:

$$\sum_{\mu} (N_i, \bar{N}_{i \pm a_\mu}). \tag{4}$$

The following transform as spinors in SU(4):

$$A_1 = (a_1 + a_2 + a_3)/2 \tag{5}$$
$$A_2 = (a_1 - a_2 - a_3)/2 \tag{6}$$
$$A_3 = (-a_1 + a_2 - a_3)/2 \tag{7}$$
$$A_4 = (-a_1 - a_2 + a_3)/2 \tag{8}$$

Thus, the fermions in the theory fall into the representations

$$\sum_{\lambda} (N_i, \bar{N}_{i \pm A_\lambda}). \tag{9}$$

In [7] it was found that the background gauge β-function vanishes to second loop order. As we go from $\mathscr{N} = 4$ to $\mathscr{N} = 0$, $\beta_g^{(1)}$ is the same term-for-term, and the first, third, and fifth terms of $\beta_g^{(2)}$ mutually cancel. The only $\mathscr{N} = 4$ case in which the remaining terms vanish is the case in which all $a_\mu \neq 0$, one $A_\lambda = 0$, and the remaining three $A_\lambda \neq 0$. For this case, all scalars are in fundamental representations, whereas the fermions are in both fundamental and adjoint representations.

Without loss of generality, we can set each a_i to be in the range $0 \leq a_i \leq (p-1)$. Furthermore, since permutations of the a_i are equivalent, we may set $a_1 \leq a_2 \leq a_3$. We define $\nu_k(p)$ to be the number of possible $\mathscr{N} = 0$ theories with k non-zero a_i $(1 \leq k \leq 3)$.

The simplest scenario is for which only one $a_i \neq 0$, ie. $a_i = (0, 0, a_3)$. Since this is clearly equivalent to $a_i = (0, 0, p - a_3)$ the value of $\nu_1(p)$ is

$$\nu_1(p) = \lfloor p/2 \rfloor \tag{10}$$

where $\lfloor x \rfloor$ is the largest integer not greater than x.

For $\nu_2(p)$ we likewise observe that $a_i = (0, a_2, a_3)$ is equivalent to $a_i = (0, p - a_3, p - a_2)$. Then we may derive, taking into account Eq.(2) that, for p *even*

$$\nu_2(p) = 2 \sum_{r=1}^{\lfloor \frac{p-2}{2} \rfloor} r = \frac{1}{4} p(p-2) \tag{11}$$

while, for p *odd*

$$\nu_2(p) = 2 \sum_{r=1}^{\lfloor \frac{p-2}{2} \rfloor} r + \lfloor \frac{p}{2} \rfloor = \frac{1}{4}(p-1)^2 \tag{12}$$

For $\nu_3(p)$, the counting is slightly more complicated. Again, we have the equivalence of $a_i = (a_1, a_2, a_3)$ with $(p - a_3, p - a_2, p - a_1)$ as well as Eq.(2) to contend with. With regard to this condition, let $\nu_p(p)$ be the number of theories with $\sum a_i = p$ and $\nu_{2p}(p)$ be the number with $\sum a_i = 2p$. Then because of the equivalence of (a_1, a_2, a_3) with $(p - a_3, p - a_2, p - a_1)$, it follows that $\nu_p(p) = \nu_{2p}(p)$. In particular, also note that the theory $a_i = (a_1, p/2, p - a_1)$ is a self-equivalent (SE) one; let the number of such theories be $\nu_{SE}(p)$. Then it can be seen that $\nu_{SE}(p) = p/2$ for p even, and $\nu_{SE}(p) = 0$ for p odd. The value will be calculated below; in terms of it $\nu_3(p)$ is given by

$$\nu_3(p) = \frac{1}{2}[\bar{\nu}(p) - 2\nu_p(p) + \nu_{SE}(p)] \tag{13}$$

where $\bar{\nu}(p)$ is the number of unrestricted (a_1, a_2, a_3) satisfying $1 \leq a_i \leq (p-1)$ and $a_1 \leq a_2 \leq a_3$. Its value is given by

$$\bar{\nu}(p) = \sum_{a_3=1}^{p-1} \sum_{a_3=1}^{p-1} a_2 = \frac{1}{6} p(p^2 - 1) \tag{14}$$

It remains only to calculate $\nu_p(p)$ given by

$$\nu_p(p) = \sum_{a_1=1}^{\lfloor \frac{p}{3} \rfloor} \left(\left\lfloor \frac{p - a_1}{2} \right\rfloor - a_1 + 1 \right) \tag{15}$$

The value of $\nu_p(p)$ depends on the remainder when p is divided by 6. Specifically, consider $p = 6k$ where k is an integer. Then

$$\nu_p(p) = \sum_{a_1=odd}^{2k-1} \left(3k + \frac{1}{2} - \frac{3a_1}{2} \right) + \sum_{a_1=even}^{2k} \left(3k + \frac{3a_1}{2} + 1 \right) = 3k^2 = \frac{1}{12} p^2 \tag{16}$$

Hence from Eq.(13)

$$\nu_3(p) = \frac{1}{2}\left[\frac{1}{6}p(p^2-1) - \frac{1}{6}p^2 + \frac{p}{2}\right] = \frac{p}{12}(p^2-p+2) \tag{17}$$

Taking $\nu_1(p)$ from Eq.(10) and $\nu_2(p)$ from Eq.(12) we find for $p = 6k$

$$\nu_{TOTAL}(p) = \nu_1(p) + \nu_2(p) + \nu_3(p) = \frac{p}{12}(p^2+2p+2) \tag{18}$$

For $p = 6k+1$ or $p = 6k+5$ one finds similarly

$$\nu_3(p) = \frac{1}{12}(p-1)^2(p+1) \quad (p = 6k+1 \;\; or \;\; 6k+5) \tag{19}$$

$$\nu_{TOTAL} = \frac{1}{12}(p-1)(p+1)(p+2) \quad (p = 6k+1 \;\; or \;\; 6k+5) \tag{20}$$

For $p = 6k+2$ or $p = 6k+4$

$$\nu_3(p) = \frac{1}{12}(p+1)(p^2-2p+4) \quad (p = 6k+2 \;\; or \;\; 6k+4) \tag{21}$$

$$\nu_{TOTAL} = \frac{1}{12}(p^3+2p^2+2p+4) \quad (p = 6k+2 \;\; or \;\; 6k+4) \tag{22}$$

and finally for $p = 6k+3$

$$\nu_3(p) = \frac{1}{12}(p^3-p^2-p-3) \quad (p = 6k+3) \tag{23}$$

$$\nu_{TOTAL} = \frac{1}{12}(p^3+2p^2-p-6) \quad (p = 6k+3) \tag{24}$$

The values of $\nu_1(p)$, $\nu_2(p)$, $\nu_3(p)$, $\nu_{TOTAL}(p)$ and $\Sigma_{p'=2}^{p} \nu_{TOTAL}(p')$ for $2 \leq p \leq 41$ are listed in Table 1.

But of all these candidates for conformal $\mathcal{N} = 0$ theories, how many if any are actually conformal? The criterion found in[7] for the vanishing of the two-loop RGE $\beta-$function $\beta_g^{(2)} = 0$, for the gauge coupling, is that $a_1 + a_2 = a_3$. We will denote the number of theories fulfilling this by $\nu_{alive}(p)$.

If p is odd there is no contamination by self-equivalent possibilities and the result is

$$\nu_{alive} = \sum_{r=1}^{\frac{p-1}{2}} (p-2r) = \frac{1}{4}(p-1)^2 \quad (p = odd) \tag{25}$$

For p even some self equivalent cases must be subtracted. The sum in Eq. (25) is $\frac{1}{4}p(p-2)$ and the number of self-equivalent cases to remove is $\lfloor p/4 \rfloor$ with the results

$$\nu_{alive} = \frac{1}{4}p(p-3) \quad (p = 4k) \tag{26}$$

$$\nu_{alive} = \frac{1}{4}(p-1)(p-2) \quad (p = 4k+2) \tag{27}$$

Table 1. Values of $v_1(p)$, $v_2(p)$, $v_3(p)$, $v_{TOTAL}(p)$, $\Sigma_{p'=1}^{p} v_{TOTAL}(p')$, $v_{alive}(p)$ and $\Sigma_{p'=2}^{p} v_{alive}(p')$ for $2 \leq p \leq 41$.

p	$v_1(p)$	$v_2(p)$	$v_3(p)$	$v_{TOTAL}(p)$	Σv_{TOTAL}	$v_{alive}(p)$	$\Sigma v_{alive}(p)$
2	1	0	1	2	2	0	0
3	1	1	1	3	5	1	1
4	2	2	5	9	14	1	2
5	2	4	8	14	28	4	6
6	3	6	16	25	53	5	11
7	3	9	24	36	89	9	20
8	4	12	39	55	144	10	30
9	4	16	53	73	217	16	46
10	5	20	77	102	319	18	64
11	5	25	100	130	449	25	89
12	6	30	134	170	619	27	116
13	6	36	168	210	829	36	152
14	7	42	215	264	1093	39	191
15	7	49	261	317	1410	49	240
16	8	56	323	387	1797	52	292
17	8	64	384	456	2253	64	356
18	9	72	462	543	2796	68	424
19	9	81	540	630	3426	81	505
20	10	90	637	737	4163	85	590
21	10	100	733	843	5006	100	690
22	11	110	851	972	5978	105	795
23	11	121	968	1100	7078	121	916
24	12	132	1108	1252	8330	126	1042
25	12	144	1248	1404	9734	144	1186
26	13	156	1413	1582	11316	150	1336
27	13	169	1577	1759	13075	169	1505
28	14	182	1769	1965	15040	175	1680
29	14	196	1960	2170	17210	196	1876
30	15	210	2180	2405	19615	203	2079
31	15	225	2400	2640	22255	225	2304
32	16	240	2651	2907	25162	232	2536
33	16	256	2901	3173	28335	256	2792
34	17	272	3185	3474	31809	264	3056
35	17	289	3468	3774	35583	289	3345
36	18	306	3796	4110	39693	297	3642
37	18	324	4104	4446	44139	324	3966
38	19	342	4459	4820	48959	333	4299
39	19	361	4813	5193	54152	361	4660
40	20	380	5207	5607	59759	370	5030
41	20	400	5600	6020	65779	400	5430

In the last two columns of Table 1 are the values of $v_{alive}(p)$ and $\sum_{p'=2}^{p} v_{alive}(p')$.

Asymptotically for large p the ratio $v_{alive}(p)/v_{TOTAL}(p) \sim 3/p$ and hence vanishes although $v_{alive}(p)$ diverges; the value of the ratio is *e.g.* 0.28 at p = 5 and at p = 41 is 0.066. It is currently being studied how the two-loop requirements $\beta_Y^{(2)} = 0$ and $\beta_H^{(2)} = 0$ select from such theories. That result will further indicate whether any $v_{alive}(p)$ can survive to all orders, and hence give a candidate for an ultraviolet-finite $\mathcal{N} = 0$ theory.

This work was supported in part by the US Department of Energy under Grant No. DE-FG02-97ER-41036.

REFERENCES

1. P.H. Frampton and W.F. Shively, Phys. Lett. **B454,** 49 (1999). *hep-th/9902168.*
2. J. Maldacena, Adv. Theor. Math. Phys. **2,** 231 (1998). *hep-th/9711200.*
3. S.S. Gubser, I.R. Klebanov and A.M. Polyakov, Phys. Lett. **B428,** 105 (1998). *hep-th/9802109.*
4. E. Witten, Adv. Theor. Math. Phys. **2,** 505 (1998). *hep-th/9803131.*
5. L. Susskind and E. Witten, *hep-th/9805114.*
6. S. Mandelstam, Nucl. Phys. **B213,** 149 (1983).
7. P.H. Frampton, Phys. Rev. **D60,** 041901 (1999). *hep-th/9812117.*
8. S. Kachru and E. Silverstein, Phys. Rev. Lett. **80,** 4855 (1998). *hep-th/9802183.*
9. S. Ferrara, A. Kehagias, H. Partouche and A. Zaffaroni, Phys. Lett. **B431,** 42 (1998). *hep-th/9804006.*
10. J.R. Russo, Phys. Lett. **B435,** 284 (1998). *hep-th/9808117.*
11. M. Schmaltz, Phys. Rev. **D59,** 105018 (1999). *hep-th/9805218.*
12. M. Berkooz and S.-J. Rey, JHEP **9901,** 014 (1999). *hep-th/9807200.*
13. J. Distler and F. Zamaro, Adv. Theor. Math. Phys. **2,** 1405-1439 (1999). *hep-th/9810206.*
14. J.A. Harvey, Phys. Rev. **D59,** 026002 (1999). *hep-th/9807213.*
15. I.R. Klebanov and A.A. Tsyetlin, JHEP **9903,** 015 (1999). *hep-th/9901101.*
16. J. Kakushadze, Phys. Rev. **D59,** 045007 (1999). *hep-th/9806091.*
17. A.A. Tsyetlin and K. Zarembo, Phys. Lett. **B457,** 77-86 (1999). *hep-th/9902095.*
18. M. Bershadsky, Z. Kakushadze and C. Vafa, Nucl. Phys. **B523,** 59 (1998). *hep-th/9803076.*
19. M. Bershadsky and A. Johansen, Nucl. Phys. **B536,** 141 (1998). *hep-th/9803249.*
20. A. Lawrence, N. Nekrasov and C. Vafa, Nucl. Phys. **B533,** 199 (1998). *hep-th/9803015.*
21. N. Nekrasov and S.L. Shatashvili, Phys. Rept. **320,** 127-129 (1999). *hep-th/9902110.*
22. P.H. Frampton and T.W. Kephart, Phys. Rev. **D64,** 086007 (2001). *hep-th/0011186.*